THERMODYNAMIC LIMIT TO THE EXISTENCE

of

Inanimate and Living Systems

A. PAGLIETTI
University of Cagliari, Italy

SEPCO - ACERTEN
2014

Published in United Kingdom by Sepco-Acerten Publ., 2014

Revised Printing

9 10 11 12

ISBN-13: 978-88-909437-0-6

Publisher's Cataloguing in Publication Data.

Paglietti, A., 1944–
Thermodynamic limit to the existence of inanimate and living systems
/ by A. Paglietti
p. cm
Includes bibliographical references and index
1. Thermodynamics. 2. Materials—Elastic properties. 3. Electrolytic cells.
4. Biophysics. 5. Turbulent flow.
I. Title
QC311.xx 536'-7xx
ISBN-13: 978-88-909437-0-6 (hardcover: alk. paper)

Cover background: *Quasar* by ESO/M. Kornmesser (adapted excerpt)

Preface

This book is for the brave. It opens a gate at the frontier of science to new ground that hitherto has been left unexplored. Whether you are dealing with the elastic limit of solids or with the energy that an electrochemical cell can store, measuring the energy that keeps the living cell alive, seeking a new perspective on the hardening and quenching of metals, inquiring into the origin of turbulence in fluid flow, or even exploring the capacity of muscle cells to provide the burst of energy required to break some athletic record, this book will help you. Your understanding of the natural sciences will never be the same.

Surely, some general physical law must be at work here for these bold and diverse claims to be true. In fact, they are all consequences of the first and the second laws of thermodynamics. It is the combined implications of these two laws that are considered in this book and that lead to some new and general conclusions.

In short, the first law implies that at finite temperature the internal energy that a system can store per unit volume should be finite. The second law imposes that the heat that the system absorbs cannot exceed its entropy variation times the absolute temperature. Add that there is no physical limit to the time rate at which work can be dissipated into heat. Then, the first and the second laws together will produce a general inequality that relates the time rate of internal energy to that of entropy in any process in which the system does not receive work from the surroundings.

Because internal energy and entropy are state functions, that inequality defines a range of states that are admissible for the system, granted some general accessibility hypotheses. Under broad assumptions, which are met by most natural systems, this leads to the conclusion that the maximum amount of energy that a system can store and supply is attained at the boundary of that range. In most cases, the knowledge of just one state of the boundary suffices to determine the entire admissible range of the system and the maximum amount of energy that it can store. This has a series of important implications, which this book examines in detail.

A firm understanding of the basic concepts of classical thermodynamics is an essential prerequisite to the analysis presented. The needed background is provided in Chapter 1, while Appendices B, C, and D cover more advanced topics, such as the difference between ideal and real systems, the thermodynamics of chemically reacting mixtures, and the effect of pressure on liquid solutions. This should make the book reasonably self-contained and accessible to readers from all branches of the natural sciences, regardless of their background in thermodynamics.

The concepts that form the foundation of the present approach are covered in Chapters 2 through 5. Chapters 6 through 11 deal with applications to such diverse fields as continuum mechanics, electrochemistry, biophysics and fluid dynamics. The analysis developed in this part of the book shows how the present theory can provide new insights into the particular disciplines or problems to which it is applied. It also affords answers to numerous unsolved problems and opens the way to new, unexpected applications.

The last chapter is devoted to the concept of intrinsic heat. Its aim is to provide the reader with a firm understanding of the concept of entropy, still within the framework of the phenomenological, macroscopic approach on which classical thermodynamics is based.

I am indebted to the many authors of books and papers whose works taught me the wide range of concepts that are needed to write such a book. These authors are all cited in the references. It is unavoidable, though, that the help I received in a less direct—and sometime unexpected—way from many other people remains in the background. My sincere gratitude goes to all of them as well.

My heartfelt thanks go to all those people who devoted their time to read and comment on the manuscript at various stages of its preparation. I am particularly grateful to Dr. Giorgio Carta for reading the final draft of this book. His untiring attention to detail and his comments helped me to eliminate several ambiguities and errors from the material presented here. Mr. W. Ponder corrected my English and was generous in comments that helped me to greatly improve my writing in a way that, sometimes, went well beyond the mere linguistic editing. Of course, the book may still contain lapses and errors. But those are entirely my responsibility.

A. PAGLIETTI
March 2014

Contents

1

The Basic Thermodynamic Setup

This chapter reviews the few basic concepts upon which the whole edifice of thermodynamics is founded. They are presented in their most elementary form, which is adequate for systems whose properties, conditions, and processes are the same at every point of the system. This is the realm of spatially uniform systems and processes, also referred to as homogeneous systems and homogeneous processes. It is governed by classical thermodynamics. The vast majority of thermodynamic applications can be treated within this framework, either rigorously or to a satisfactory degree of accuracy. The present chapter also introduces the main notations and the basic formulae of classical thermodynamics. Thus it provides the minimum background that is needed for this book.

In this chapter as well as in most of the book, "body" and "system" are treated as synonyms. Even so, there is a slight difference in the meanings of these words. As discussed in more detail in Chapter 2, "system" puts the emphasis on the theoretical model that we associate with the portion of matter under consideration, while "body" refers prevalently to the real-world entity. Most often, this distinction need not be made or is implicit from the context, and the two words can be interchanged freely.

1.1 Temperature

Thermodynamics cannot be conceived without temperature. It is a primitive quantity that measures the degree of "hotness" of a body or its parts. Temperature joins mass, length, and time in the physical description of the properties of any substance. Although temperature can also be introduced as a quantity derived from other physical observables, there is not much practical advantage in doing so. The notion of temperature helps to express the property of hotness or coldness of a body in quantitative terms.

Thermometers are used to measure temperature and can be made in several different ways by exploiting one or more of the many properties of the body that are sensitive to its degree of hotness. To be definite, we may refer to a gas thermometer, which is made of a mass of a gas that is sealed within a vessel with diathermal walls. When the vessel is placed in thermal contact with an otherwise isolated body, the pressure of the gas will change over time until it reaches a final, constant value. In this condition, the body is said to be in thermal equilibrium with the gas in the vessel. The change in the pressure of the gas depends on the temperature of the body in equilibrium with it.

It is a matter of experimental evidence that if the same equilibrium pressure is measured when the thermometer is in thermal equilibrium with any two bodies separately, then the two bodies are in thermal equilibrium with each other. What is meant is that no change in the properties of the two bodies is produced if they are placed in thermal contact with each other through a rigid, diathermal wall. This result expresses the so-called *zeroth law of thermodynamics*, which states that if two bodies are in thermal equilibrium with a third body (e.g., a thermometer), then they are also in thermal equilibrium with each other. It legitimates taking the reading from a thermometer (in the case of a gas thermometer, the equilibrium pressure of the gas within the vessel) as a measure of the temperature of the body.

Of course, in order to define a temperature scale, one also must choose an appropriate unit of temperature and assign a value of temperature to a given reference state. The reference state is usually taken as the triple point of pure water (the point at which vapour, liquid, and solid water coexist in equilibrium), because this is a unique condition that can be achieved easily in practice. The triple point of pure water coincides approximately with the state of melting ice at normal atmospheric pressure. To fix the size of the unit of temperature, a second reference state is needed. This is usually the boiling point of water at normal atmospheric pressure. The so-called *absolute temperature* scale is then obtained by assuming the temperature of boiling water to be 100 units above that of the triple point of water and by assigning the value of 273.16 to the latter. When measured in this scale, the temperature is called *absolute*, and its unit is the Kelvin (K).

The absolute temperature scale owes its name to the fact that it can also be established as a consequence of some basic considerations on the maximum efficiency of thermal engines, without making any reference to a particular gas or any other substance. Hence, the attribute "absolute" is attached to that scale. The details of such a procedure are standard and will not be repeated here. However, it must be stressed that the procedure leads to the conclusion that no temperature can exist below 0 K. For this reason, the temperature of 0 K is referred to as the *absolute zero* temperature.

Of course, an unlimited number of different temperature scales can be defined by similar procedures to the one described above. Each scale assigns a different temperature value to the reference points that are used. In fact, many different temperature scales have been proposed in the past, and a few of them are still in use today. They are denoted collectively as empirical temperature scales in order to distinguish them from the absolute scale defined above.

Although empirical temperatures are used extensively in practice, the absolute temperature scale is typically used in theoretical thermodynamics. The main reason is that it makes for simpler formulae when dealing with the second law of thermodynamics and its consequences. For this reason, we shall henceforth refer to it.

1.2 Heat

The notion of heat is based on the following experiment. Two otherwise isolated bodies that possess two different temperatures are put in thermal contact with each other through a rigid, diathermal wall or an equivalent device. The temperature of the hotter body will then decrease and that of the colder body will increase until a condition of thermal equilibrium is reached, in which the two bodies are at the same temperature. The process does not involve any exchange of work, since the diathermal wall is rigid, and the two bodies are isolated from the surrounding world. Due to their initial temperature difference, however, the two bodies exchange something that makes their temperatures change. That something is called *heat*. It is measured by instruments called calorimeters.

This is the most basic definition of heat, and it is referred to as the calorimetric definition of heat. It does not depend on the concept of energy or on the atomic structure of the material. The calorimetric unit of measure of heat is the *calorie* (cal). It represents the amount of heat required to increase by 1 K the temperature of 1 g of water from a given initial temperature. For the so-called *15 deg calorie*, the initial temperature of the water is 287.66 K (14.5 °C).

The calorie is seldom used today. Instead, heat is often measured by energy. This is made possible due to the well-known equivalence of heat and energy, established by the celebrated experiments conducted by Joule in the mid-1800s. According to this equivalence, 1 cal is equivalent to 4.1860 J, if the initial temperature of water is assumed to be 287.66 K (14.5 °C). Accordingly, the quantity $\mathcal{J} = 4.1860$ J/cal is referred to as the *mechanical equivalent of heat*. Slightly different values of $\mathcal{J}$ are sometimes used.

Strictly speaking, the energetic definition of heat is not always viable. Heat only becomes equivalent to energy after it is absorbed by the body and turned into internal energy. However, there are situations in which one must deal with heat before it is absorbed by a body as internal energy. For example, consider the case in which heat is flowing through a heat-conducting bar that joins two heat reservoirs that are at different temperatures. When a steady-state situation is reached, the temperature of the points of the bar and all of the bar's state variables remain constant. Thus the bar elements do not absorb any heat, while heat flows through the bar. A flowing amount of heat cannot be confused with its equivalent in energy, simply because Joule's experimental result about the equivalence between heat and energy does not apply to flowing amounts of heat, i.e. before thermal equilibrium is reached [50]. In the considered instance, the heat keeps flowing through the bar without releasing energy to it, because each cross-section of the bar is at a different temperature, which prevents the flowing heat from stopping.

In thermal equilibrium conditions both the calorimetric and the energetic definitions of heat are equally acceptable. For this reason, distinguishing between the two definitions of heat is unnecessary in classical thermodynamics; which explains why heat and thermal energy are widely considered as one and the same thing. The situation is bound to be different outside of thermodynamic equilibrium. However, the distinction between heat and thermal energy can be ignored in the rest of this book, because the analysis to be presented in what follows refers to thermal equilibrium conditions.

Before we close this section, a few words about notation are in order. We follow the almost universal convention to denote the heat absorbed or lost by the system by the letter Q. Also, we assume that Q is greater than zero when representing a quantity of heat that flows into the body from its surroundings. Therefore, a negative value of Q represents the amount of heat that leaves the body. In general, Q may vary with time; to emphasize this, we write $Q = Q(t)$, where t denotes time. The time derivative of any quantity is denoted by a superimposed dot or by the notation $\mathrm{d}\cdot/\mathrm{d}t$. Thus, the quantity $\dot{Q}$ (i.e., $\mathrm{d}Q/\mathrm{d}t$) indicates the time rate at which heat is absorbed ($\dot{Q} > 0$) or rejected ($\dot{Q} < 0$). The notation $\mathrm{d}Q$ indicates an infinitesimal amount of heat that the body absorbs ($\mathrm{d}Q > 0$) or loses ($\mathrm{d}Q < 0$). If this takes place in the infinitesimal time interval $\mathrm{d}t$, then $\mathrm{d}Q = \dot{Q}\,\mathrm{d}t$.

1.3 External Work

The work performed by a force acting upon a point of a body is defined as the product of the force times the component of the displacement of the point in the direction of the force. The kind of work that is considered in thermodynamics is *external work*. This is the work done by the forces that the surrounding world exerts upon the body under consideration.

Energy measures the capacity to perform work. It is therefore measured in work units. A body can store energy. It also can exchange energy with the surroundings by doing work in many different ways and by heat transfer. Energy exchange via heat transfer stands apart from the other energy exchanges, because it is subjected to restrictions imposed by the second law of thermodynamics.

A general definition of non-thermal energy transfer can be introduced by means of the notion of generalized force and generalized displacement. A *generalized force* is any external entity (e.g., mechanical, electrical, magnetic, and chemical entities) that acts upon a system and produces changes in its state. These changes are measured by appropriate *generalized displacements*, conjugated with the considered generalized force. They are defined such that their scalar product with the conjugated force measures the non-thermal energy (*generalized work*) that the system exchanges with the surroundings as the force is applied or removed. This definition includes, as a particular case, the notion of work as a scalar product of force and displacement that is used in classical mechanics. The concepts of generalized force and generalized displacement, however, help to describe the exchange of non-thermal energy by a system in a more general framework.

Of course, different systems have different generalized forces and generalized displacements associated with them. In gases, the most frequently considered systems in classical thermodynamics, the generalized forces and the conjugate displacement reduce to pressure (p) and volume (V). No matter how generalized forces and generalized displacements are defined, their scalar product must have the physical dimensions of energy and is therefore measured in work units.

Symbolically, all generalized forces acting on a system can be denoted collectively by $\boldsymbol{F}$. In general, $\boldsymbol{F}$ is a vector of appropriate dimensions. Likewise, a vector $\boldsymbol{\xi}$ of the same geometrical dimensions as $\boldsymbol{F}$ denotes the generalized displacements conjugated to $\boldsymbol{F}$. In homogeneous systems, the variable $\boldsymbol{\xi}$ is the same at every point of the system. This eliminates any ambiguity about the value of $\boldsymbol{\xi}$ at the points at which the generalized forces are applied. If W denotes work, then the scalar product defined as

$$\mathrm{d}W = \boldsymbol{F} \cdot \mathrm{d}\boldsymbol{\xi} \tag{1.3.1}$$

represents the generalized work done by the generalized forces $\boldsymbol{F}$ as the system undergoes the infinitesimal, generalized displacement specified by $\mathrm{d}\boldsymbol{\xi}$.

The notation introduced in the above equation does not make clear whether a positive value for W indicates the work done by the system or on the system. For this reason, we specify the notation further by indicating the work done on the system as W_{in}. This is assumed to be positive when it represents the amount of non-thermal energy that the body receives from its surroundings. Accordingly, a negative value of W_{in} indicates the transfer of non-thermal energy from the body to its surroundings. The subscript "in" appended to W reminds us that W_{in} is an input of energy into the body. Likewise, the notation W_{out} indicates the work done by the forces that the body exerts upon its surroundings. Here, W_{out} is positive when these forces do work on the surroundings, which means that the body is losing energy. A negative value for W_{out}

indicates that the body is receiving non-thermal energy from the surroundings.

Of course, if a body exerts a force upon the surroundings, the surroundings exert the opposite force upon the body. Both forces are applied to the same point, so the displacement of their common point of application is the same. From the above convention concerning the signs of W_{in} and W_{out}, it follows that these two quantities are always opposite to each other:

$$W_{in} = - W_{out} \, . \tag{1.3.2}$$

Generally, the quantity W, whether it is W_{in} or W_{out}, will vary over time during a process. Therefore, it makes sense to speak of their time derivatives, $\dot{W}_{in} \equiv \mathrm{d}W_{in}(t)/\mathrm{d}t$ and $\dot{W}_{out} \equiv \mathrm{d}W_{out}(t)/\mathrm{d}t$, at any time of a process or of the infinitesimal amounts of work, $\mathrm{d}W_{in} = \dot{W}_{in}\,\mathrm{d}t$ and $\mathrm{d}W_{out} = \dot{W}_{out}\,\mathrm{d}t$, exchanged by the body in an infinitesimal time interval $\mathrm{d}t$. From the present conventions, it follows that positive values of $\dot{W}_{in}$ and $\mathrm{d}W_{in}$ denote inputs of non-thermal energy into the body, while positive values of $\dot{W}_{out}$ and $\mathrm{d}W_{out}$ indicate losses of energy from the body.

1.4 The First Law

In order to formulate the first law of thermodynamics, the notion of *state* must be introduced. Broadly speaking, we say that a system is in a given physicochemical state when all of its variable properties assume given values. Usually, the most important variable properties of a system can be expressed in terms of the system's *state variables*. The state of the system is defined as a set of values for its state variables. A more detailed definition of state will be given in the next chapter.

A *process* is an ordered sequence of states. A body is said to undergo a cyclic process, or a *cycle*, if its state at the end of the sequence is the same as it was at the beginning. It is a well-established, experimental fact that no system or body is capable of absorbing or producing a net amount of energy while operating in a cycle. This is referred to as *the first law of thermodynamics*.

There are several ways to express the first law of thermodynamics. They range from the statement that the energy of an isolated body cannot change to the impossibility of producing energy from nothing (inexistence of a perpetual motion machine of the first kind). However, a mathematical formulation of this law is needed both for the theory and the applications. To determine it, we turn our attention to any restricted portion of matter that is separated from its surroundings by a definite boundary. The boundary may either be fixed or variable over time. In any case, it is assumed to be impervious to matter, so that the system cannot exchange matter with its surroundings. A system with this property is referred to as a *closed system*.

Consider a closed system undergoing a cyclic process. Let A denote the initial state of the system, and assume that the cycle takes the system to another state B and then back to the initial state A. In general, the two branches of the cycle, namely the process from A to B and the process from B to A, will pass through different states and need not share any other common state (Fig. 1.4.1). During any infinitesimal part of this cycle, the system will absorb heat, $\mathrm{d}Q$, and non-thermal energy, $\mathrm{d}W_{\mathrm{in}}$. However, since the cycle brings the system back to its initial state, the first law of thermodynamics requires that no net energy be absorbed or produced by the system as a result of the cycle. This is expressed mathematically by the following relation:

$$\oint \mathcal{J}\mathrm{d}Q + \oint \mathrm{d}W_{\mathrm{in}} = 0\,, \tag{1.4.1}$$

where the circle over the integration symbol indicates that the integration is performed along the whole cycle. The above equation can also be written as

$$\oint (\mathrm{d}Q + \mathrm{d}W_{\mathrm{in}}) = 0\,, \tag{1.4.2}$$

where we dropped the symbol $\mathcal{J}$ with the understanding that Q is measured in energy units.

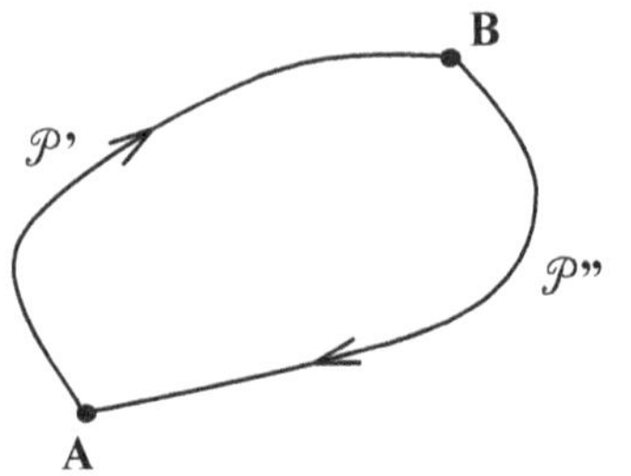

Fig. 1.4.1. A cycle passing through states A and B and the decomposition of the cycle into two branches (𝒫' from A to B and 𝒫" from B to A)

Equation (1.4.2) applies to any cycle between states A and B. Let $\mathcal{P}'$ be any process from A to B and $\mathcal{P}''$ be another process from B to A. The sequence $\mathcal{P}'$, $\mathcal{P}''$ forms a cycle (Fig. 1.4.1). Therefore, we obtain

$$\underset{\mathcal{P}'}{\int_{\mathrm{A}}^{\mathrm{B}}}(\mathrm{d}Q+\mathrm{d}W_{\mathrm{in}})+\underset{\mathcal{P}''}{\int_{\mathrm{B}}^{\mathrm{A}}}(\mathrm{d}Q+\mathrm{d}W_{\mathrm{in}})=0\,, \tag{1.4.3}$$

by applying eq. (1.4.2) to this cycle. This equation can also be written as

$$\underset{\mathcal{P}'}{\int_{\mathrm{A}}^{\mathrm{B}}}(\mathrm{d}Q+\mathrm{d}W_{\mathrm{in}})=\underset{\mathcal{P}''}{\int_{\mathrm{A}}^{\mathrm{B}}}(\mathrm{d}Q+\mathrm{d}W_{\mathrm{in}})\,, \tag{1.4.4}$$

because $\int_{\mathrm{B}}^{\mathrm{A}} = -\int_{\mathrm{A}}^{\mathrm{B}}$.

Since $\mathcal{P}'$ and $\mathcal{P}''$ are any two processes from state A to state B and from state B to state A, eq. (1.4.4) means that the quantity ΔU, defined by

$$\Delta U=\int_{\mathrm{A}}^{\mathrm{B}}(\mathrm{d}Q+\mathrm{d}W_{\mathrm{in}})\,, \tag{1.4.5}$$

has the same value no matter the process joining state A to state B (although, of course, the value of ΔU may be different for different A and B). This is equivalent to saying that, for every system, there must be a state function, U, with the following property: as the system passes from any state A to any other state B, the increase in U is given by

$$\Delta U=U_{\mathrm{B}}-U_{\mathrm{A}}=Q+W_{\mathrm{in}}\,, \tag{1.4.6}$$

irrespective of the process from A to B. In eq. (1.4.6), the quantities Q and W_{in} are the amounts of thermal and non-thermal energy, respectively, that the system absorbs in the process. Their actual values depend on the process, but, according to eq. (1.4.6), the sum of Q and W_{in} does not.

The function U introduced above is known as the *internal energy* of the system. Actually, eq. (1.4.6) only defines the variation in the system's internal energy from its initial value U_{A}. This is acceptable, because the variation in U, rather than the actual value of U, is what matters in thermodynamics. The above arguments show that the first law of thermodynamics requires that every system possess an internal energy function. This function generally differs for each system.

Clearly, eq. (1.4.6) is also a sufficient condition to ensure that no net energy is absorbed or produced by the system during a cyclic process. Thus, eq. (1.4.6) is equivalent to the first law of thermodynamics and therefore can be considered as a mathematical formulation of that law.

Equation (1.4.6) is usually referred to as the *energy balance equation*, because it expresses a balance between the energy that a system absorbs and the corresponding variation of its internal energy. The same equation can also be expressed in a number of equivalent forms. For instance, by applying eq. (1.4.6) to an infinitesimal time interval, d*t*, of any given process, we obtain

$$\mathrm{d}U = \mathrm{d}Q + \mathrm{d}W_{\mathrm{in}} \,, \tag{1.4.7}$$

which expresses the energy balance equation in differential form. This equation confirms that the sum of $\mathrm{d}Q$ and $\mathrm{d}W_{\mathrm{in}}$ is an exact differential, although neither $\mathrm{d}Q$ nor $\mathrm{d}W_{\mathrm{in}}$ are generally such when considered separately. After dividing both sides of eq. (1.4.7) by d*t*, the same equation can also be written in time-rate form:

$$\dot{U} = \dot{Q} + \dot{W}_{\mathrm{in}} \,. \tag{1.4.8}$$

Of course, all of the above equations can be expressed in terms of W_{out} rather than W_{in} simply by substituting $-W_{\mathrm{out}}$ for W_{in}, according to eq. (1.3.2).

External energy

In the above derivation, the energy that the system possesses due to its motion or its position relative to the surrounding world is ignored. The energy due to the system's motion is its *kinetic energy*. The *potential energy* is the energy that the system possesses if any field of forces (e.g., gravitational, electric, magnetic) acts on it. Both energies are sometimes referred to collectively as *external energy* to distinguish them from internal energy, *U*, which is stored in the mass of the system and is not related to the system's position or its motion in space. A general energy balance should include both internal and external energy. In general, therefore, eq. (1.4.6) should be expressed as

$$\Delta U + \Delta U^{\mathrm{e}} = Q + W_{\mathrm{in}} \,, \tag{1.4.9}$$

where U^{e} denotes the external energy of the system.

However, for most of the processes addressed in this book, the changes in the external energy of the system can either be neglected or treated separately. Often, they are due to a non-dissipative process. In that case, the so-called mechanical energy conservation theorem of classical mechanics ensures that the change in external energy, ΔU^{e}, is equal to an appropriate part of the total external work W_{in}. Under these conditions, the energy balance can still be written in the form of eq. (1.4.6), provided that the quantity W_{in} that appears in that equation is understood to be the net work done on the system in addition to the work that produces ΔU^{e}. With this proviso, we ignore the external energy contribution to the energy balance unless otherwise indicated. Accordingly, we refer to eq. (1.4.6) as the mathematical expression of the first law of thermodynamics.

1.5 Reversible and Quasi-Static Processes

We defined a process as an ordered sequence of states. The process is *reversible* if the following two conditions are met:

(1) The process can be performed according to a given sequence of states (*direct process*) and according to the inverse sequence (*inverse process*).

(2) During the inverse process, both the system and the surroundings recover, in the inverse sequence, the same physical conditions that they had in the direct process.

A system is said to be in an *equilibrium state* if all time rates of its state variables as well any state variable that represents a time rate are equal to zero. An ordered sequence of equilibrium states is called an *equilibrium process* or, equivalently, a *quasi-static process*. It should be clear that a system can never perform an equilibrium process in a finite time interval, apart from the trivial process that maintains the system fixed at a given equilibrium state.

From the above definition of reversibility, it follows that a reversible process of any real system must be an equilibrium process. Otherwise, in the inverse process, the non-vanishing time rates of its state variables would assume values opposite to those they assumed in the direct process, making it impossible to meet condition (2). The notion of reversible process involves both the system and its surroundings. It plays a central role in thermodynamics. However, it is of little or no relevance to theories, like classical mechanics, where the change in the world that surrounds the system is not considered.

Because a reversible process must be quasi-static, it cannot be executed in a finite interval of time. Therefore, reversible processes are not real processes; rather, they are mere theoretical abstractions that can never be realized in practice no matter how useful they may be for theoretical considerations.

It should also be observed that an equilibrium process need not be reversible either. Consider a bar of heat-conducting material in a steady-state condition at a non-uniform temperature with its ends kept in contact with two heat reservoirs at different temperatures. The bar is in an equilibrium state, because the time rates of all of its state variables vanish in steady-state conditions. Therefore, the process of the bar is quasi-static. It is also trivially invertible since the state of the bar remains constant (Appendix A). It is not reversible, though, because it does not meet the reversibility condition (2) above. This is due to the flow of heat through the bar, which produces changes in the surroundings even if the state of the bar remains unchanged.

The preceding example shows that there are equilibrium states that cannot belong to any reversible process. In the example, this was due to the presence of a non-vanishing, stationary, temperature gradient. More generally, a given equilibrium state of a system cannot belong to a reversible process if the system's surroundings cannot be prevented from undergoing changes in their states while the system is kept at the equilibrium state. This important point has to be considered appropriately in continuum thermodynamics, where a general macroscopic approach to non-homogeneous systems is sought. The same point, however, can be ignored in classical thermodynamics, because the latter is concerned with homogeneous systems undergoing homogeneous processes. In that case, the lack of spatial gradients in the state variables of the system makes any equilibrium state of the system compatible with the equilibrium of its surroundings. In other words, every equilibrium state of a system in spatially uniform conditions, i.e., every equilibrium homogeneous state of the system, can belong to a reversible process.

1.6 The Second Law

The second law of thermodynamics is closely related to the way in which spontaneous heat transfer takes place in nature. When two bodies at different temperatures are placed in thermal contact, heat spontaneously flows from the hotter body to the colder body. Moreover, any process that brings the heat back to the hotter body invariably requires the expenditure of some non-thermal energy. The first law of thermodynamics does not impose any privileged direction to the flow of heat. It only is concerned with the balance of the energy that enters or leaves the system, quite apart from the kind of energy that is considered and from where it comes or to where it goes. In particular, the first law of thermodynamics holds true irrespective of the temperature of the body that absorbs or supplies energy. This means, for instance, that the first law does not exclude that a cold body may become colder simply as a result of being put in thermal contact with a hotter body and that the hotter body may become still hotter as a result.

Surely, some other fundamental physical principle must be at work here to prevent such an unphysical occurrence. This principle is the *second law of thermodynamics*. In the words of Clausius:

> No process is possible whose sole result is the transfer of heat from a colder to a hotter body.

An alternative way to state the same law is the so-called Kelvin-Planck postulate:

> It is impossible to extract heat from a heat reservoir at uniform temperature and transform it entirely into work by means of an engine that operates in a cycle.

A hypothetical engine capable of operating in defiance of this postulate is usually referred to as a perpetual motion machine of the second kind. If it existed, such machine would be as useful as a perpetual motion machine of the first kind (Section 1.4), since it could transform the almost unlimited thermal energy that is freely available at ambient temperature from the surrounding world into useful work. For proof that the Clausius postulate and the Kelvin-Planck postulate are equivalent, see, e.g., [17, Sect. 1.10] and [74, Sect. 7.7].

To express the above postulates mathematically, let $\mathrm{d}Q$ denote the amount of heat that the system exchanges with its surroundings in any infinitesimal part of any given process. Moreover, let T be the absolute temperature of the part of the system's surface through which the heat $\mathrm{d}Q$ is exchanged. In the absence of thermal equilibrium, T will differ from the temperature of other points of the system. However, T will coincide with the system's temperature if the latter is uniform. Whether or not the temperature of the system is uniform, it can be shown that a condition that is both necessary and sufficient to satisfy the above postulates is that the following relation:

$$\oint \frac{\mathrm{d}Q}{T} \leq 0 \tag{1.6.1}$$

should be met by every system undergoing a cyclic process (see, e.g., Sects 1.11-1.14 of [17]). It is shown, moreover, that the condition:

$$\oint \frac{\mathrm{d}Q}{T} = 0 \tag{1.6.2}$$

is both necessary and sufficient for a cycle to be reversible. Note that the temperature of the system may vary during the cycle. However, for the cycle to be reversible, T must be uniform throughout the system (recall that every state of a reversible process must be an equilibrium, homogeneous state and, hence, a uniform-temperature state). A more general condition for the reversibility of any process, whether a cycle or not, is given later in this section.

Although every process can be part of a cycle, the expression of the second law of thermodynamics in the form (1.6.1) is impractical for use with non-cyclic processes. An alternative expression that applies to every infinitesimal part of any process can be obtained introducing a new state function of the system, called *entropy*. The main steps of the procedure are described below.

To begin, let us confine our attention to reversible cycles. In this case, eq. (1.6.2) holds true. It follows that there must exist a function S of the state variables of the system with the property that

$$\Delta S = S_{\mathrm{B}} - S_{\mathrm{A}} = \int_{\mathrm{A}}^{\mathrm{B}} \frac{\mathrm{d}Q}{T}, \tag{1.6.3}$$

where A and B are any two states belonging to the cycle. In classical thermodynamics, function S is referred to as the entropy of the system. The proof of eq. (1.6.3) is analogous to that leading to eq. (1.4.5) from eq. (1.4.2), and it is not repeated here. However, since eq. (1.6.2) only applies to reversible cycles, an important difference is that now A and B must be two states that can be joined by a reversible process, i.e., two equilibrium, homogeneous states. This contrasts with the case of internal energy, where A and B can be any two states of the system.

Because the entropy thus introduced is defined for equilibrium, homogeneous states, it can be applied to determine the entropy changes in any process, whether reversible or not, provided that the process takes place between such states. This definition of entropy is perfectly adequate to classical thermodynamics. The same definition must be generalized appropriately if we want to make it applicable to non-equilibrium and/or spatially non-uniform states. However, much of the analysis of this book is based on classical thermodynamics; therefore, the entropy definition given above is adequate for the present purpose.

Equation (1.6.3) applies, in particular, when states A and B are the initial and the final states, respectively, of an infinitesimal part of any reversible cycle passing through them. In that case, it reduces to

$$\mathrm{d}Q = T\,\mathrm{d}S\,. \tag{1.6.4}$$

Since any infinitesimal part of a reversible process can always be a part of a reversible cycle, it follows that eq. (1.6.4) is a general condition for the reversibility of any process. On the other hand, if the process between two equilibrium states is performed irreversibly, from relations (1.6.1) and (1.6.3) it is not difficult to infer that

$$\int_{\mathrm{A}}^{\mathrm{B}} \frac{\mathrm{d}Q}{T} < \Delta S\,. \tag{1.6.5}$$

In particular, if states A and B are in the same infinitesimal neighbourhood, then relation (1.6.5) yields

$$\mathrm{d}Q < T\,\mathrm{d}S\,, \tag{1.6.6}$$

which applies instead of eq. (1.6.4) when the process is irreversible.

From eqs (1.6.4) and (1.6.6), it is concluded that any process that takes place between any two equilibrium, homogeneous states that belong to the same infinitesimal neighbourhood must always satisfy the following relation:

$$\mathrm{d}Q \le T\,\mathrm{d}S\,, \tag{1.6.7}$$

which is usually referred to as the *entropy inequality*. In this relation, the equality sign applies when the process is reversible.

Conversely, in the restricted range of the states that can be joined by a reversible process, the entropy inequality is both a necessary and sufficient condition for relation (1.6.1) to be fulfilled. Therefore, inequality (1.6.7) represents a mathematical expression of the second law of thermodynamics in a form that applies directly to any infinitesimal part of any process. It applies whether or not the process is reversible, provided that the starting and the ending states of the considered part of the process are two equilibrium, homogeneous states; i.e., two states that, at least in principle, can be joined reversibly.

1.7 The Third Law and the Calorimetric Measurement of Entropy

As was the case with the first law and internal energy, eq. (1.6.3) defines an entropy increase ΔS rather than the entropy S itself. There is an important difference between the two cases, though. Equation (1.4.5) can be used to calculate ΔU in any process, while eq. (1.6.3) can only give the correct value of ΔS when applied to a reversible process. This may pose some problems in the practical determination of ΔS, since it may not be easy or feasible to make the system follow a reversible process joining any two given states, because the process must be quasi-static (Section 1.5). The problem is solved by the *third law of thermodynamics*. The third law can be expressed in a number of different ways, but the following is the most suitable for the present purposes:

> In equilibrium conditions, the entropy of a system approaches to zero as the system's temperature approaches to absolute zero, irrespective of the value of the remaining state variables.

This law has several important consequences. In particular, it enables us to determine the value of the entropy at any equilibrium state of the system. To illustrate this, let $\boldsymbol{\xi}$ collectively denote the state variables of the system other than T. Therefore, any state of the system will be defined by the pair$\{\boldsymbol{\xi}, T\}$, so that we can write:

$$S = S(\boldsymbol{\xi}, T). \tag{1.7.1}$$

Let us then take state $\{\boldsymbol{\xi}, 0\text{ K}\}$ as A and state $\{\boldsymbol{\xi}, T\}$ as B. From eq. (1.6.3) we obtain

$$S(\boldsymbol{\xi}, T) = \int_{0\text{ K}}^{T} \frac{\mathrm{d}Q}{T}, \tag{1.7.2}$$

because $S(\boldsymbol{\xi}, 0) = 0$ for every value of $\boldsymbol{\xi}$, according to the third law. Here, the integral refers to a reversible process joining the two end states, as this is the condition under which eq. (1.6.3) applies. In the considered case, the two end states share the same values of $\boldsymbol{\xi}$, which means that the process that joins them can be a reversible heating process at constant $\boldsymbol{\xi}$ from $T = 0$ K to T. In practice, such a process is always feasible with any degree of accuracy, provided that the system is heated slowly enough to keep its temperature uniform during the process.

Alternatively, a reversible cooling process from T to 0 K could be used. This changes the sign of the integral on the right-hand side of eq. (1.7.2), which is then rewritten as

$$S(\boldsymbol{\xi}, T) = -\int_{T}^{0\,\mathrm{K}} \frac{\mathrm{d}Q}{T} . \tag{1.7.3}$$

The quantity $\mathrm{d}Q$ relevant to a heating/cooling process can be calculated from the values of the specific heat capacity (specific heat) of the system for constant values of the variables $\boldsymbol{\xi}$. We denote this specific heat as $c_{\boldsymbol{\xi}}$:

$$c_{\boldsymbol{\xi}} = c_{\boldsymbol{\xi}}(T) = \frac{1}{M} \frac{\mathrm{d}Q}{\mathrm{d}T}\bigg|_{\boldsymbol{\xi}=\mathrm{const}} , \tag{1.7.4}$$

where M is the mass of the system. For $\boldsymbol{\xi}$=const, we have that $\mathrm{d}Q = \mathrm{d}U$, as evident from eqs (1.3.1) and (1.4.7). Accordingly, we can also express $c_{\boldsymbol{\xi}}$ as

$$c_{\boldsymbol{\xi}} = c_{\boldsymbol{\xi}}(T) = \frac{1}{M} \frac{\partial U}{\partial T}\bigg|_{\boldsymbol{\xi}=\mathrm{const}} . \tag{1.7.5}$$

From eq. (1.7.4) the amount of heat absorbed in any infinitesimal interval of the considered heating/cooling process can be expressed as:

$$\mathrm{d}Q = M\, c_{\boldsymbol{\xi}}\, \mathrm{d}T . \tag{1.7.6}$$

In ideal gases, the variables $\boldsymbol{\xi}$ reduce to just one variable. This can be the pressure of the gas, p, or the volume of the gas, V. In this case, the specific heat at constant $\boldsymbol{\xi}$ usually is denoted as c_p or c_v, respectively.

However, not all of the heat absorbed by the system in a heating process can be expressed in the form of eq. (1.7.6). At certain temperatures, the system may undergo phase changes, such as solid-to-liquid transitions, liquid-to-gas transitions, or allotropic transitions between different solid phases. In a phase change, the system absorbs a definite amount of heat, say Q_t, at a constant temperature T_t. The heat absorbed depends on the system and the phase transition under consideration. Their overall contribution to the integral in the right-hand side of eq. (1.7.2) can be expressed as:

$$\sum_{t=1}^{n} \frac{Q_t}{T_t}, \qquad (1.7.7)$$

where n is the total number of phase transitions that the system undergoes in the heating process from 0 K to T. The index t in the above formula runs from 1 to n and it refers to the phase transition that occurs at temperature T_t. From eqs (1.7.6) and (1.7.7), it follows that, in the presence of phase transitions, eq. (1.7.2) becomes:

$$S(\boldsymbol{\xi}, T) = M\int_{0\,\mathrm{K}}^{T} \frac{c_{\xi}}{T}\mathrm{d}T + \sum_{t=1}^{n} \frac{Q_t}{T_t}. \qquad (1.7.8)$$

This equation determines the *absolute entropy, S*, of the system at any equilibrium state{$\boldsymbol{\xi}$, T}. If the reference is to a cooling process rather than a heating process, the equation is similar, but the right-hand side of the equation obtained from eq. (1.7.3) has the opposite sign.

This method of calculating entropy is referred to as the *calorimetric method.* It is the only method that is fully consistent with the macroscopic approach at the basis of classical thermodynamics. The statistical method is another method frequently used to calculate the entropy of a system. It is based on statistical mechanics and quantum theory. Although it generally gives entropy values that are surprisingly close to those obtained by the calorimetric method, there are some exceptions. In the case of a discrepancy, both methods should be considered in order to determine why they fail to give the same answer and to determine which of the two answers should be taken as the correct one in the theory under consideration.

The practical value of the calorimetric method stems from the fact that specific heat data for almost all known materials are available in the literature. Often, these data refer to the specific heat at constant pressure (typically, 1 atm) and for a discrete set of temperatures. From well-established theoretical or semi-empirical formulae, these values can be extrapolated to other states of interest. In particular, the extrapolation procedure eliminates the difficulty of determining the ratio dQ/T when the temperature is near absolute zero, thereby making the calorimetric determination of absolute entropy feasible in practice.

It should be noted that the third law of thermodynamics provides a guarantee that any two states of a system that are homogeneous and at equilibrium, say state $\{\boldsymbol{\xi}_1, T_1\}$ and state $\{\boldsymbol{\xi}_2, T_2\}$, always can be joined by at least one reversible process. Conceptually, this process can be constructed as follows. First, a reversible cooling process at constant $\boldsymbol{\xi}$ brings the system from state $\{\boldsymbol{\xi}_1, T_1\}$ to state $\{\boldsymbol{\xi}_1, 0\text{ K}\}$. This cooling process is followed by an isothermal process at 0 K, in which $\boldsymbol{\xi}$ changes from $\boldsymbol{\xi}_1$ to $\boldsymbol{\xi}_2$, leading the system to state $\{\boldsymbol{\xi}_2, 0\text{ K}\}$. Such isothermal process at 0 K is reversible because the third law requires that $\mathrm{d}S = 0$ in any process at $T = 0$ K and because $\mathrm{d}Q = 0$. The last condition follows from relation (1.6.7) and from the fact that $\mathrm{d}Q$ cannot be less than zero, since no heat can be extracted from a body at $T = 0$ K. Finally, a reversible heating process from state $\{\boldsymbol{\xi}_2, 0\text{ K}\}$ to state $\{\boldsymbol{\xi}_2, T_2\}$ completes the reversible process between state $\{\boldsymbol{\xi}_1, T_1\}$ and state $\{\boldsymbol{\xi}_2, T_2\}$.

1.8 Upper Bound to the Heat Absorption Rate

Following directly from the second law of thermodynamics, there is a limit to the maximum time rate at which any real system can absorb heat. To show this let us divide both sides of relation (1.6.7) by $\mathrm{d}t$ to obtain:

$$\dot{Q} \leq T\dot{S}\,. \tag{1.8.1}$$

That is, the heat input rate, $\dot{Q}$, can never exceed the entropy rate, $\dot{S}$, multiplied by the absolute temperature at which the heat is being absorbed by the system. Therefore, the product $T\dot{S}$ sets an upper bound to the maximum time rate at which heat can be absorbed by the system from the surroundings.

Consider then the set of all isothermal processes that join any two given states of the system at a given temperature T. Time integration of relation (1.8.1) along any such process yields

$$Q \leq T\Delta S\,, \tag{1.8.2}$$

where Q represents the amount of heat absorbed during the process, and ΔS is the entropy difference between the final and the initial state of the system. Of course, Q will be different for different processes. The quantity $T\Delta S$, however, is the same for all processes in the considered set, because it depends only on the initial and final states of the system. The quantity $T\Delta S$ represents, therefore, the upper bound to the maximum amount of heat that the system can absorb in any isothermal process that joins the two given states.

A similar restriction holds true when non-isothermal processes are considered. In that case, however, the system temperature changes during the process and the above upper bound to the maximum absorbable heat should then be referred to each infinitesimal part of the process. No limit whatsoever is imposed by inequality (1.8.1) or, equivalently, by inequality (1.6.7), on the negative values of $\dot{Q}$. Therefore, the time rate at which a system can give off heat is not limited *a priori* by any physical law.

The above discussion provides an additional physical reason that helps explain why entropy—or some other equivalent quantity—is indispensable to thermodynamics. To wit, nature sets no limit on the rate at which a system can turn non-thermal energy into heat (i.e., to the negative values of $\dot{Q}$) as long as, of course, energy is available for that purpose. The same is not true for the heat absorption rate. The rate at which a system absorbs heat—which thus becomes part and parcel of the system's internal energy—cannot exceed the upper limit $T\dot{S}$. This interpretation of the second law in the form (1.6.7) is, in all respects, equivalent to and as fundamental as the condition that no heat can flow spontaneously from cold to hot or that it is impossible to build a perpetual motion machine of the second kind.

As far as the second law is concerned, the actual value of the above upper limit is immaterial. In fact, each system has its own limit, which is completely defined by its entropy function once the process is specified. As discussed in more detail in Chapter 11, this interpretation of the second law gives the entropy a clear physical meaning that is entirely within the scope of macroscopic thermodynamics. It also explains why no thermodynamic system can be physically realistic if the notion of entropy is ignored.

Admittedly, to set a bound to $\dot{Q}$ through the time derivative of an abstract function, such as S, times the absolute temperature, may not be the most direct way to proceed. However, the path of science is not always straightforward, and historical reasons exert a strong bias. In the present case, the formal analogy of eq. (1.6.2) with eq. (1.4.2) was undoubtedly crucial to the birth of entropy. That came at a price, however, because entropy was defined for equilibrium states only and appeared in the theory without a definite physical connotation. Its introduction was as formal as the mathematical analogy between eqs (1.6.2) and (1.4.2) from which it originated. This gave rise to all sorts of conjectures and interpretations about the meaning of entropy, most of which are still alive and well today, much to the bewilderment of the unaware student. A more direct approach to the upper bound of the heat absorption rate is presented in Chapter 11.

The asymmetry in the capacity of a system to absorb and supply heat stands at the very root of the irreversibility of natural phenomena. For a process to be reversible, the amount of heat absorbed at each time of the process must be opposite to the corresponding amount of heat absorbed in the inverse process. Thus, if $\dot{Q}_{\mathrm{D}}$ and $\dot{Q}_{\mathrm{I}}$ are the time rates of the heat absorbed by the system at corresponding points of the direct process and the inverse process, respectively, then the reversibility of the process requires that

$$\dot{Q}_{\mathrm{I}} = -\dot{Q}_{\mathrm{D}} . \qquad (1.8.3)$$

Of course, reversibility also implies that an analogous condition should be met by all of the other time rates involved in the process. Thus, the following equation concerning the entropy rate should apply to a reversible process:

$$\dot{S}_{\mathrm{I}} = -\dot{S}_{\mathrm{D}} . \qquad (1.8.4)$$

From inequality (1.8.1), it follows that

$$\dot{Q}_{\mathrm{D}} \leq T\dot{S}_{\mathrm{D}} \qquad (1.8.5)$$

and

$$\dot{Q}_{\mathrm{I}} \leq T\dot{S}_{\mathrm{I}} \qquad (1.8.6)$$

should be met in the direct process and the inverse process, respectively. Therefore, if in the direct process we have that

$$\dot{Q}_{\mathrm{D}} < T\dot{S}_{\mathrm{D}} , \qquad (1.8.7)$$

that is

$$-\dot{Q}_{\mathrm{D}} > -T\dot{S}_{\mathrm{D}} , \qquad (1.8.9)$$

then, from relations (1.8.4), (1.8.6) and (1.8.9) we can conclude that

$$-\dot{Q}_{\mathrm{D}} > T\dot{S}_{\mathrm{I}} \geq \dot{Q}_{\mathrm{I}} . \qquad (1.8.10)$$

This shows that the reversibility condition (1.8.3) cannot be met if inequality (1.8.7) holds true in the direct process. Hence, any process that satisfies inequality (1.8.7) or, more generally, any process in which

$$\dot{Q} < T\dot{S} \tag{1.8.11}$$

cannot be reversible.

By the same arguments, it can be shown that conditions (1.8.3) and (1.8.4) are compatible with the second law (1.8.1) only if the latter is met as an equation:

$$\dot{Q} = T\dot{S}\,. \tag{1.8.12}$$

It must be concluded that eq. (1.8.12) is a necessary and sufficient condition for the reversibility of a process, in agreement with what is stated in Section 1.6.

1.9 Dissipation

Let A = $\{\boldsymbol{\xi}, T\}$ and B = $\{\boldsymbol{\xi} + \mathrm{d}\boldsymbol{\xi}, T + \mathrm{d}T\}$ be any two states of the system that belong to the same infinitesimal neighbourhood in state space. Any process from A to B will produce the following changes in U and S:

$$\mathrm{d}U = U(\boldsymbol{\xi} + \mathrm{d}\boldsymbol{\xi}, T + \mathrm{d}T) - U(\boldsymbol{\xi}, T) \tag{1.9.1}$$

and

$$\mathrm{d}S = S(\boldsymbol{\xi} + \mathrm{d}\boldsymbol{\xi}, T + \mathrm{d}T) - S(\boldsymbol{\xi}, T). \tag{1.9.2}$$

Conversely, any process from B to A will produce the changes

$$\mathrm{d}U' = U(\boldsymbol{\xi}, T) - U(\boldsymbol{\xi} + \mathrm{d}\boldsymbol{\xi}, T + \mathrm{d}T) = -\,\mathrm{d}U \tag{1.9.3}$$

and

$$\mathrm{d}S' = S(\boldsymbol{\xi}, T) - S(\boldsymbol{\xi} + \mathrm{d}\boldsymbol{\xi}, T + \mathrm{d}T) = -\,\mathrm{d}S\,. \tag{1.9.4}$$

Every process, whether from A to B or from B to A, must comply with the first and second laws. If the process from A to B takes place reversibly, we have that

$$\mathrm{d}Q = T\,\mathrm{d}S. \tag{1.9.5}$$

From the second law (1.6.7), it then follows that the system absorbs (supplies, if $\mathrm{d}S < 0$) the maximum (minimum) amount of heat possible if the process is performed reversibly.

On the other hand, from the first law (1.4.7), we obtain

$$\mathrm{d}W_{\text{in}} = \mathrm{d}U - \mathrm{d}Q. \tag{1.9.6}$$

From this equation and from eq. (1.9.5) we can conclude that, if $\mathrm{d}S > 0$, the amount of work absorbed by the system in the process from A to B reaches its minimum value when the process takes place reversibly. Conversely, if in going from A to B we have that $\mathrm{d}S < 0$, then the system supplies the largest amount of work to the surroundings when the process is performed reversibly.

If the inverse process (i.e., the process from B to A) is not performed reversibly, the amount of heat that the system absorbs in the same process is

$$\mathrm{d}Q' < T\,\mathrm{d}S' = -\,T\,\mathrm{d}S. \qquad (1.9.7)$$

Therefore, from this and from eq. (1.4.7) we calculate that the corresponding amount of work absorbed by the system is

$$\mathrm{d}W'_{\mathrm{in}} = \mathrm{d}U' - \mathrm{d}Q' > -\,\mathrm{d}U + T\,\mathrm{d}S = -\,\mathrm{d}W_{\mathrm{in}}\,, \qquad (1.9.8)$$

as immediately follows from eqs (1.9.3), (1.9.4), and (1.9.6). In view of this and considering eq. (1.3.2), the work produced by the system if the process from A to B takes place irreversibly is given by

$$\mathrm{d}W'_{\mathrm{out}} = -\,\mathrm{d}W'_{\mathrm{in}} < \mathrm{d}W_{\mathrm{in}}\,. \qquad (1.9.9)$$

However, if the process is performed reversibly, then

$$\mathrm{d}W'_{\mathrm{out}} = \mathrm{d}W_{\mathrm{in}}\,. \qquad (1.9.10)$$

In this case, $\mathrm{d}Q' = -\,\mathrm{d}Q = -\,T\mathrm{d}S$.

The quantity $\mathrm{d}D$ defined by:

$$\begin{aligned}\mathrm{d}D &= \mathrm{d}W_{\mathrm{in}} - \mathrm{d}W'_{\mathrm{out}} = \mathrm{d}U - T\,\mathrm{d}S + \mathrm{d}U' - \mathrm{d}Q' \\ &= \mathrm{d}U - T\,\mathrm{d}S - \mathrm{d}U - \mathrm{d}Q' = T\,\mathrm{d}S' - \mathrm{d}Q'\end{aligned} \qquad (1.9.11)$$

is the work lost to perform the infinitesimal cycle that goes from A to B reversibly and then from B back to A irreversibly. If the entire cycle were performed reversibly, then eq. (1.9.10) would apply, and no work would be lost ($\mathrm{d}D = 0$). It can then be concluded that the quantity $\mathrm{d}D$ represents the work dissipated in the irreversible part of the cycle, i.e., from B to A.

Since the role of A and B can be interchanged, we can remove the prime signs from the last equation (1.9.11) and conclude that the amount of non-thermal energy dissipated in any process that joins any two states in the same infinitesimal neighbourhood is given by

$$\mathrm{d}D = T\,\mathrm{d}S - \mathrm{d}Q. \tag{1.9.12}$$

Of course

$$\mathrm{d}D \geq 0, \tag{1.9.13}$$

as immediately follows from relations (1.9.12) and (1.6.7). The equality sign in relation (1.9.13) holds true if the process is performed reversibly.

The quantity D entering eq. (1.9.11) is referred to as *dissipation*. We also can rewrite relation (1.9.12) as

$$\dot{D} = T\dot{S} - \dot{Q} \geq 0\,. \tag{1.9.14}$$

This can be integrated along a finite process to yield

$$D = \int \dot{D}\,\mathrm{d}t = \int (T\dot{S} - \dot{Q})\,\mathrm{d}t \geq 0\,, \tag{1.9.15}$$

which enables us to calculate the dissipation that is produced by any finite process that joins any two states of the system. For reversible processes, $D = 0$. In any irreversible process, $D > 0$. For this reason, these processes are also called dissipative.

1.10 Heat Absorbed at Constant Pressure and Constant Volume

When dealing with compressible fluids or chemical reactions, some mechanical work supplied to or by the system is usually done by the external pressure, p, against the change of volume of the system (*volume work*). Any infinitesimal work input, $\mathrm{d}W_{\mathrm{in}}$, can accordingly be written as follows:

$$\mathrm{d}W_{\mathrm{in}} = \mathrm{d}W_{\mathrm{in}}^{\mathrm{e}} - p\,\mathrm{d}V\,, \tag{1.10.1}$$

where $W_{\mathrm{in}}^{\mathrm{e}}$ denotes the work input in excess of volume work. The negative sign in the front of $p\,\mathrm{d}V$ on the right-hand side of this equation stems from the fact that p is assumed to be positive when representing compression. Thus, a negative value of $\mathrm{d}V$ is required for p to make a positive contribution to $\mathrm{d}W_{\mathrm{in}}$.

By means of eqs (1.10.1) and (1.4.7), the heat that the system absorbs in each infinitesimal part of a process can be expressed as:

$$\mathrm{d}Q = \mathrm{d}U - \mathrm{d}W_{\mathrm{in}}^{\mathrm{e}} + p\mathrm{d}V. \qquad (1.10.2)$$

It is often convenient to introduce a function H, called *enthalpy,* which is defined as

$$H = U + p\,V. \qquad (1.10.3)$$

Enthalpy is clearly a state function of the system since U is a state function and p and V are state variables. In terms of H, eq. (1.10.2) can be written as

$$\mathrm{d}Q = \mathrm{d}H - \mathrm{d}W_{\mathrm{in}}^{\mathrm{e}} - V\mathrm{d}p. \qquad (1.10.4)$$

Constant pressure processes

Many processes occur at constant pressure, in particular at atmospheric pressure. For constant pressure processes, eq. (1.10.4) reduces to

$$\mathrm{d}Q = \mathrm{d}H\big|_p - \mathrm{d}W_{\mathrm{in}}^{\mathrm{e}}. \qquad (1.10.5)$$

Here, the subscript p attached to $\mathrm{d}H$ emphasizes that the latter quantity is intended to be calculated by keeping p constant. There are many situations in which $\mathrm{d}W_{\mathrm{in}}^{\mathrm{e}} \neq 0$. However, the case in which $\mathrm{d}W_{\mathrm{in}}^{\mathrm{e}} = 0$ is the most common condition at which a spontaneous chemical reaction usually occurs. When $\mathrm{d}W_{\mathrm{in}}^{\mathrm{e}} = 0$, eq. (1.10.5) reduces to

$$\mathrm{d}Q = \mathrm{d}H\big|_p. \qquad (1.10.6)$$

For a finite process, this integrates to

$$Q = \Delta H\big|_p, \qquad (1.10.7)$$

where Q is the heat absorbed in the process.

When applicable, eq. (1.10.7) provides a way to measure the change in the system's enthalpy (ΔH) from the heat that the system absorbs. A negative value of ΔH means that a negative value of heat was absorbed, indicating that the system supplied heat to the surroundings. Thus, a reaction that produces a decrease in the enthalpy of the reacting material must be exothermic, meaning that it must release heat to the surroundings ($Q < 0$).

Since H is a state function, ΔH is independent of the process that joins the initial state to the final state of the system. From eq. (1.10.7), it follows that every constant-pressure process between any two given states makes the system absorb the same amount of heat if $W_{\text{in}}^{\text{e}} = 0$ in the process, i.e., if no work other than $-p\text{d}V$ is done on the system. If the process is performed reversibly, then dQ is given by eq. (1.9.5). Then, from eq. (1.10.7), we infer that the following relation

$$\text{d}H\big|_p = T\,\text{d}S\big|_p \tag{1.10.8}$$

must apply to every reversible, constant-pressure process in which $\text{d}W_{\text{in}}^{\text{e}}$ is zero. If, in addition to being isobaric, the process is also isothermal, eq. (1.10.8) integrates to yield

$$\Delta H\big|_{p,T} = T\,\Delta S\big|_{p,T}\,. \tag{1.10.9}$$

Processes at p = 0

When the process occurs at zero pressure, $pV = 0$. By differentiating eq. (1.10.3), it then follows that

$$\text{d}H\big|_{p=0} = \text{d}U\big|_{p=0}\,. \tag{1.10.10}$$

Constant volume processes

If the process takes place at a constant volume and $\text{d}W_{\text{in}}^{\text{e}} = 0$, we obtain the following from eq. (1.10.3):

$$\text{d}H = \text{d}U + V\text{d}p. \tag{1.10.11}$$

In view of eq. (1.10.4), this means that

$$Q = \Delta U\big|_V\,, \tag{1.10.12}$$

since $W_{\text{in}}^{\text{e}} = 0$. In this case, Q depends only on the end states of the process, because U is a state function. This is also obvious from eq. (1.4.6), because no work is performed by the process.

1.11 Isothermal Thermodynamics

Many processes of practical interest take place at constant temperature. In biology, isothermal processes are the rule rather than the exception. They also are quite common in many branches of chemistry and engineering. It is useful, therefore, to record here some of the main consequences of the laws of thermodynamics when it comes to constant temperature processes.

From the first two laws, eqs (1.4.7) and (1.6.7), respectively, it generally follows that

$$\mathrm{d}U - T\,\mathrm{d}S \leq \mathrm{d}W_{\mathrm{in}}\,. \tag{1.11.1}$$

The equality sign in this relation applies to reversible processes only. For isothermal processes, the same relation can be written as

$$\mathrm{d}(U - TS)_T \leq \mathrm{d}W_{\mathrm{in}}\,. \tag{1.11.2}$$

Since both U and S are state functions and T is a state variable, the difference, $U - TS$, is a state function and is usually referred to as *Helmholtz free energy*. We denote it by Ψ:

$$\Psi = U - TS\,. \tag{1.11.3}$$

In terms of Ψ, inequality (1.11.2) is read as:

$$\mathrm{d}\Psi\big|_T \leq \mathrm{d}W_{\mathrm{in}}\,. \tag{1.11.4}$$

For a finite process, it is rewritten as

$$\Delta\Psi\big|_T \leq W_{\mathrm{in}}\,. \tag{1.11.5}$$

This means that more work must be done on a system if the isothermal process takes place irreversibly rather than reversibly.

In view of eq. (1.3.2), the same relation also means that, for isothermal processes,

$$\mathrm{d}W_{\mathrm{out}} \leq -\mathrm{d}\Psi\big|_T\,, \tag{1.11.6}$$

and hence

$$W_{\mathrm{out}} \leq -\Delta\Psi\big|_T\,. \tag{1.11.7}$$

That is, the maximum work that a system can produce in an isothermal process equals the corresponding loss in its Helmholtz free energy. Such an amount of work can be produced only in a reversible process.

When all external forces vanish, or when p is the only agent that does work but the process takes place at a constant volume, then $\mathrm{d}W_{out}$ vanishes. In this case, we infer the following from relation (1.11.4)

$$\mathrm{d}\Psi|_T \le 0. \tag{1.11.8}$$

This condition must be met at any time during the process. This means, in particular, that in isothermal conditions a state of equilibrium of a system that it is free of external forces (which implies that $W_{out} = 0$, no matter the process) must be a state of a minimum free energy, Ψ. In the absence of external forces the system cannot move from that state, according to relation (1.11.8).

Isothermal constant pressure processes

More generally, for any isothermal process that occurs at constant pressure we infer from eqs (1.11.1), (1.10.1), and (1.10.3) that:

$$\mathrm{d}G\big|_{p,T} = \mathrm{d}(H - T\,S)_{p,T} \le \mathrm{d}W_{in}^{e}. \tag{1.11.9}$$

Here G is the so-called *Gibbs free energy* function (also referred to as the *free enthalpy* function), defined by any of the following equalities:

$$G = H - TS = U + p\,V - TS = \Psi + p\,V. \tag{1.11.10}$$

For reversible processes, relation (1.11.9) gives

$$\mathrm{d}G\big|_{p,T} = \mathrm{d}W_{in}^{e}. \tag{1.11.11}$$

For a finite process and in terms of the work done by the system, we obtain

$$\Delta G\big|_{p,T} = -\,\Delta W_{out}^{e}, \tag{1.11.12}$$

thanks to eq. (1.3.2). Equation (1.11.12) is used extensively in chemistry to measure the change in the Gibbs free energy that occurs as a result of a chemical reaction. This involves making the reaction occur reversibly in a galvanic cell and measuring the electric energy output, W_{out}^{e}.

Isothermal processes at constant pressure and volume

If an isothermal process takes place both at constant p and at constant V, then from eqs (1.11.10), (1.10.1) and (1.10.3) we obtain

$$\mathrm{d}G = \mathrm{d}\Psi. \tag{1.11.13}$$

Hence,

$$\Delta G = \Delta \Psi. \tag{1.11.14}$$

For the same process, we also obtain

$$\mathrm{d}H = \mathrm{d}U. \tag{1.11.15}$$

Therefore,

$$\Delta H = \Delta U. \tag{1.11.16}$$

Isothermal processes at constant pressure when $W_{\text{in}}^{\text{e}} \equiv 0$

When p is the only agency that does work on the system, then the quantity W_{in}^{e} must vanish identically, as follows from eq. (1.10.1). In this case, from relation (1.11.9) we infer that

$$\mathrm{d}G\,|_{p,T} \leq 0. \tag{1.11.17}$$

For a finite process,

$$\Delta G\,|_{p,T} \leq 0. \tag{1.11.18}$$

Again, the equality sign in the above relations is only applicable when the process takes place reversibly.

Relation (1.11.17) implies that, at constant pressure and under isothermal conditions, the equilibrium state of any system must be a state of minimum Gibbs free energy if no external forces other than p act on the system. In particular, if $p = 0$, then $G \equiv \Psi$. In this case, condition (1.11.17) coincides with condition (1.11.8). However, if W_{in}^{e} does not identically vanish, then condition (1.11.17) is no longer applicable. In that case, the state of equilibrium is controlled by the more general relation (1.11.9).

When dealing with systems in which chemical reactions take place, the difference, a, between the Gibbs free energy of the reactants and that of the products in a balanced chemical equation is called *chemical affinity*. This quantity is taken as a measure of the tendency of a chemical reaction to occur spontaneously when the reacting material is at constant p and T. If a happens to be negative, the reaction will tend to proceed in the reverse direction, i.e., from products to reactants, as inferred from the fact that the process must meet relations (1.11.17) and (1.11.18), see, e.g., [17, Sect. 3.4] or [23, Sect. 9b]. The *reaction Gibbs free energy* is more widely used than a for the same purposes. It is usually denoted as $\Delta_{\text{r}}G$ and is defined as the difference between the Gibbs free energy of the reaction products and that of the reactants

(Appendix C.4.) Clearly, $\Delta_r G = -a$. Thus, the more negative $\Delta_r G$ is, the greater the tendency of a reaction to occur. Note that $\Delta_r G$ is nothing more than a different way to call $\Delta G|_{p,T}$ in the case of a chemical reaction.

1.12 A Note on the Prerequisite of Homogeneity

To some extent, all real systems are non-homogeneous. Living systems are not an exception. Be they plants, animals, or even single cells, all living systems are an assemblage of different organs with different properties and functions at different points. Inanimate systems can also be highly non-homogeneous. Even when they are made of one single material, their state may differ from point to point, which effectively makes them non-homogeneous.

Classical thermodynamics refers to homogeneous systems in homogeneous states. Does that mean that its validity is confined to ideal systems and ideal engines? No, any real system can be treated as homogeneous either by looking at it on an appropriately large scale or by breaking it into appropriately small parts. The main point here is that the laws of thermodynamics discussed in the previous sections apply to any homogeneous system, irrespective of its size and complexity. By looking at a real system on a large enough scale, we can blur out much of its inhomogeneities. Likewise, by breaking the real system into appropriately small parts, we can regard it as consisting of homogeneous elements and apply classical thermodynamics to those elements.

On the other hand, there is no need to include in the model every detail of the real body, nor would this be possible due to the unlimited number of details that even the simplest, real-world system exhibits. Invariably, we are interested in just a limited number of its properties. Thus, we construct an idealized system to model those properties and ignore all other details. The considered properties may be the same throughout the system, which makes it homogeneous even if the real-world object it represents is not.

However, when modelling a spatially non-uniform system as a homogeneous system or when breaking it into homogeneous parts, we may implicitly neglect the presence of non-local effects. In fact, the state of an element of a system may be affected by the state of the neighbouring elements, which, in mathematical terms, means that some spatial derivatives of the state variables of the element may be state variables themselves. This possibility should not be overlooked when a non-homogeneous system is modelled as an assemblage of homogeneous elements. For instance, take the case of a non-uniform temperature system made of a material whose response depends on the temperature gradient. Even if each of its infinitesimal elements can be considered as having a well-defined temperature, the temperature gradient at the element should nevertheless be included among its state variables.

Finally, the mathematical expressions of the laws of thermodynamics (though by no means their physical meanings) may be different from the expressions given here if they refer to non-homogeneous systems. In particular, when dealing with non-uniform temperature systems, the classical expressions of the first law and the second law must be generalized to account for the fact that the system is not in thermal equilibrium (see [48] and [49]). It turns out that the discrepancy between the classical and the non-equilibrium expressions can only be significant in the presence of large temperature gradients, which is unlikely in most practical applications. The issue, however, is crucial for a better understanding of non-equilibrium thermodynamics.

2

The Admissible Range of Inanimate and Living Systems

A system is an idealized model to describe a definite portion of matter or body, whether inanimate or living. A bar of metal, a living cell, and any combination of inanimate and living materials are just some of the infinite instances of the real-world bodies that we may want to model. The system represents the behaviour of the body when external actions are acting upon it. It focuses on the properties of the real-world object that we choose to describe and provides a means of predicting and controlling their evolution in time.

Throughout this book, we shall consistently consider systems that refer to a finite portion of matter. Accordingly, the word "system" will be treated as synonymous with "finite system". Infinite systems are often introduced to indicate large reservoirs of heat or of other forms of energy. Their thermodynamics, however, falls outside the scope of this book.

In this chapter, we consider why living systems behave so differently from inanimate systems and why such a difference is physically plausible. In doing so, we must distinguish between two kinds of state variables: physical and subjective. Both kinds of variables are controlled by physics; however, subjective variables are related to so many subtle changes in the microscopic structure of the system that a complete physical model of them would be hopelessly complicated.

The analysis presented in this book is concerned mainly with the limits that the laws of thermodynamics impose on the maximum amount of energy that a system can store or release. These limits are independent of the way in which the system uses its energy or how it reacts to external actions. Thus, although a living system may respond quite differently to an external input depending on its subjective variables, the same variables do not influence the physical limits we are looking for. In other words, when it comes to the maximum amount of energy that a system can store or release, it makes no difference whether the system is inanimate or living. This simple fact discloses a hitherto little-explored opening of physics into the field of biology.

2.1 State Variables and Constitutive Equations

The most basic step in defining a system is to choose the set of variables that specify its state. As mentioned in the previous chapter, these are the *state variables* of the system, a set of variables that suffice to determine all variable properties of the system at any of its points. The state variables provide all of the information that we need to know about the assigned system. This is, of course, quite different from all of the possible information about the real-world object that is modelled by the system. The mere fact that a variable describes a property of the system does not, in itself, qualify it as a state variable. Position and time, for instance, can hardly be taken as state variables for a system that can be in the same state at different times and in different places.

Generally speaking, the state variables of a system form a finite set of physical quantities of various dimensions and different mathematical nature, such as scalars, vectors, and tensors of various orders. When their explicit identification is not required, we denote them simply as $\boldsymbol{\xi}^{(i)}$ ($i = 1, 2, \ldots, n$). Otherwise, a specific symbol is introduced for each variable. In general, the state variables assume different values at different times and at different points of the system. That is, $\boldsymbol{\xi}^{(i)} = \boldsymbol{\xi}^{(i)}(t, \boldsymbol{X})$, where t is time and $\boldsymbol{X}$ indicates the position vector of the points of the system. When the dependence on $\boldsymbol{X}$ can be excluded (i.e., in homogeneous systems), we have $\boldsymbol{\xi}^{(i)} = \boldsymbol{\xi}^{(i)}(t)$. In general, we find it convenient to single out temperature, T, from the other state variables of the system. The other variables are collectively referred to as $\boldsymbol{\xi}$. Therefore, by following a notation already used in the previous chapter, the state of the system is specified as $\{\boldsymbol{\xi}, T\}$.

The choice of the state variables of a system is, to some extent, a question of convenience. Once these variables are chosen, all variable properties of the system are expressed as appropriate functions of them, usually referred to as *state functions*. Different systems are characterized by different state functions. By borrowing terminology used in the mechanics of continuous media, we shall also refer to such functions as the *constitutive functions* (or *constitutive equations*) of the system.

The constitutive equations of a system may or may not depend on time. If they do, we speak of a system in evolution or in formation. However, it should be observed that this kind of time dependence is a relative concept, since it depends on the time scale we use to observe the system. Over a large enough time scale, every system will probably exhibit time-dependent, constitutive equations. Aging, phase changes, and the transformation of inert matter into a living organism are just a few examples of situations that involve time-dependent constitutive equations. However, when considered over an appropriately short time span (which, depending on the system, may run from a fraction of a second to eons), any system can be considered as possessing time-independent constitutive features.

Some properties of the system have a constant value, independent of the state variables. In general, the constitutive equations themselves will involve constant factors. All of these constants will be referred to as *constitutive constants* or simply *system constants*.

2.2 Limits to the Capacity for Storing Energy

Energy conservation implies that every system should somehow accommodate the energy it receives. How this is done can dramatically affect the system, especially if the energy input cannot be stored as internal energy. In that case, the system must move to a place of higher potential energy, accelerate to acquire kinetic energy, or even break into pieces or change its state of aggregation. For instance, a rigid body cannot deform to accommodate the energy it receives from an impact with another body. If the impact does not break it, the rigid body must drastically change its state of rest or motion. The fate of an ice cube provides another common instance. If the energy that the ice cube receives from the surroundings exceeds the energy that the ice can accommodate while remaining in the solid state, the ice cube liquefies and eventually disappears altogether.

Many systems can absorb energy from their surroundings without appreciable change in their position, state of motion, or state of aggregation. They do so by storing the external energy input as internal energy, which they then release under the right circumstances. This enables the system to keep their constitutive identity in spite of all the interactions with the surrounding world. This capacity is an essential prerequisite for a system's existence.

There is a limit, however, to the internal energy that, at a finite temperature, a system can store and release per unit volume. This limit is a direct consequence of the first law of thermodynamics. If a finite volume system contained an infinite amount of internal energy per unit volume, any finite part of that system could act as an unlimited source of energy. Likewise, if the system could store an infinite amount of energy per unit volume, any finite part of it could act as an unlimited sink of energy. Both occurrences would contradict the energy conservation principle, i.e., the first law of thermodynamics. It must be concluded that there is a limit to the maximum amount of internal energy that, at a finite temperature, a system can store and release per unit volume and that, moreover, this limit is finite.

The physical evidence of this conclusion is so obvious that the same conclusion could also be introduced as a very reasonable hypothesis, without even mentioning its connection with the first law of thermodynamics. If that standpoint is adopted, the analysis of the present book should be considered as referring to the systems that can only store a finite amount of internal energy per unit volume at a finite temperature. Appeal to the first law, however, helps us to point out that such systems are actually quite general.

Energy exchanges between a system and its surroundings occur frequently in the real world. A large enough limit to the internal energy that the system can store and release allows the system to withstand large external disturbances without breaking or changing its state of rest or motion at the slightest input of energy. Also, a large enough limit to the internal energy that the system can store helps the system to withstand large external actions when appropriate. Living systems are like inanimate systems in this respect. Their capacity to survive relates directly to their capacity to store internal energy and to spend it when needed. Contrary to what applies to inanimate systems, however, a living system can, to some extent, control how and when the energy exchange occurs.

Of course, the limit to the internal energy that a system can store per unit volume is generally different at different temperatures. The value of this limit depends on the system's constitutive equations and represents one of the most basic—though seldom considered—properties of the system.

2.3 Subjective Variables: The Hallmark of Life

In defiance of the universal law of gravitation, whispering "psst, psst" at your cat may be all you need to take her/him upstairs. The same trick will not work with the stones in your lawn, however. Yet gravity applies to cats and stones equally, as you would readily see if you were to throw both of them out the window. Foolish though these arguments may appear, they contain all the ingredients that are needed to distinguish between living (or biological) systems and inanimate (or physical) systems. Doing this appropriately is of utmost importance for applying thermodynamics to living systems.

No matter how complex an inanimate system is, all of its state variables are physical quantities. They will be referred to as *physical state variables*. Their evolution can, at least in principle, be predicted from the laws of physics. So much so that one would feel compelled to challenge the validity of the constitutive equations of the systems—or even the laws of physics themselves—should the system not behave as predicted.

Living systems are different. There should be little question as to whether their state can be described by a large enough set of physical variables. In the case of a living system, however, some of these variables can only be defined in a scale that is too small compared to the scale of the system itself. The number of variables needed to make the response of a living system fully predictable from the laws of physics would, moreover, be prohibitively large.

When dealing with living systems, a better approach is to distinguish between physical variables and *subjective variables*. In doing so, we give up the possibility of predicting every aspect of the behaviour of these systems from physics alone. At the macroscopic scale that we use to describe the behaviour of a living system, subjective variables make the system's response dependent on the qualitative properties of the system itself, on the qualitative properties of its surroundings and/or on the system's choice of how to react on each particular occasion. Subjective variables relate to the system's state of hunger, fear, alertness, previous exposure to certain kinds of phenomena or situations, and training. Together with the physical variables, subjective variables complete the description of the state of a living system and are therefore essential for predicting how it will respond to external actions.

Consequently, there are many more external actions that influence the response of a living system than those that are of significance to inanimate systems. In a living system, these actions include the information that the system receives about certain qualitative properties of the surrounding world, such as its lightness and colour, whether it is comfortable or hostile, how it smells, and so forth. Such properties do not influence the behaviour of inanimate systems unless the system has been designed specifically to sense them. And, if this is the case, these properties must be represented by quantities that are physically measurable, while a qualitative appraisal of them may be enough for a living system to change its response drastically. For instance, the white and black stripes of a zebra crossing on the road may condition the response of a pedestrian (or even that of a dog), although the same stripes do not exert any measurable physical force on the pedestrian (or dog).

The subjective variables of a living system, and hence its state, are not controlled by the system's physical variables—at least not completely. This fundamental difference between inanimate and living systems makes it impossible to predict the response of a living system on the basis of physics only. However, no matter how whimsical a living system might be, all of its energy exchanges with the surrounding world must comply with the laws of thermodynamics. In particular, the thermodynamic limits that we are going to consider in this book concerning the maximum amount of free energy that the system can store or supply are independent of the subjective variables. As a consequence, these limits apply to any system, whether inanimate or living.

2.4 Primary State Variables

In the absence of specific constraints, the physical state variables of a system generally may assume any value that is compatible with their definition. For some of these variables, however, this is not true—not even in principle. The second law of thermodynamics may require that the values of certain state variables fall within a certain definite range. It will be shown in the next chapter that the state variables whose range of variation may be restricted by thermodynamics are those that enter the constitutive equations of the internal energy and entropy of the system. These physical variables will be referred to as *primary state variables* and will be further discussed in the next chapter.

Because we shall be concerned almost exclusively with the primary state variables of the system, we shall refer to them simply as state variables. Any departure from this usage will be mentioned explicitly, should any risk of ambiguity arise.

2.5 Admissible Range

The way in which a body responds to external actions is regulated by the laws of physics and by the constitutive properties of the body itself. The constitutive properties are specified by appropriate constitutive equations and are different for different bodies. In principle, they must be determined experimentally. Consider, then, any real-world object with given constitutive properties. It is a matter of experimental evidence that the states that the object can attain must fall within a definite range. Beyond that range, the object ceases to respond according to the considered constitutive properties. In other words, it ceases to exist as an object with those properties.

The existence of such a range is a fundamental fact concerning any real-world system. For instance, if we strain an elastic metal spring beyond a certain limit, the spring will cease to respond elastically and it will deform plastically or it will break. Even more drastic changes, if possible, may take place in a living organism. Increasing one or more of its state variables further and further may kill the organism, cause it to metamorphose into a different organism, or even split it into two or more separate living organisms, to quote just a few of the possibilities. Therefore, beyond certain values of the state variables, the system that models the body ceases to represent it realistically. When this happens, a new system with different properties must be introduced to describe the real object under consideration. The determination of the range in which a physical system can behave according to a given set of constitutive properties is the main concern of the chapters that follow.

Many processes taking place in this world are isothermal or nearly so. The set of all of the states that a system can cover through every possible process at a constant temperature will be referred to as the *admissible range* of the system at that temperature. This is the set of all the states that, at the considered temperature, are compatible with the system's constitutive equations and the laws of physics. In the next chapter, this range will be determined under rather broad hypotheses from the laws of thermodynamics.

No change in a system's constitutive equations can be produced by a process that does not cross the boundary of the admissible range. This is almost a truism. Since any change in the properties of the system transforms the system into a new system with different constitutive equations, we have that as soon as a constitutive change has occurred the state of the system will pertain to the admissible range of a new system. This means that any constitutive change must necessarily take place at the frontier of the original range, because that is the only part of the original range that the original system has in common with the new system.

The boundary to the admissible range of a system is referred to as the *limit surface* of the system (see Section 3.3 for a detailed definition). The limit surface need not be one single surface. Depending on the system, it may consist of separate surfaces, which may or may not be bounded in certain directions.

2.6 Admissible Actions and Processes

Each physical state variable of the system has an associated action (or generalized force) that does work as that variable changes. These actions are collectively referred to as *admissible actions*. Any action not included in this set will either have no effect on the system or activate variables other than the ones that define the state of the considered system.

Every action applied to the body produces changes in the state of the body itself by driving it along an appropriate process. All processes that a system can undergo under all possible combinations of admissible actions, in every possible time sequences, constitute the set of the *admissible processes* of the system.

If we confine our attention to isothermal processes, the set of all states that the system can cover through all admissible isothermal processes coincides with the admissible range of the system, as defined in the previous section.

2.7 Incidental Restraints and Activity Range

The constitutive properties of a system may impose limitations on the number of state variables that can be assigned independently. For instance, incompressibility is a constitutive property imposing that the deformation of the body should occur at constant volume. This reduces the number of variables that are needed to specify the deformation of the system. The *free state variables* of a system are defined as the state variables that specify the system's state and are independent of each other.

Whether expressed in terms of free state variables or not, the admissible range of a system is subjected to the restrictions that come from the laws of physics. However, apart from any physical law, the range of states that the system can actually access may also be reduced by the presence of *incidental restraints*. The latter are limitations to the system's behaviour that are not imposed by any physical law or by the system's constitutive equations. An incidental restraint can impede access to some states that would otherwise be available to the system. Also, it may force an otherwise reversible process to proceed one way only by imposing on the system a kind of pawl-and-ratchet constraint.

We can distinguish between different types of incidental restraints. Many of them are equivalent to the application of an external device that restricts the values that one or more state variables (or their increments) can assume. Other incidental restraints may be more subtle in nature. For instance, consider the effect of a microscopic crack in a section of an otherwise homogeneous steel wire. Suppose that the wire is acted upon by two tensile forces applied to the ends of the wire. As the forces exceed a certain value, the crack will almost instantly grow, and the wire will break. This prevents the wire from reaching the maximum bearing capacity compatible with the material of which it is made. Being microscopic, one single crack does not affect the constitutive equations of the wire significantly. In particular, the crack hardly affects the load/displacement response of the wire before rupture occurs. This crack, however, restricts the maximum elongation of the wire by causing rupture before the wire's maximum elongation, compatible with the laws of physics and the constitutive properties of the material, is reached.

Another example of an incidental restraint is provided by a body constrained to glide on a board of finite dimensions. As long as it moves within the board, the body exhibits the same response to external forces, irrespective of its position. This state of affairs ends abruptly as the body crosses the border of the board. The displacements that are compatible with the constitutive equations of the gliding body and the physical laws may well be unlimited, yet a board of finite dimensions restricts the body's displacements to finite values depending on the board size.

An incidental restraint does not modify the system's constitutive equations and thus is not a part of the system. If the incidental restraint is removed, the system will gain full access to all states that are compatible with its constitutive equations and the laws of physics. For instance, a fly is represented by the same physical system whether it is free to move in space or confined within a room. The walls of the room set a restraint to the coordinates of the fly, but they do not affect the fly's constitutive properties. The admissible range for the fly's position is much larger than the space within the room. It depends, among other things, on the energy that the fly can expend. Such a range is determined, at least in principle, by the laws of physics and by the constitutive equations of the fly. It has nothing to do, however, with the room in which the fly is confined.

To distinguish between the admissible range of a system and the range of states that the system can actually access, we shall refer to the latter range as the *activity range* of the system. In the presence of incidental restraints, the activity range is only a part of the system's admissible range. If no incidental restraint acts on the system, the activity range coincides with the admissible range.

Of course, by removing the incidental restraints, we can expand the activity range of the system up to its admissible range. However, we cannot remove the restrictions that come from physical principles, nor can we eliminate a constitutive restraint without changing the system. Thus, the admissible range will be the largest of all possible activity ranges that a system of given constitutive equations can ever exhibit.

3

Thermodynamic Limit to the Existence of a System

As discussed at length in the previous chapter, every real system has its own range of states in which it is physically admissible. The system cannot exist outside of that range, not even in principle. Once the constitutive equations of a system are given, the laws of thermodynamics dictate what the admissible range should be. In this chapter, we show how.

The arguments presented here are quite general. They apply to inanimate and living systems alike, and they lead to the determination of the surface that bounds the region of the admissible states of the system in the space of its primary state variables. This region is the system's *admissible range* that was introduced in Section 2.5. Beyond that range, the system ceases to be physically admissible, and its constitutive equations must change if the boundary of that range is crossed. As a result, the system becomes a different system with a new admissible range.

Knowledge of the admissible range of a system is essential for determining the maximum amount of energy that the system can store compatibly with its constitutive equations. The greater this amount, the better the capacity of the system to withstand the energy demands that are imposed on it by an ever-changing environment. For living systems, this capacity may well relate to their fitness for survival.

3.1 Loading, Unloading, and Neutral Processes

If not otherwise stated, we henceforth confine our attention to uniform temperature systems undergoing isothermal processes. All processes considered in this chapter are assumed to be driven by admissible actions, as defined in Section 2.6. Processes that are driven by non-admissible actions are responsible for the formation of a new system. They are discussed at length in Chapter 5.

The following three special classes of isothermal processes have a fundamental role in the analysis that follows.

Loading processes

> An isothermal processes in which $\dot{W}_{\text{in}} > 0$ at any time of the process is a *loading process*.

In this kind of process, the system consistently *absorbs* non-thermal energy from its surroundings.

Unloading processes.

> An isothermal processes in which $\dot{W}_{\text{in}} < 0$ at any time of the process is an *unloading process*.

In this kind of process, the system consistently *supplies* non-thermal energy to its surroundings.

Neutral processes

> An isothermal processes in which $\dot{W}_{\text{in}} = 0$ at any time of the process is a *neutral process*.

In this kind of process, the system does not exchange any non-thermal energy with its surroundings.

3.2 Exhaustion States, Saturation States, and Limit Surface

As observed in Sect. 2.2, the maximum amount of internal energy that a finite system can store at any given temperature is finite. It follows, in particular, that the amount of energy that any finite system can absorb or release while operating isothermally must be finite. There must then be states in which the system becomes incapable of absorbing any more energy from the surroundings as well as states in which the system has no energy to supply. These simple observations justify the definitions of exhaustion and saturation states given below. Again, isothermal conditions are assumed

Exhaustion states

> A state of the system from which no unloading process can start without violating the system's constitutive equations is an *exhaustion state*.

If the constitutive equations of the system are smooth enough, then in an exhaustion state the system's free energy should be at a minimum or at least have the lowest value with respect to all of its neighbouring states. This conclusion is an immediate consequence of relation (1.11.4).

Once a system reaches an exhaustion state, it has no more free energy to spend while operating isothermally. This means that the system can only leave an exhaustion state through a loading process or a neutral process. However, if the system leaves an exhaustion state through a neutral process, then its free energy cannot increase. In this case, the process can only be a sequence of exhaustion states.

An exhaustion state has different meanings in different systems. In an elastic body, for instance, the exhaustion state is the unstrained state, also referred to as the stress-free state. This is the state in which all elastic energy of the body vanishes. The body cannot move from that state to a different state unless some external energy is supplied to deform it.

The free energy mentioned above is the internal free energy of the system. When considering the relationship between exhaustion state and free energy, it is important to distinguish between internal energy and external energy. The expressions of internal energy and entropy of many systems of practical interest often include terms that represent an external contribution to the system's free energy to all effects. As such these contributions do not affect the free energy that can be stored within the system and, therefore, they must be ignored when the thermodynamic limit to the system's internal energy is sought. This question is of utmost importance to the analysis presented in this book. It is dealt with in detail in Section 3.11 to Section 3.14.

Saturation states

> A state of the system from which no loading process can start without violating the system's constitutive equations is a *saturation state*.

A system can leave a saturation state either through an unloading process or a neutral process. However, once in a saturation state, the system cannot absorb any more non-thermal energy from the surroundings because no loading process can start from that state. At any given temperature, most natural systems possess just one exhaustion state. However, the same systems can usually store energy in a number of different ways. This endows them with a set of different saturation states at that temperature.

The difference between the free energy in any given state and the free energy in the exhaustion state is a measure of the maximum amount of energy that the system can supply isothermally to its surroundings as work and/or heat through a process that starts from the considered state. In a space in which the distance between two states is measured by the difference between their free energies, the saturation states are the farthest states from the exhaustion state. In that space, the locus of all saturation states represents the boundary of the admissible range, i.e., the *limit surface* of the system. This justifies the following definition:

Limit surface

> At any given temperature, the locus of the saturation states of a system in the space of its state variables is the *limit surface* of the system at the considered temperature.

Because no loading process can start from a saturation state, the system's free energy must reach an upper limit in that state. This is an immediate consequence of relation (1.11.4) and of the fact that every process leaving a saturation state must do so through a process at the beginning of which $dW_{in} \leq 0$. The same applies to every point of the limit surface, since that surface is made of saturation states. Thus, in the state space, the points of the limit surface represent limits to the free energy that a system can store isothermally.

3.3 Accessibility Hypotheses

To gain some insight into the thermodynamics of the admissible range, we need to know whether any two given states in that range can be linked by a process, and, if so, which type of process. To keep the analysis simple and general, we focus on systems that possess a unique exhaustion state. Most real-world systems, both inanimate and living, can be modelled as systems possessing this property.

Systems with multiple exhaustion states may be of practical interest. Typically, systems that have bumps and wells in their free energy function possess multiple exhaustion states. If a system has multiple exhaustion states, all of these states must either have the same internal energy or be devoid of any unloading process that links them. Otherwise, the system could unload from an exhaustion state at a higher energy to another exhaustion state at a lower energy, which would make the higher energy exhaustion state contradict the very definition of an exhaustion state. Although much of the analysis of this book could be generalized to cover a system with multiple exhaustion states, we make no such attempt.

By referring to the admissible range of systems that possess just one single exhaustion state, we introduce the following hypotheses:

(i) *From any state in the admissible range of the system, there is at least one unloading process (i.e., a process where $\dot{W}_{in} < 0$) that leads to the exhaustion state and belongs to the admissible range. The possibility that $\dot{W}_{in} = 0$ at the end of the process (i.e., at the exhaustion state) is not excluded.*

(ii) *Any state in the admissible range of the system can be reached by at least one loading process (i.e., a process where $\dot{W}_{in} > 0$) that starts from the exhaustion state and belongs to the admissible range. Again, the possibility that $\dot{W}_{in} = 0$ at the beginning of the process (i.e., at the exhaustion state) is not excluded.*

(*iii*) *Let* A *and* B *denote any two states of the admissible range of the system at an infinitesimal distance from each other (i.e., two points that belong to the same infinitesimal neighbourhood in the state space). These states can always be linked by either a neutral or a loading (unloading) process. Moreover, if a neutral process links* A *to* B, *then there is at least one neutral process that links* B *to* A; *if a loading (unloading) process links* A *to* B, *there is at least one unloading (loading) process that links* B *to* A.

All of the processes considered in the above hypotheses are understood to be isothermal, since the admissible range is defined under isothermal conditions.

Hypothesis (*i*) must hold for the system to be capable of recovering the exhaustion state through its own means (i.e., without absorbing non-thermal energy from its surroundings). The same hypothesis excludes that, in the absence of external forces, the system could reach equilibrium at a state other than the exhaustion state. This is an immediate consequence of the virtual work principle of classical mechanics and of the fact that the hypothesis in question assumes that from every state other than the exhaustion state there is at least one unloading process to the exhaustion state. In some applications, an appropriate convexity hypothesis concerning the free energy function is usually introduced or understood in order to exclude that the system could reach equilibrium at a state other than the exhaustion state when no external force acts upon the body. Hypothesis (*i*), however, is more general than that.

Hypothesis (*ii*) makes the system capable of moving from any state in the admissible range and returning to that state, provided that it is allowed to absorb or lose non-thermal energy, as appropriate.

Hypothesis (*iii*) rules out systems that are so dissipative that they cannot perform a process—not even an infinitesimal one—without absorbing energy from their surroundings. Such systems always dissipate energy, irrespective of the kind of process they undergo. A rigid body, sliding with friction over a horizontal plane, is the simplest instance of such a system. Although systems of this kind are a physical reality, they do not fall within the present theory.

The hypotheses introduced above are general and reasonable enough to cover most real-world systems of practical interest. The processes to which they refer may well be irreversible, as is the case with real-world processes.

3.4 Equipotential Property of the Limit Surface

In Section 3.2, we defined the limit surface as the locus of the saturation states of the system. We observed that to leave a saturation state a system cannot absorb non-thermal energy from the surroundings. We consider here the particular case of a limit surface made of just one single continuous surface. In the more general, but still practically important, case of a limit surface made of two or more separate surfaces, the arguments presented below apply to each separate continuous part of that surface. In every case, the functions U and S are assumed to be regular enough for the present arguments to apply.

Under these assumptions, let A and B be any two points on the limit surface of a system at an infinitesimal distance from each other. The state variables corresponding to these points are defined by values that differ by infinitesimal quantities at most. From accessibility Hypothesis (*iii*) of the previous section, we can exclude that there is any unloading process from A to B, since this would be equivalent to admitting the existence of a loading process from B to A. The latter process would not be admissible, since B is on the limit surface, which means that no loading process can start from it. From the same accessibility hypothesis, it then follows that states A and B can be joined by at least two neutral processes, one from A to B and the other from B to A. In these processes, $\dot{U}=\dot{Q}$, as follows immediately from the first law of thermodynamics and from the fact that $\dot{W}_{\text{in}}=0$ in a neutral process. Of course, the same processes must also comply with the second law. This requires that both processes should satisfy the following equation:

$$\dot{U}=T\,\dot{S}\,. \tag{3.4.1}$$

Otherwise, one of them would violate inequality (1.8.1) and thus the second law. Equation (3.4.1) means that in going from A to B, the following relation must hold true:

$$\mathrm{d}(U-TS)=0. \tag{3.4.2}$$

In view of definition (1.11.3), this means that

$$\mathrm{d}\Psi=0. \tag{3.4.3}$$

That is,

$$\Psi=\text{const.} \tag{3.4.4}$$

It can therefore be concluded that the boundary of the admissible range must be equipotential for Ψ.

If the limit surface is composed of separate parts, then the above arguments imply that each part must be equipotential for Ψ. In that case, however, the value of Ψ may be different in each part.

Obviously, the same conclusions do not generally apply to the boundary of the activity range if the latter does not coincide with the boundary of the admissible range (see Section 2.7). In that case, it cannot be excluded that two points at the boundary of the activity range can be joined by a loading process, which makes the above arguments leading to equations (3.4.3) and (3.4.4) inapplicable. In other words, the boundary to the activity range of a system need not be equipotential for Ψ.

3.5 Isothermal Production of Non-Thermal Energy

The maximum amount of non-thermal energy that an inanimate system can supply in an isothermal process has long been recognized as a property of the system. Classical thermodynamics relates this property to the so-called Helmholtz free energy of the system and teaches us how to compute it. The same property appears to be of paramount importance also for living systems because of its obvious bearing upon their fitness to survive in an energy-demanding environment.

The maximum amount of non-thermal energy that a system can supply while undergoing an isothermal process from any state A to another state B can be calculated from eq. (1.11.7), which can be rewritten more explicitly as:

$$W_{\text{out}}\Big|_{T=\text{const}}^{\text{AB}} \leq \Psi_{\text{A}} - \Psi_{\text{B}} \,. \tag{3.5.1}$$

The quantity on the left-hand side of this relation is the non-thermal energy that the system supplies to the surroundings in the process. It is a process-dependent quantity. The quantities Ψ_{A} and Ψ_{B} on the other side of the relation are the values of Ψ in state A and state B, respectively. These values, and thus their difference $\Psi_{\text{A}}-\Psi_{\text{B}}$, are the same irrespective of the process that joins A to B, because Ψ is a state function. By setting

$$\Delta_{\text{AB}}\Psi = \Psi_{\text{A}} - \Psi_{\text{B}}, \tag{3.5.2}$$

relation (3.5.1) can be written as:

$$W_{\text{out}}\Big|_{T=\text{const}}^{\text{AB}} \leq \Delta_{\text{AB}}\Psi \,. \tag{3.5.3}$$

Classical thermodynamic arguments concerning the free energy of a system apply to living systems, because free energy represents the capacity of a system for doing work. It is immaterial how, when, and if the system exploits that capacity. Therefore, it makes no difference whether the system that has that capacity is a living or an inanimate system.

3.6 Determination of the Admissible Range

We now apply the second law to determine the admissible range and limit surface of any given system, once the system's constitutive equations are known.

To begin, let us consider any unloading process starting from any state, other than the exhaustion state, belonging to the system's admissible range. The existence of such a process is granted by the accessibility hypotheses introduced in Section 3.3. In an unloading process, $\dot{W}_{\text{in}}$ must consistently be less than zero. It follows that, at any time during this process, the system supplies non-thermal energy to its surroundings. Of course, the process must meet the first and second laws of thermodynamics. Since $\dot{W}_{\text{in}} < 0$, we infer from eq. (1.4.8) that

$$\dot{U} < \dot{Q} . \qquad (3.6.1)$$

From this and from inequality (1.8.1), it follows that

$$\dot{U} < T\,\dot{S} \qquad (3.6.2)$$

during any unloading process.

Inequality (3.6.2) does not generally apply to a loading process. In such a process $\dot{W}_{\text{in}} > 0$, which from eq. (1.4.8) means that $\dot{U} > \dot{Q}$. This cannot lead to any definite conclusion about whether $\dot{U}$ should be greater than, equal to, or less than $T\dot{S}$. The same inequality (3.6.2) need not apply to neutral processes either. In that case, from similar arguments as those leading to inequality (3.6.2), we can only infer that the weaker inequality $\dot{U} \leq T\dot{S}$ should apply to any neutral process, the equality sign holding true if the process is reversible.

Result (3.6.2) has profound implications for the system. Recall that, according to hypothesis (*i*) of Section 3.3, from any state in the admissible range, there is always at least one unloading process that belongs to that range and that leads to the exhaustion state. Therefore, at any time during such a process, the following equation must hold according to relations (3.6.2) and (1.11.3):

$$\dot{\Psi} = \dot{U} - T\dot{S} < 0. \tag{3.6.3}$$

It follows that the free energy of the system must decrease as the system unloads isothermally from any state in the admissible range. For systems that possess a unique exhaustion state, this means that Ψ, and hence the difference $U - TS$, reaches its lowest value at the exhaustion state.

No loading process can start from a point at the boundary of the admissible range, because that boundary is the locus of the saturation states. However, there is always an unloading process starting from any point of the said boundary due to the accessibility hypothesis (*i*). It follows that the difference $U - TS$, i.e. the system's free energy, Ψ, must reach its largest value at a point on that boundary. This value is denoted as $\Psi_{\max}$. That is,

$$\Psi_{\max} = max\,(U - TS\,), \tag{3.6.4}$$

where $\Psi_{\max}$ is a characteristic constant of the system.

If the boundary of the admissible range is one single continuous surface, then $\Psi_{\max}$ is also the value of Ψ at every point on that surface, because the latter is equipotential for Ψ as shown in Section 3.4. In this case, the following relation determines the admissible range of the system:

$$U - TS \leq \Psi_{\max}\,. \tag{3.6.5}$$

This inequality is valid for every state in the admissible range of the system. It applies in the equality form only to the points of the boundary of that range.

Result (3.6.5) involves the values of U and S rather than their time rates, $\dot{U}$ and $\dot{S}$, respectively. The time rates are different for different processes; U and S are independent of the process, because they are state functions, namely $U = U(\boldsymbol{\xi}, T)$ and $S = S(\boldsymbol{\xi}, T)$. For constant T, relation (3.6.5) defines a region in the space of the $\boldsymbol{\xi}$ variables. This is the space of the primary state variables of the system, because the only state variables that are involved in the above arguments are those entering the expressions of U and S. The region defined by relation (3.6.5) is composed of all states that the system can access through an isothermal process complying with the system's constitutive equations and the laws of thermodynamics. Therefore, it is the system's admissible range. Its boundary is expressed by

$$U - TS = \Psi_{\max} \tag{3.6.6}$$

and is the system's limit surface.

In terms of the Helmholtz free energy, the admissible range and admissible surface of the system can be expressed as

$$\Psi \leq \Psi_{\max} \tag{3.6.7}$$

and

$$\Psi = \Psi_{\max}, \tag{3.6.8}$$

respectively. These relations are the immediate consequence of eqs (3.6.4)-(3.6.6) and definition (1.11.3).

Quite often in practice, functions U and S are continuous. In this case, the entire surface (3.6.6) is fully determined by the constitutive equations for U and S once a point of that surface is known, because $\Psi_{\max}$ is determined by eq. (3.6.6) once the values of U and S at that point are calculated. Accordingly, if functions U and S are given and continuous, the experimental determination of the entire limit surface reduces to the determination of just one point of that surface.

On the other hand, if U and S are piecewise continuous functions, then a relation like eq. (3.6.5) applies separately to each sub-domain where U and S are continuous. In this case, a different value of $\Psi_{\max}$ will generally apply to each sub-domain, which will produce a different limit surface for different sub-domains, according to eq. (3.6.6).

The role of the second law of thermodynamics in the derivation of the results of this section should not be downplayed. This law imposes that every unloading process should satisfy the entropy inequality (1.6.7) and thus inequality (3.6.2). This implies that the difference $U - TS$, i.e., the system's free energy, must attain its greatest value at the boundary of the admissible domain—a conclusion that cannot be reached in this generality without making an appeal to the second law. In particular, the assumption that a finite system can only store a finite amount of energy under isothermal conditions is not enough to justify the existence of the limit surface.

We touch here on a very important but seldom exploited point concerning the role of the second law of thermodynamics. Its traditional role in physics is to control the maximum amount of heat that a system can absorb in a process, quite apart from the domain of states that are accessible to the system. The above arguments show that the same law also restricts the states that a system of given constitutive equations can attain, granted some general accessibility conditions.

3.7 Formally Simpler Expressions of Admissible Range and Limit Surface

Without introducing any further restriction, we can express inequality (3.6.5) in an even more practical form. The internal energy of a system is defined to within an arbitrary additive constant that depends on the reference value of internal energy. By choosing the reference value appropriately, we can make the difference $U - TS$, and thus Ψ, assume any given value at a given state. The reference value of U does not affect the admissible range and limit surface of the system, because, in view of eq. (3.6.4), a change in the reference value of U is equivalent to adding a same constant to both sides of relation (3.6.5).

We often assume that the reference value of U is chosen so that $\Psi_{max} = 0$. With this choice, relation (3.6.5) reduces to

$$U \leq TS, \tag{3.7.1}$$

which represents a formally simpler expression of the *admissible range*. Accordingly, the *limit surface* is expressed as

$$U = TS\,. \tag{3.7.2}$$

Notice, however, that the function U that appears into relations (3.7.1) and (3.7.2) is not the same as the function U that appears in the more general relation (3.6.5). The two functions differ by a constant. This is not a problem, because the change in $(U - TS)$, not its actual value, is what matters in the definition of the admissible range. This is easily shown by rewriting relation (3.6.5) as

$$U - TS \leq max\,(U - TS), \tag{3.7.3}$$

thanks to eq. (3.6.4). As observed, a change in the reference value of U adds the same constant to both sides of relation (3.7.3), which is invariant under such a change. The admissible range that the same relation defines is therefore left unchanged.

Relations (3.7.1) and (3.7.2) can also be expressed in terms of the system's Helmholtz free energy (1.11.3). In that case, they become

$$\Psi \leq 0 \tag{3.7.4}$$

and

$$\Psi = 0, \tag{3.7.5}$$

respectively. These relations provide further alternative expressions of the admissible range and the admissible surface of the system. Again, the expression of Ψ here is different from the expression of Ψ in eqs (3.6.7) and (3.6.8) due to the different values of the internal energy constant.

To put it differently, relations (3.7.4) and (3.7.5), as well as relations (3.7.1) and (3.7.2) from which they are derived, are valid provided that $U - TS$ and thus Ψ vanish at the boundary of the admissible range. To meet this condition, the arbitrary constant in the general expression of U, and thus of Ψ, is to be chosen appropriately. With this choice, Ψ becomes negative throughout the admissible range and attains its most negative value at the exhaustion state. (Remember that the free energy of the system cannot increase during unloading.)

3.8 Crossing the Limit Surface

No process can bring a system outside of the limit surface without producing changes in the system's constitutive equations. The reason is the following. The free energy Ψ defines a scalar field in the space of the state variables of the system. The limit surface is a level surface of that field, because it is equipotential for Ψ. As a consequence, at the points of the limit surface, the vector grad Ψ is normal to that surface and oriented in the direction of the increase of Ψ. From relation (3.6.7) it then follows that the admissible range of the system lies on the side of the limit surface opposite to the direction of grad Ψ. Under these conditions, the crossing of the limit surface would cause the system to store more free energy than the maximum amount of free energy that it can store. This is impossible, as it would violate relation (3.6.7) and, therefore, would not be compatible with the system's constitutive equations for U and S. To go beyond the limit surface, therefore, the system must change these constitutive equations and thus its own admissible range.

A consequence of this is that, once the limit surface is crossed, the system cannot come back by itself (i.e., through an unloading process) to the state it had before crossing the limit surface. No change in the system's constitutive equations can be brought about by a process that occurs within the admissible range (see Sect. 2.5). Therefore, to come back to the original state, and hence recover its original constitutive equations, the system must of necessity cross its new limit surface, which cannot be done through an unloading process.

In other words, a process that brings a system outside of its limit surface produces constitutive changes in the system that cannot be reversed by an unloading process ($\dot{W}_{in} < 0$). To reverse these changes, a process where $\dot{W}_{in} > 0$ is needed. This means that some external agency must spend non-thermal energy on the system, granted that a process that reverses the above constitutive changes is feasible. In want of such an external agency, the crossing of the limit surface produces permanent—and usually major—changes in the system.

3.9 Non-Interacting Homogeneous Parts of Non-Homogeneous Systems

The hypothesis that the system should be homogenous is essential to the analysis of the previous sections. No concept of classical thermodynamics applies rigorously unless referred to a homogeneous system. This point was already considered in Section 1.12, where we discussed how a non-homogeneous system can, for certain purposes, be modelled as a homogeneous system.

In order to better illustrate the role of this hypothesis as far as the thermodynamic limit is concerned, consider the example of the lead-acid battery, commonly found in automobiles. The battery is obviously a non-homogeneous system. It is made of a number of different parts. The most external part is the casing. As any other elastic body, the casing can store internal energy in the form of strain energy. The maximum amount of strain energy that an elastic material can store is considered in detail in Chapter 6. Of course, that energy is of little interest if we are concerned with the electric energy that the battery can supply. That energy is stored by an assembly of different components inside the casing. The main components are the electrolyte (a dilute sulphuric acid solution) and two sets of electrodes or plates (actually, a set of lead plates and a set of lead-oxide plates, packed together via appropriate separators). Together, all these components result in a highly non-homogeneous system that possesses different properties at different points.

We can disregard much of the above details, however, if we are only concerned with the electric energy that the battery can store. In that case, the system's temperature and the amount of ions available to the battery's electrodes are the main state variables of the system. They suffice to build a simple but effective thermodynamic model of the battery, the details of which are presented in Chapters 7 and 8. This model is a homogeneous system, because its state variables are not space-dependent.

Of course, it does not make sense to treat the electric energy stored in the battery as equivalent to the strain energy stored by its casing. The two energies do not interact. In the next section, we introduce the notion of thermodynamically orthogonal sets of state variables, which can coexist independently in the same physical system without interacting with each other.

3.10 Thermodynamically Orthogonal State Variables

In some cases, the set $\boldsymbol{\xi}$ of all state variables of the system can be subdivided into two or more subsets, say $\boldsymbol{\xi}^{\mathbf{A}}$, $\boldsymbol{\xi}^{\mathbf{B}}$, ..., $\bar{\boldsymbol{\xi}}$, that decompose the internal energy and entropy of the system into the following sums of functions:

$$U = U(\boldsymbol{\xi}, T) = U^{\mathbf{A}}(\boldsymbol{\xi}^{\mathbf{A}}, \bar{\boldsymbol{\xi}}, T) + U^{\mathbf{B}}(\boldsymbol{\xi}^{\mathbf{B}}, \bar{\boldsymbol{\xi}}, T) + \ldots + \bar{U}(\bar{\boldsymbol{\xi}}, T) \quad (3.10.1)$$

and

$$S = S(\boldsymbol{\xi}, T) = S^{\mathbf{A}}(\boldsymbol{\xi}^{\mathbf{A}}, \bar{\boldsymbol{\xi}}, T) + S^{\mathbf{B}}(\boldsymbol{\xi}^{\mathbf{B}}, \bar{\boldsymbol{\xi}}, T) + \ldots + \bar{S}(\bar{\boldsymbol{\xi}}, T). \quad (3.10.2)$$

In the above equations, each subset $\boldsymbol{\xi}^{\mathbf{I}}$ only appears in the corresponding functions $U^{\mathbf{I}}$ and $S^{\mathbf{I}}$ (**I**=**A**, **B**, ...), while both the subset $\bar{\boldsymbol{\xi}}$ and the variable T may enter some or all of the above functions $U^{\mathbf{I}}$ and $S^{\mathbf{I}}$.

It will be assumed that the subsets of variables $\boldsymbol{\xi}^{\mathbf{A}}$, $\boldsymbol{\xi}^{\mathbf{B}}$, ..., $\bar{\boldsymbol{\xi}}$ are separate and do not include temperature. In some cases, subset $\bar{\boldsymbol{\xi}}$ may be empty. Moreover, it is not excluded that one of the two functions in some of the pairs of functions $\{U^{\mathbf{I}} = U^{\mathbf{I}}(\boldsymbol{\xi}^{\mathbf{I}}, \bar{\boldsymbol{\xi}}, T),\ S^{\mathbf{I}} = S^{\mathbf{I}}(\boldsymbol{\xi}^{\mathbf{I}}, \bar{\boldsymbol{\xi}}, T)\}$ may be independent of $\boldsymbol{\xi}^{\mathbf{I}}$ or may vanish identically. For instance, both the pair of functions $\{U^{\mathbf{I}} = U^{\mathbf{I}}(\boldsymbol{\xi}^{\mathbf{I}}, \bar{\boldsymbol{\xi}}, T),\ S^{\mathbf{I}} = S^{\mathbf{I}}(T)\}$ and the pair $\{U^{\mathbf{I}} = U^{\mathbf{I}}(\boldsymbol{\xi}^{\mathbf{I}}, \bar{\boldsymbol{\xi}}, T),\ S^{\mathbf{I}} \equiv 0\}$ fall within the scope of the decomposition described by eqs (3.10.1)-(3.10.2).

The variables $\boldsymbol{\xi}^{\mathbf{I}}$ of each subset of variables are assumed to be independent of those of the other subsets of variables and of their time rates. Also, decomposition (3.10.1)-(3.10.2) is understood to be maximal. In other words, functions $U^{\mathbf{I}}(\boldsymbol{\xi}^{\mathbf{I}}, \bar{\boldsymbol{\xi}}, T)$, $S^{\mathbf{I}}(\boldsymbol{\xi}^{\mathbf{I}}, \bar{\boldsymbol{\xi}}, T)$ cannot be decomposed further into sums of separate sub-subsets of the variables $\boldsymbol{\xi}^{\mathbf{I}}$ according to formulae similar to (3.10.1) and (3.10.2).

The subsets $\boldsymbol{\xi}^{\mathrm{A}}$, $\boldsymbol{\xi}^{\mathrm{B}}$, ..., $\bar{\boldsymbol{\xi}}$ possessing the above mentioned properties will be referred to as *thermodynamically orthogonal*.

In the rest of this section, we focus on isothermal processes at constant values of $\bar{\boldsymbol{\xi}}$. From eqs (3.10.1) and (3.10.2), it can be verified immediately that, for constant $\bar{\boldsymbol{\xi}}$ and T, both the internal energy and the entropy of the system are the sum of independent contributions, namely U^{A}, U^{B}, ... and S^{A}, S^{B}, ... Each contribution comes from an independent set of variables $\boldsymbol{\xi}^{\mathrm{A}}$, $\boldsymbol{\xi}^{\mathrm{B}}$, ..., respectively. Under the same conditions, the terms $\bar{U}(\bar{\boldsymbol{\xi}}, T)$ and $\bar{S}(\bar{\boldsymbol{\xi}}, T)$ are constants.

Now, let $\dot{W}^{\mathrm{A}}$, $\dot{W}^{\mathrm{B}}$, ... be the (non-thermal) power associated with the time rates of the variables $\boldsymbol{\xi}^{\mathrm{A}}$, $\boldsymbol{\xi}^{\mathrm{B}}$, ..., respectively. That is,

$$\dot{W}^{\mathbf{I}} = \boldsymbol{X}^{\mathbf{I}} \cdot \dot{\boldsymbol{\xi}}^{\mathbf{I}}, \qquad (\mathrm{I} = \mathrm{A, B}, \ldots), \qquad (3.10.3)$$

where vector $\boldsymbol{X}^{\mathbf{I}}$ represents the conjugate force to the variables $\dot{\boldsymbol{\xi}}^{\mathbf{I}}$, and the dot product denotes an appropriate scalar product between $\boldsymbol{X}^{\mathbf{I}}$ and $\dot{\boldsymbol{\xi}}^{\mathbf{I}}$. By introducing eqs (3.10.3) and (3.10.1) into the first law (1.4.8), we obtain

$$\dot{U}^{\mathbf{A}} + \dot{U}^{\mathbf{B}} + \ldots = \dot{Q} + \dot{W}^{\mathbf{A}} + \dot{W}^{\mathbf{B}} + \ldots . \qquad (3.10.4)$$

The term corresponding to the time rate of the variables $\bar{\boldsymbol{\xi}}$ is missing from this expression, since the process takes place at constant $\bar{\boldsymbol{\xi}}$.

Starting from any state of the system, the time rates of the pair $\dot{U}^{\mathbf{I}}$ and $\dot{W}^{\mathbf{I}}$ are independent of those of pairs $\dot{U}^{\mathbf{J}}$ and $\dot{W}^{\mathbf{J}}$ ($\mathrm{I} \neq \mathrm{J}$), because the variables $\boldsymbol{\xi}^{\mathrm{I}}$ are independent of $\boldsymbol{\xi}^{\mathrm{J}}$. Under these conditions, eq. (3.10.4) implies that

$$\dot{Q} = \dot{Q}^{\mathbf{A}} + \dot{Q}^{\mathbf{B}} + \ldots , \qquad (3.10.5)$$

where

$$\dot{Q}^{\mathbf{I}} = \dot{U}^{\mathbf{I}} - \dot{W}^{\mathbf{I}} \qquad (\mathrm{I} = \mathrm{A, B}, \ldots). \qquad (3.10.6)$$

This means that, in addition to the global energy balance equation (1.4.8), the following partial energy balance equations should also apply:

$$\dot{U}^{\mathbf{I}} = \dot{Q}^{\mathbf{I}} + \dot{W}^{\mathbf{I}} \qquad (\mathrm{I} = \mathrm{A, B}, \ldots). \qquad (3.10.7)$$

Likewise, by introducing eqs (3.10.2) and (3.10.5) into the entropy inequality (1.8.1), we obtain

$$\dot{Q}^{\mathbf{A}} + \dot{Q}^{\mathbf{B}} + \ldots \leq T\dot{S}^{\mathbf{A}} + T\dot{S}^{\mathbf{B}} + \ldots . \qquad (3.10.8)$$

Again, since the variables in each subset $\boldsymbol{\xi}^{\mathrm{I}}$ are independent of the variables in the other subsets, we infer from inequality (3.10.8) that the following partial entropy inequalities should apply to the considered processes:

$$\dot{Q}^{\mathbf{I}} \leq T\dot{S}^{\mathbf{I}} \qquad (\mathrm{I} = \mathrm{A}, \mathrm{B}, \ldots). \tag{3.10.9}$$

We can conclude that the subsets of variables $\boldsymbol{\xi}^{\mathrm{A}}$, $\boldsymbol{\xi}^{\mathrm{B}}$, ..., $\overline{\boldsymbol{\xi}}$, enable us to decompose the energy balance equation and the entropy inequality into a set of simpler, uncoupled relations, (3.10.7) and (3.10.9), respectively, with each relation being relevant to a different subset of variables $\boldsymbol{\xi}^{\mathrm{I}}$. Therefore, when U and S are decomposed as in eqs (3.10.1) and (3.10.2), the system can be represented as an assemblage of simpler, independent systems, each of which is described by one particular subset of state variables $\boldsymbol{\xi}^{\mathrm{I}}$. This justifies calling these subsets thermodynamically orthogonal. Such a representation only applies as long as the processes at constant $\overline{\boldsymbol{\xi}}$ and T are considered.

If the state variables of the system can be grouped into thermodynamically orthogonal subsets, the admissible range of the system will decompose into separate sub-ranges, one for each different set of variables $\boldsymbol{\xi}^{\mathrm{I}}$. Of course, these ranges apply at constant $\overline{\boldsymbol{\xi}}$. Clearly, in the case where subset $\overline{\boldsymbol{\xi}}$ is empty, the above sub-ranges define the only admissible ranges of the system.

More explicitly, from the analysis developed in Section 3.6, we can infer that, at constant $\overline{\boldsymbol{\xi}}$, the admissible range relevant to the set of state variables $\boldsymbol{\xi}^{\mathrm{I}}$ is defined by

$$U^{\mathrm{I}} - TS^{\mathrm{I}} \leq C^{\mathrm{I}} \qquad (\mathrm{I} = \mathrm{A}, \mathrm{B}, \ldots). \tag{3.10.10}$$

This particularizes relation (3.6.5) to the present case. Similar to what we observed in Section 3.7, the values of constant $C^{\mathbf{I}}$ that appear in the above relations depend on the reference value assigned to the relevant internal energy $U^{\mathbf{I}}$. Note that because the problem is now split into a set of separate problems, in determining the admissible range of each set of variables, we may choose a different reference value for each component $U^{\mathbf{I}}$ in order to make each constant $C^{\mathbf{I}}$ disappear.

State variables that belong to different thermodynamically orthogonal sets are referred to as *thermodynamically orthogonal variables*. Conversely, state variables that are not thermodynamically orthogonal are referred to as *thermodynamically coupled*. A change in a thermodynamically coupled variable will affect the internal energy that is available to all the other variables in the same set. The same is not true for variables that belong to different thermodynamically orthogonal sets.

The above considerations are useful, in particular, when determining the admissible range of a linear thermoelastic material (see Chapter 6). Also, in complex systems, such as living organisms, it is not unusual for different organs or different set of cells to be controlled by different sets of thermodynamically orthogonal state variables. Each organ will therefore possess its own thermodynamic range and limit. In this case, the attainment of the thermodynamic limit for the whole organism will be conditioned by the organ that reaches its own limit first.

In order to test whether any two given variables, ξ_h^A and ξ_k^B, are thermodynamically orthogonal, we check whether the partial derivatives $\partial U/\xi_h^A$ and $\partial S/\xi_h^A$ depend on ξ_k^B. If not, then the two variables belong to two different thermodynamically orthogonal sets of variables. This follows immediately from eqs (3.10.1) and (3.10.2) and from the definition of thermodynamically orthogonal sets of variables. In other words, if the change in U and S due to a change in ξ_h^A is independent of the value of ξ_k^B, then the variables ξ_h^A and ξ_k^B belong to two different thermodynamically orthogonal sets.

3.11 Non-Thermodynamic Systems

Consider the special case of a system whose internal energy depends on some state variables $\boldsymbol{\xi}$ but not on temperature. That is,

$$U = U(\boldsymbol{\xi}), \tag{3.11.1}$$

or $\partial U/\partial T \equiv 0$. In view of eq. (1.7.5), this implies that the system's specific heat is equal to zero no matter the value of $\boldsymbol{\xi}$. As apparent from eq. (1.7.6), the system is incapable of exchanging heat with the surroundings, regardless of the process it undergoes. From eq. (1.7.8) it then follows that the entropy of the system should vanish identically, i.e., $S(\boldsymbol{\xi}, T) \equiv 0$. The following equation

$$\dot{S} = 0 \tag{3.11.2}$$

will, therefore, apply to every process of such a system.

Any system with an internal energy of the form (3.11.1) is referred to as a *non-thermodynamic system*. The inability of non-thermodynamic systems to exchange heat with the surroundings means that, for these systems,

$$\dot{Q} = 0\,. \tag{3.11.3}$$

From this equation and eq. (3.11.2), it follows that these systems satisfy the second law of thermodynamics (1.8.1) identically, in the form 0 = 0. Non-thermodynamic systems, therefore, are unconditionally consistent with the second law. Moreover, all the processes that they undergo are reversible, because they meet relation (1.8.1) as equality.

When applied to a non-thermodynamical system, the energy balance equation (1.4.8) reduces to

$$\dot{U} = \dot{W}_{\text{in}}\,, \tag{3.11.4}$$

as evident from eq. (3.11.3).

Systems that possess an internal energy function of the form (3.11.1) are considered routinely in various classical branches of physics, ranging from mechanics to electricity. Temperature has no citizenship there. The reason behind the success of these theories is that gravity and electric forces are insensitive to temperature. A kettle of water has the same weight whether it is hot or cold. Likewise, the trajectory of a comet is not affected by the comet's temperature, nor does temperature have any effect on the forces that two electric charges exert upon each other.

From eq. (3.11.2) and from definitions (1.11.3) and $(1.11.10)_1$, we infer that the following equations should apply to every non-thermodynamic system:

$$\mathrm{d}U \equiv \mathrm{d}\Psi \tag{3.11.5}$$

and

$$\mathrm{d}H \equiv \mathrm{d}G. \tag{3.11.6}$$

From these two equations and from eqs (1.11.13) and (1.11.15), we also infer that the relations

$$\mathrm{d}U = \mathrm{d}\Psi = \mathrm{d}G = \mathrm{d}H \tag{3.11.7}$$

must apply to every non-thermodynamic system that undergoes a process at constant pressure and volume.

Because its internal energy does not depend on temperature, a non-thermodynamic system cannot produce heat by cooling itself. Any heat deriving from the system must result from the dissipation of the work that the system produces. The dissipation must take place in the environment surrounding the system, due to the inability of a non-thermodynamic system to supply heat.

It should be clear by now that the behaviour of non-thermodynamic systems is unaffected by temperature, heat, or entropy. For this reason, the thermodynamic notions of admissible range and limit surface developed in the previous sections do not apply to them. This does not necessarily mean that a finite non-thermodynamic system can store an infinite amount of energy. It simply means that any limit to the states that such a system can attain while complying with its constitutive equations derives from conditions or constraints that are of a different nature –possibly geometric or mechanical– than the thermo-dynamical one considered in this book.

Non-thermodynamic components of a thermodynamic system

When determining the admissible range of a thermodynamic system, we should first ascertain if the system includes any non-thermodynamic component. As observed above, such a component is not restricted by the laws of thermodynamics and therefore cannot affect the admissible range of the system to which it belongs. When applying relation (3.6.5) or (3.7.1) to determine the system's admissible range, therefore, the contribution to U from any non-thermodynamic component of the system should be ignored. Failure to do so would mean introducing into relations (3.6.5) and (3.7.1) an internal energy term that, being not controlled by thermodynamics, would unduly affect the prediction of the system's admissible range.

The presence of a non-thermodynamic component in a system can be detected by inspecting the expression of its internal energy. It suffices to ascertain whether that expression can be split into the sum of two terms, one of which has the form (3.11.1). In other words, an internal energy function that can be expressed as

$$U(\boldsymbol{\xi}, T) = U'(\boldsymbol{\xi}') + U''(\boldsymbol{\xi}'', T) \qquad (3.11.8)$$

indicates that the system contains a non-thermodynamic component whose internal energy is

$$U' = U'(\boldsymbol{\xi}'). \qquad (3.11.9)$$

The sets of variables $\boldsymbol{\xi}'$ and $\boldsymbol{\xi}''$ introduced here are understood to be such that their union coincides with $\boldsymbol{\xi}$. One possibility is that $\boldsymbol{\xi}' \equiv \boldsymbol{\xi}'' \equiv \boldsymbol{\xi}$.

Once U is decomposed as in eq. (3.11.8), the admissible range and the limit surface of the system can be calculated by substituting $U''(\boldsymbol{\xi}'', T)$ for $U(\boldsymbol{\xi}, T)$ into relations (3.7.1) and (3.7.2). Thus, the admissible range of the system is determined by

$$U'' \leq T\,S, \tag{3.11.10}$$

and its limit surface is

$$U'' = T\,S. \tag{3.11.11}$$

Note that there is no need to distinguish between the entropy of the whole system and that of the system without its non-thermodynamic part. As shown above, the entropy of the non-thermodynamic part is a constant, possibly equal to zero. As far as relations (3.11.10) and (3.11.11) are concerned, such a constant is of little relevance. The reason is that function U'' contains an arbitrary additive constant resulting from the choice of the reference value of internal energy. That constant can always be chosen in such a way as to cancel the effect of any constant term coming from S in the above relations.

Pure thermo-entropic components

A similar situation occurs when the system contains a *pure thermo-entropic* component. This occurs when the entropy, S', of this component depends on temperature only. That is,

$$S' = S'(T). \tag{3.11.12}$$

Equation (3.11.2) and its consequences (3.11.3) through (3.11.7) apply to this component, provided that isothermal processes are considered. Therefore, under isothermal conditions, a pure thermo-entropic component of the system is equivalent to a non-thermodynamic one. Accordingly, its contribution to Ψ should be ignored when seeking the thermodynamic admissible range of the system at the considered temperature.

The same conclusion holds true if S' also depends on some state variables, say $\boldsymbol{\xi}'$, which remain constant in the considered process. Therefore, any system component that possesses an entropy function of the form

$$S' = S'(T, \boldsymbol{\xi}') \tag{3.11.13}$$

behaves as a pure thermo-entropic component of the system as long as isothermal processes at constant $\boldsymbol{\xi}'$ are considered. A particular case, which is often of relevance to chemistry, is when variables $\boldsymbol{\xi}'$ reduce to just the external pressure, p.

The presence of a pure thermo-entropic component in a system can be detected by examining the expression of its entropy function $S = S(T, \boldsymbol{\xi})$. Because entropy, like internal energy, is an extensive quantity, any pure thermo-entropic component should appear in that expression as a separate additive term of the form (3.11.12) or (3.11.13).

In conclusion, let U'' and S'' be the contributions to U and S, respectively, in addition to the contributions that come from any non-thermodynamic or pure thermo-entropic component that the system may contain. That is,

$$U'' = U - U' \tag{3.11.14}$$

and

$$S'' = S - S'. \tag{3.11.15}$$

For systems that contain non-thermodynamic or pure thermo-entropic components, we must substitute U'' and S'' for U and S, respectively, when we seek the admissible range and limit surface of the system via relations (3.7.1) and (3.7.2).

Alternatively, if relations (3.7.4) and (3.7.5) are used for the same purpose, then the quantity Ψ should be replaced by

$$\Psi'' = U'' - TS''. \tag{3.11.16}$$

This is the free energy of the system without the contribution from the non-thermodynamic or pure thermo-entropic components. Accordingly, the system's admissible range is expressed as

$$U'' - TS'' \leq 0 \tag{3.11.17}$$

or

$$\Psi'' \leq 0. \tag{3.11.18}$$

The expression of the limit surface follows by taking the equality case in the same relations.

Recall that relations (3.7.1) and (3.7.2) as well as relations (3.7.4) and (3.7.5) imply a choice of the arbitrary constant that enters the internal energy function of the system. For systems in which U'' and S'' do not coincide with U and S, respectively, the above constant must be chosen so that $U'' = TS''$ at the limit surface. Alternatively, one may refer to relations (3.6.5) through (3.6.8), provided that U, S, and Ψ are replaced by U'', S'', and Ψ'', respectively.

3.12 Pure Thermo-Entropic Components of Open and Chemically Reacting Systems

In the previous section, we implicitly assumed that the amount of material making up the system did not change in amount or type. This is not true if the system is open and exchanges material with the surroundings or if the system, although closed, contains chemically-reacting species.

The amount of a species, i, in a system is measured by the number of moles, n_i, of that species. This number may change if the system is open (due to material entering and leaving the system) or if chemical reactions take place in the system. In the last case, the composition of the system will vary as a reaction proceeds, because the reacting material disappears, and new material is produced as a result. Even if the state of a species in the system remains the same, the overall contribution of that species to the system's free energy may vary, simply because its amount, n_i, changes.

In particular, the contribution to S from any pure thermo-entropic component that the system may contain depends on the amount of that component, quite apart from how the entropy of the same component depends on the variables that define the component's state. This may hinder recognition of the pure thermo-entropic character of that component, because its contribution to S may appear to have a different form than (3.11.12) or (3.11.13). For this reason, when looking for the presence of a pure thermo-entropic component in an open system or a reacting mixture, it is better to refer to the molar entropies of each single species in the system. These entropies refer to a fixed quantity (one mole) of the considered species and therefore are independent of their actual amount. In the expression of molar entropies, the presence of a pure thermo-entropic term appears unambiguously in the form (3.11.12) or (3.11.13).

For open and/or chemically-reacting systems, the pure thermo-entropic terms in the expressions of S may be detected and eliminated as follows. We start by observing that the total Gibbs free energy of the system can be expressed as the sum of the contributions from each species it contains. That is,

$$G = \sum_{i} n_i \mu_i , \tag{3.12.1}$$

where μ_i denotes the molar Gibbs free energy (or chemical potential) of component *i* (see Appendix C.1, for details and notations). The quantities μ_i can, in turn, be expressed as

$$\mu_i = \mu_i^*(T, p) + f_i(T, \boldsymbol{\xi}_i), \tag{3.12.2}$$

where p is the ambient pressure and $\boldsymbol{\xi}_i$ are the state variables that refer to species *i* under consideration (Appendix C.5). For liquid solutions, the dependence on p can be ignored, and the above equation can be written as

$$\mu_i = \mu_i^*(T) + f_i(T, \boldsymbol{\xi}_i). \tag{3.12.3}$$

This expression also applies to mixtures in general, provided that we limit our attention to constant pressure processes. By applying eq. (C.6) to the molar Gibbs free energies, μ_i, of each species in the system, we obtain their partial molar entropies s_i. By referring to eq. (3.12.3), we can write them as

$$s_i(T, \boldsymbol{\xi}_i) = s_i'(T) + s_i''(T, \boldsymbol{\xi}_i), \tag{3.12.4}$$

where

$$s_i' = s_i'(T) = -\frac{\partial \mu_i^*(T)}{\partial T} \tag{3.12.5}$$

and

$$s_i'' = s_i''(T, \boldsymbol{\xi}_i) = -\frac{\partial f_i(T, \boldsymbol{\xi}_i)}{\partial T}. \tag{3.12.6}$$

The same equations also apply when μ_i is given by eq. (3.12.2) if we restrict our attention to constant pressure processes.

Equations (3.12.4) and (3.12.5) show that the molar entropy s_i of species *i* contains a pure thermo-entropic component s_i'. Therefore, as observed in the previous section, this component, and thus the Gibbs free energy term $\mu_i^*(T)$ relevant to it, should be ignored when calculating the admissible range of the system.

If required, the internal energy contribution, u_i', from one mole of that component can be obtained from

$$u_i' = \mu_i^*(T) + T\, s_i'(T), \tag{3.12.7}$$

to within an additive constant. This equation follows from definition (1.11.10) as applied to the molar quantities considered here and in light of the fact that both pressure and volume are constant (an acceptable assumption for non-gaseous phases). Again, the contribution u_i' to the internal energy of the whole system should be ignored when determining the admissible range of the system.

The notions of a non-thermodynamic component and a pure thermo-entropic component are essential to a correct determination of the admissible ranges of electrochemical cells (Chapter 7) and biological systems (Chapter 9).

3.13 Non-Thermodynamic Processes

There are several ways in which the internal energy and entropy of a complex system can change, and not all of them are thermodynamic in character. When determining the admissible range of a system, we must exclude any contribution to U and S that is not regulated by thermodynamics. The mere transport of material to and from the system, for instance, need not be a thermodynamic process. For instance, refuelling an automobile changes its total internal energy and entropy by adding new material to the system (i.e. the automobile). However, refuelling can hardly be considered a thermodynamic process. A similar situation applies to living organisms when they feed themselves or expel waste material.

These kinds of processes do not affect the internal energy and entropy of the active part of the system. For this reason, the state variables that describe the energy reserves should not be considered as belonging to the set of variables that determine the admissible range of the system. It should also be noted that a fictitious change in the value of the internal energy of the system may occur when the internal energy is being defined relative to a reference state that varies during the process.

To elaborate on the last point, let us observe that the internal energy of any system can be expressed as

$$U(p, T, \boldsymbol{\xi}) = U^\circ + f(p, T, \boldsymbol{\xi}\,;\, U^\circ), \tag{3.13.1}$$

where the quantity U° denotes the internal energy of the system at the chosen reference state. The function $f = f(p, T, \boldsymbol{\xi}; U^\circ)$ introduced above is the excess internal energy with respect to U°. It depends on the reference state used, which justifies the presence of the parameter U° (which is not a state variable) among the independent variables of function f in the above equation. Sometimes, the reference state is assumed to vary with the value of some specified state variables of the system, usually p and T. In that case, we obtain

$$U^\circ = U^\circ(p, T), \tag{3.13.2}$$

and eq. (3.13.1) can be written more explicitly as

$$U(p, T, \boldsymbol{\xi}) = U(p, T) + f(p, T, \boldsymbol{\xi}; U^\circ(p, T)). \tag{3.13.3}$$

For chemically-reacting mixtures, the reference state of each component species is referred to as the *standard state* of that species and may be defined differently for different species (Appendix C.5). Usually, the standard state depends on p and T, so an equation such as (3.13.3) applies to each species in the mixture.

Because the reference state can be chosen arbitrarily, one may also treat U° as an independent variable. In that case, from eqs (3.13.1) or (3.13.3), the differential of U can be written as

$$\mathrm{d}U = \mathrm{d}U^\circ + \mathrm{d}f\big|_{U^\circ} + \frac{\partial f}{\partial U^\circ}\mathrm{d}U^\circ, \tag{3.13.4}$$

where

$$\mathrm{d}f\big|_{U^\circ} = \frac{\partial f}{\partial p}\mathrm{d}p + \frac{\partial f}{\partial T}\mathrm{d}T + \frac{\partial f}{\partial \boldsymbol{\xi}}\mathrm{d}\boldsymbol{\xi}. \tag{3.13.5}$$

Expression (3.13.4) shows that $\mathrm{d}U$ can be split into the sum of two contributions, according to the following relation:

$$\mathrm{d}U = \mathrm{d}U' + \mathrm{d}U''. \tag{3.13.6}$$

The first contribution,

$$\mathrm{d}U' = \mathrm{d}U^\circ + \frac{\partial f}{\partial U^\circ}\mathrm{d}U^\circ, \tag{3.13.7}$$

is due to a change in the assumed reference state. It represents the increase in U due to the change in the reference state. As such, it is not due to a thermodynamic process.

In contrast, the other contribution to $\mathrm{d}U$,

$$\mathrm{d}U'' = \mathrm{d}f\big|_{U^\circ}, \tag{3.13.8}$$

represents the change in internal energy brought about by a change in the system's state variables. Therefore, if we are interested in determining the change in the system's internal energy produced by a thermodynamic process, we should ignore $\mathrm{d}U'$ and consider $\mathrm{d}U$ as

$$\mathrm{d}U = \mathrm{d}U'' \tag{3.13.9}$$

rather than as eq. (3.13.6). Failure to do so would mean including a term that does not depend on the process in the energy change produced by the process. Of course, $\mathrm{d}U' = 0$ if the reference state is kept constant, in which case eqs (3.13.6) and (3.13.9) coincide.

3.14 The Free Energy that Determines the Admissible Range

As observed in Section 3.11, the free energy, Ψ'', defined by eq. (3.11.16) is the part of the system's total free energy, Ψ, that is not due to non-thermodynamic and/or pure thermo-entropic components. As apparent from relation (3.11.18), Ψ'' determines the admissible range of the system, if the system contains non-thermodynamic and/or pure thermo-entropic components. When this occurs, the system's limit surface is equipotential for Ψ'', and, in general, will not be equipotential for Ψ.

While Ψ'' reaches its largest value at the limit surface and its lowest value at the exhaustion state, the same is not true for Ψ if the system contains non-thermodynamic and/or pure thermo-entropic components. In that case, when the system is set free from external actions, it may head to a state that is different from the exhaustion state of the system (i.e., the state where Ψ'' reaches its lowest value), because that state may not coincide with the state at which the system's total free energy reaches its lowest value.

Note that the maximum amount of free energy that a system can store/supply isothermally is determined by the difference between the highest and the lowest values of Ψ in the admissible range, i.e., Ψ_{max} and Ψ_{min}, respectively. This quantity may be quite different from the difference between the value, Ψ''_{max}, which Ψ'' assumes at the limit surface and the value, Ψ''_{exh}, which Ψ'' assumes at the exhaustion state. Note also that the states at which the total free energy attains the values Ψ_{max} and Ψ_{min} are generally different from those at which Ψ'' reaches the values Ψ''_{max} and Ψ''_{exh}.

Of course, Ψ'' coincides with Ψ for systems that do not contain non-thermodynamic and/or pure thermo-entropic components. However, when Ψ'' is different from Ψ, the system's admissible range and limit surface are determined by Ψ'', no matter how small the latter function is in comparison with the total free energy of the system.

4

The Geometric Interpretation

This chapter presents a geometric interpretation of the results of the previous chapter. It provides invaluable insight into how U and S combine to determine the admissible range of a system. It also illustrates why the capacity of a system to store and supply energy is crucially dependent on the difference between U and TS and how a change in these functions affects the system's admissible range. The analysis presented here should provide a better understanding of the physics involved in the present theory and facilitate its application.

4.1 Graphical Representation in Three or Two Dimensions

For most systems of practical interest, the number of independent state variables in the constitutive equations of U and S (i.e., the primary state variables, as defined in Section 2.4) is greater than two. Accordingly, the geometric representation of U and TS should be made in a (hyper-)

space of more than three dimensions. One dimension represents the values of U or TS, and the other dimensions accommodate the values of the state variables upon which U and S depend. However, the essentials of the geometric relationship between U and TS can also be discussed in a simpler space of three or even two dimensions. In this case, an axis of that space is used to represent the values of U and TS. The other two axes, or just one single axis in the case of a representation in a two-dimensional space, will serve to represent the state of the system as a point in the plane of the axes, or as a point of the axis, respectively.

Figure 4.1.1 shows an example of such a representation in a three-dimensional space. Each point of the $\{\boldsymbol{\xi}\}$-plane is assumed to represent a set of values for the (primary) state variables $\boldsymbol{\xi}$ of the system. The corresponding values of U and TS are measured on the vertical axis. Temperature is excluded from $\boldsymbol{\xi}$, since we are concerned with the admissible range of the system, which is defined at constant temperature.

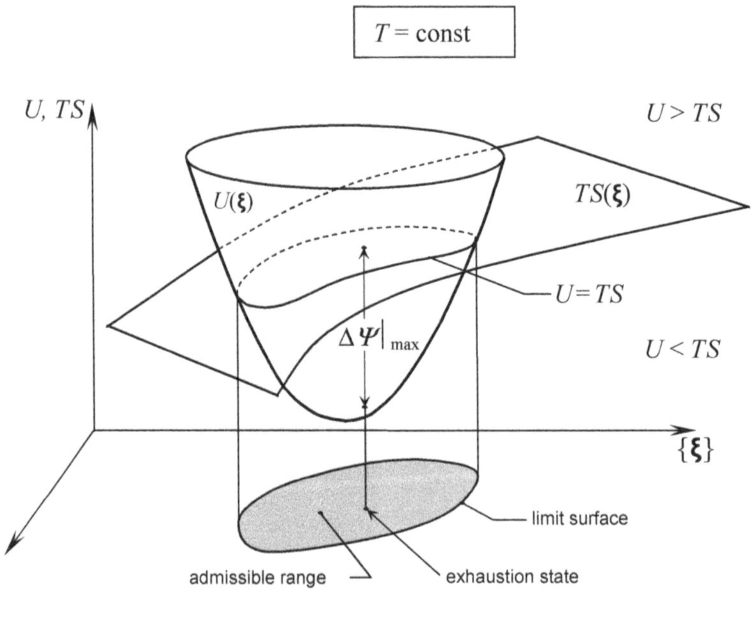

Fig. 4.1.1. A graphical representation of the admissible range in a symbolic space of three dimensions.

The above figure refers to a system whose entropy does not depend on ξ as strongly as U does. Such a feature is likely to be frequently met in practice, because the entropy of many real systems depends weakly on state variables other than temperature, while the internal energy depends strongly on them. Similar diagrams, however, can be drawn for any kind of system.

In the considered representation, the internal energy surface will move up or down depending on the reference value that we adopt for the internal energy. In the above figure, we assumed that the energy reference value was chosen so as to make constant Ψ_{max} defined by eq. (3.6.4) equal to zero. In this case, the analytical expressions of the admissible range and the limit surface assume the forms (3.7.1) and (3.7.2), respectively.

At variance with what applies to surface U, the position of surface TS is fixed. The reason is that T is constant, while the reference value of S is fixed by the third law of thermodynamics (Section 1.7). If the reference of U is understood to be chosen so that $\Psi_{max} = 0$, the values of functions U and TS must coincide at the boundary of the admissible range, in agreement with relation (3.6.6) or (3.7.2). Thus surface U and surface TS must intercept at the boundary of the admissible range. This enables us to fix their relative position once one point of that boundary is known. The graphic representation of the limit surface is easily obtained by projecting on the $\{\xi\}$-plane the intersection line of surface U and surface TS.

From relation (3.7.1) it follows that the admissible range of the system is the region of states of the $\{\xi\}$-plane where surface U does not exceed TS. Figure 4.1.1 gives a geometrical representation of that region. The exhaustion state is the point in that region where the gap between TS and U reaches its greatest value. In general, this point is different from the point where U is minimum. The symbol $\Delta\Psi|_{max}$ indicates the maximum gap between TS and U. This symbol should not be confused with Ψ_{max}, which represents the maximum value of Ψ, and is attained at the boundary of the admissible range. At the exhaustion state, Ψ reaches its minimum value. This is actually the more negative value of Ψ, if the internal energy constant is so chosen that $\Psi_{max} = 0$.

An example of a graphical representation in two dimensions is provided by Fig. 4.1.2(a). In the figure, $U(\xi)$ and $TS(\xi)$ are represented as two curves, while the points of the ξ-axis represent the states of the system. Again, in drawing this diagram we assumed that the reference value of U is chosen so that $\Psi_{max} = 0$. No point of the U-curve above curve TS can be accessed by the system, because this would contradict relation (3.7.1). The admissible range of the system is obtained by projecting on the {ξ}-axis the points where curves U and TS intercept, namely, H and K. In this representation, the admissible range appears as a segment. Its end points, A and B, represent the limit surface.

Figure 1.4.2(b) plots the free energy, $\Psi(\xi)$, of the system. This plot is obtained by taking the difference between curves U and TS of the diagram above. The choice $\Psi_{max} = 0$ makes $U < TS$ and thus $\Psi < 0$ within the admissible range. The maximum free energy, $\Psi_{max} = 0$, is attained at the ends of the admissible range. Consequently, the maximum gap between U and TS is given by $\Delta\Psi|_{max} = (0 - \Psi_{exh}) = -\Psi_{exh}$, where Ψ_{exh} is the value of Ψ at the exhaustion state. The meaning of $\Delta\Psi|_{max}$ is discussed further in Section 4.3.

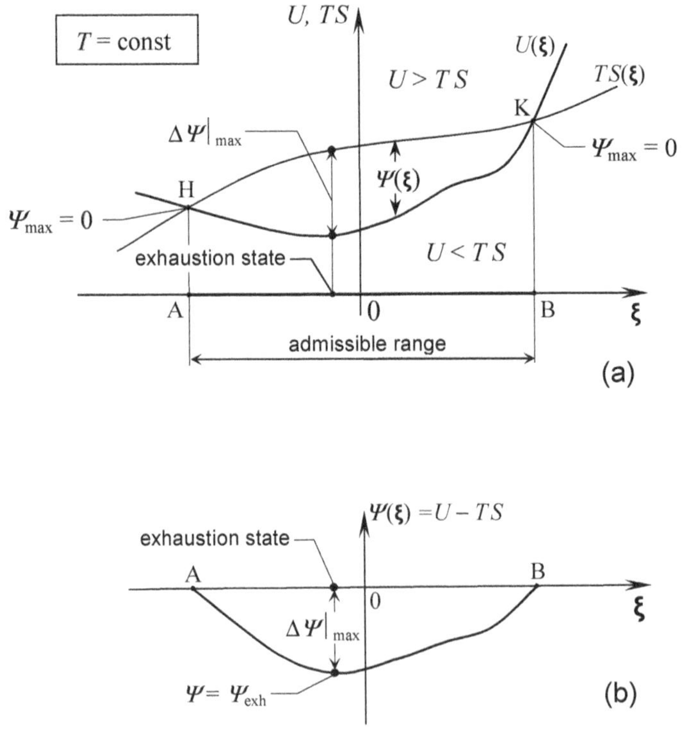

Fig. 4.1.2 (a) Two-dimensional representation of the admissible range. (b) Plot of the free energy of the system, as obtained by taking the distance between U and TS from diagram (a).

4.2 Independence of the Admissible Range from Internal Energy Reference

As variously observed, the system's internal energy, U, is defined to within an additive constant. The latter can be fixed arbitrarily or by assigning a reference value for U, which can be done arbitrarily. This arbitrariness in the value of U extends to free energy, Ψ, which is defined by eq. (1.11.3). It follows that Ψ_{max} in relation (3.6.5) is defined to within the same additive, arbitrary constant as U. This is of no consequence for the admissible range of the system, since the arbitrary constant that appears in U and Ψ_{max} cancel out from relation (3.6.5), because U and Ψ_{max} are on opposite sides of that relation. In other words, the reference value of U, and thus of Ψ, does not affect the admissible range of the system.

The situation is illustrated in Fig. 4.2.1 with reference to the two-dimensional representation. Fig. 4.2.1(a) refers to the case in which the internal energy constant is chosen so that $\Psi_{max} = 0$. The admissible range is then given by relation (3.7.1), and curves U and TS intersect at the boundary of that range. As already observed, this makes Ψ negative throughout the admissible range. The sign of Ψ is of little relevance, however, because the free energy change—not the free energy value—is what matters.

The problem of choosing a reference value for U that makes $\Psi_{max} = 0$ can be solved quite simply in practice. We only need to know a state on the limit surface and impose that U should coincide with TS at that state. In the present graphical representation, this fixes the position of the U-curve with respect to the TS-curve and hence defines the entire admissible range.

A different choice in the reference value of U will make $\Psi_{max} \neq 0$. Figure 4.2.1(b) refers to the case in which $\Psi_{max} > 0$. In that case, the U-curve turns out to be displaced upward with respect to the U-curve of Fig. 4.2.1(a). In Fig. 4.2.1(b), the new U-curve is in bold, while the old U-curve is dashed. The admissible range of the system is now given by relation (3.6.5) rather than relation (3.7.1). The boundary of this range is represented by the intersection of the new U-curve with the $(TS + \Psi_{max})$-curve. The latter is obtained by moving curve $TS(\xi)$ upward by the same amount Ψ_{max} by which the new U-curve was displaced with respect to the U-curve of Fig. 4.2.1(a). It follows that the relative position of the new U-curve and the $(TS + \Psi_{max})$-curve is the same as that of the U-curve and the TS-curve that were considered in Fig.

4.2.1(a). As a result, the admissible range of the system remains unaltered.

In Fig. 4.2.1(b), the free energy of the system is positive throughout the admissible range. However, the difference in free energy of any two given states within that range is the same as the analogous difference obtained by referring to Fig. 4.2.1(a). In particular, the difference between the maximum value of Ψ and the value of Ψ at the exhaustion state, i.e. $\Delta\Psi|_{max}$, is unaffected by the change in the reference value assumed for U. This is as it should be, since $\Delta\Psi|_{max}$ represents the maximum amount of free energy that the system can store and supply. Clearly, this quantity is independent of our choice of the reference value for U.

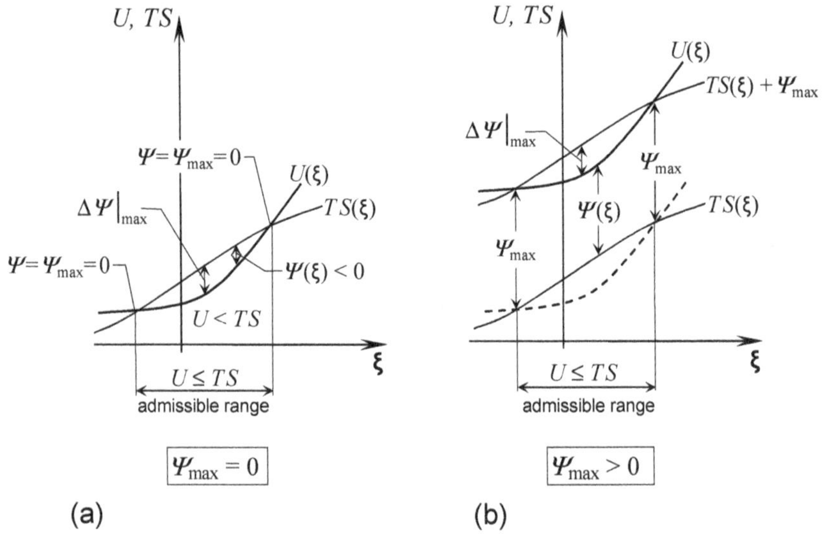

Fig. 4.2.1. Independence of the admissible range from the reference value of internal energy. (a) The reference value of U is such that $\Psi = \Psi_{max} = 0$ at the boundary of the admissible range. (b) The reference value of U makes $\Psi = \Psi_{max} > 0$ at the same boundary; this leaves the admissible range unaltered.

4.3 The Gap Between *U* and *TS*

At any state within the admissible range, the values of U can never exceed those of $(TS+\Psi_{max})$. This is the immediate consequence of relation (3.6.5). In the graphical representations considered in this chapter, this means that the admissible range of any system is the region in state space where curve/surface U is lower than curve/surface $(TS+\Psi_{max})$. As stated by eq. (3.6.6), the boundary of that region is the locus of states at which curve/surface U and curve/surface $(TS+\Psi_{max})$ intersect. The maximum gap, $\Delta\Psi|_{max}$, that the two curves/surfaces attain in the admissible range can be calculated as follows:

$$\begin{aligned}\Delta\Psi|_{max} &= [(TS+\Psi_{max})-U]_{max} \\ &= [\Psi_{max}-(U-TS)]_{max} \\ &= \Psi_{max} - min[\Psi(\boldsymbol{\xi})] = \Psi_{max}-\Psi_{exh}\,. \end{aligned} \tag{4.3.1}$$

Here, $min[\Psi(\boldsymbol{\xi})]$ denotes the smallest value that Ψ attains in that range. This is the value of the free energy of the system at the exhaustion state, already denoted as Ψ_{exh}. That is

$$min[\Psi(\boldsymbol{\xi})] = \Psi_{exh}\,, \tag{4.3.2}$$

which justifies the last equation in formula (4.3.1).

If the reference value of U is chosen so that $\Psi_{max}=0$, the previous expression of $\Delta\Psi|_{max}$ reduces to

$$\Delta\Psi|_{max} = -\Psi_{exh}\,. \tag{4.3.3}$$

For systems with a non-vanishing admissible range, this formula produces a positive value for $\Delta\Psi|_{max}$, since the choice $\Psi_{max}=0$ makes $\Psi<0$ within the admissible range and thus, in particular, $\Psi_{exh}<0$.

No matter how it is expressed, the gap $\Delta\Psi|_{max}$ represents the maximum amount of non-thermal energy that the system can supply isothermally. This energy is given off by the system in any process that starts from the boundary of the admissible range and ends at the exhaustion state.

Interestingly enough, the gap $\Delta\Psi|_{max}$ is not necessarily related to the size of the admissible range. There may be systems with a small admissible range and a large gap [Fig. 4.3.1(a)] and systems with a wide admissible range and a small gap [Fig. 4.3.1(b)]. The reason is that $\Delta\Psi|_{max}$ depends on the relative position of curves U and TS as well as their shapes. Of course, if the shapes of U and TS are fixed, the gap $\Delta\Psi|_{\max}$ relates directly to the size of the admissible range, as Fig. 4.3.1(c) shows. All diagrams of Fig. 4.3.1 refer to the case in which the reference value of U is chosen so as to make $\Psi_{\max}$ equal to zero.

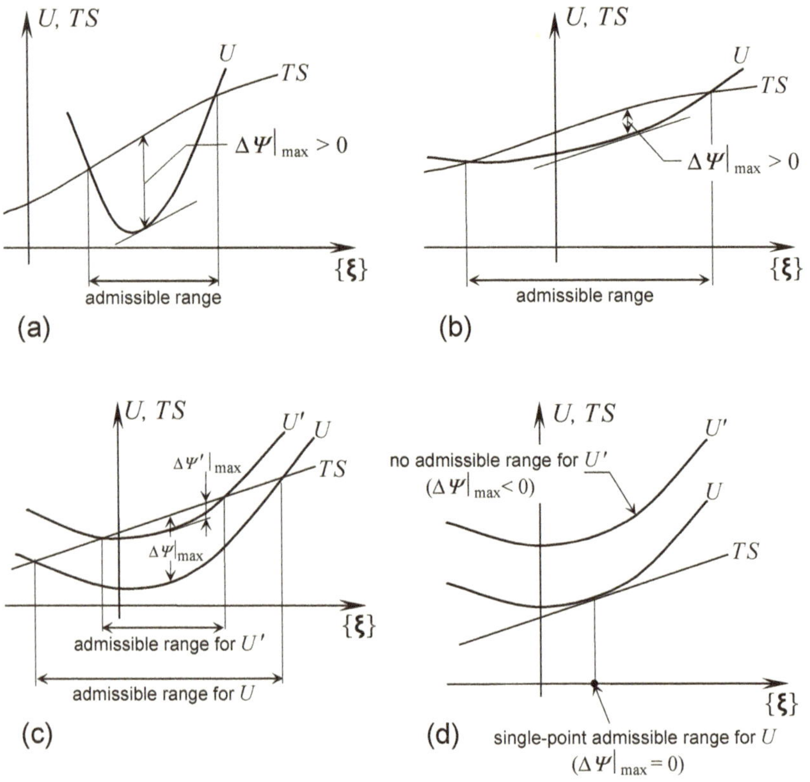

Fig. 4.3.1. The gap, $\Delta\Psi|_{\max}$, between U and TS is, in general, unrelated to the size of the admissible range of the system, insets (a) and (b). However, if the shape of curves U and TS is fixed, then $\Delta\Psi|_{\max}$ determines the size of the admissible range, inset (c). In particular, the admissible range reduces to a single point if $\Delta\Psi|_{\max} = 0$ and it vanishes altogether if $\Delta\Psi|_{\max} < 0$, inset (d).

For a system to posses a non-vanishing admissible range, the gap $\Delta\Psi|_{\max}$ must be greater than zero. This is the direct consequence of definition (4.3.1) and relation (3.6.5), or definition (4.3.3) and relation (3.7.1). If $\Delta\Psi|_{\max}$ is less than zero, there are no admissible states for the system. In this case the admissible range vanishes. On the other hand, if $\Delta\Psi|_{\max} = 0$, the admissible range reduces to just a single point [Fig. 4.3.1(d)]. In the present discussion, we exclude the very special and rather trivial case in which curves/surfaces U and TS coincide or differ by a constant. This would represent a system whose free energy is constant, independent of the state of the system. Clearly, such a system would possess an infinite admissible range. However, it would be incapable of storing or producing any non-thermal energy while operating isothermally.

4.4 Interpreting the Effect of Constitutive Changes Geometrically

The effect of a change in the constitutive equations of U and S can be followed effectively in the graphical representation introduced above. Any process that modifies the constitutive equations of a system is a *formation process*. The general features of these processes are presented at length in Chapter 5. As already discussed in Section 2.5, no formation process can take place within the admissible range of the system. Therefore, any process that changes the system's constitutive equations must bring the system outside its actual admissible range.

As an example, let us consider a steel bar subjected to tension or compression loads by two axial forces on the ends of the bar. The bar is made of a linear thermoelastic material, which is a good approximation for steel in the elastic range. Though referring to a particular case, this example shows how the effects of a change in the constitutive properties of a system can be anticipated with the help of the present graphical representation. A similar problem was considered in [51]. The general case is treated in detail in Chapter 6.

Since the bar is in a uniaxial state of stress, we can take the axial elastic strain, ε_e, as the state variable that defines the deformation of the bar. This scalar is sufficient to fully describe the state of deformation of the bar in this example, and it should be distinguished from the more general strain tensor, $\boldsymbol{\varepsilon}^e$, which is considered in Chapter 6. Besides temperature and ε_e, the state of the bar is described by other variables that are related to the previous inelastic (plastic) deformation of the bar or, more generally, to its state of preparation. These variables are denoted collectively as $\boldsymbol{\xi}_p$. We need not specify them, because we confine our attention to purely elastic deformation processes, i.e., processes at constant $\boldsymbol{\xi}_p$. Under isothermal conditions, therefore, the elastic strain, ε_e, is the only state variable that can change. The admissible range for ε_e can accordingly be referred to as the *elastic range* of the bar at the considered temperature and considered state of preparation.

The constitutive equations of the bar's internal energy and entropy are the following. The specific internal energy u (per unit mass) is

$$u = u(\varepsilon_e, \boldsymbol{\xi}_p, T) = \frac{1}{\rho_0}\left(\frac{E}{2}\varepsilon_e^{\,2} + \alpha E \varepsilon_e T\right) + f(\boldsymbol{\xi}_p, T). \qquad (4.4.1)$$

The specific entropy s (per unit mass) has the form:

$$s = s(\varepsilon_e, \boldsymbol{\xi}_p, T) = \frac{1}{\rho_0}\alpha E \varepsilon_e + g(\boldsymbol{\xi}_p, T). \qquad (4.4.2)$$

More details on these equations are provided in Chapter 6. In these equations, ρ_0 is the mass density relevant to the reference state from which ε_e is measured. This state is assumed to be the stress-free state at temperature T_0. The quantities E and α are the linear elastic modulus (Young modulus) and the thermal expansion coefficient of the material, respectively. In the same equations, the two functions $f(\boldsymbol{\xi}_p,T)$ and $g(\boldsymbol{\xi}_p,T)$ incorporate all of the contributions to u and s, respectively, that do not depend on ε_e. Both functions remain constant as long as the bar remains in its elastic range at the considered temperature. Function $f(\boldsymbol{\xi}_p,T)$ is also assumed to incorporate the arbitrary constant that enters the definition of internal energy.

Since the bar is a homogeneous system, the values of U and S of the entire bar are obtained by multiplying the specific values u and s by the total mass of the bar. Thus, the graphical representation of U and S coincides with that of u and s except for a scale factor.

As previously observed, the admissible range of a system can be determined once a point at its boundary is known. Since the internal energy constant can be fixed arbitrarily, no restriction is introduced if the values of u and Ts are assumed to coincide at that point. Then, according to relation (3.7.1), the elastic range of the bar is represented by the states for which $u \leq Ts$. The equality sign in this relation holds true at the boundary of that range, i.e., at the points where the bar reaches the elastic limit. Observe that by imposing that $u = Ts$ at one end of the elastic range, say at the end that corresponds to the elastic limit in tension, ε_e', the position of the elastic limit in compression, ε_e'', and hence the whole elastic range, is determined (Fig. 4.4.1).

Consider now a process that brings the bar beyond the elastic range. For a material such as steel, this can be done by straining the material beyond its elastic limits. The process produces plastic deformation and makes the material entrap internal energy due to the microscopic distortions that are generated in the crystal grains of the material and at the crystal grain interfaces. Although these microscopic distortions are elastic in character, they cannot be released simply by removing the macroscopic stress. As a consequence, they generally result in a permanent elastic strain, say Δe, which adds to the purely plastic strain and makes the material entrap the elastic energy Δu_e. This phenomenon is studied extensively in [52].

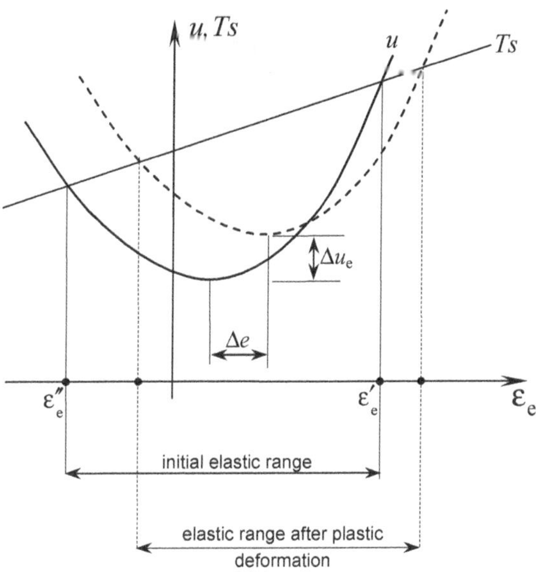

Fig. 4.4.1. The elastic range of a linear, elastic steel bar before and after plastic deformation. The solid parabola refers to the bar's internal energy before plastic deformation. The dashed curve is the internal energy after plastic deformation. The *Ts* line is assumed to be unaffected by plastic deformation.

Plastic deformation of iron and steel is known to leave both E and α nearly unchanged. In view of eq. (4.4.1), this means that the shape of the u-curve in the $[u, \varepsilon_e]$-plane remains unaltered. However, the change in the value of $f(\boldsymbol{\xi}_p,T)$ due to the change in $\boldsymbol{\xi}_p$ and the change in the stress free state of the material produced by Δe result in a rigid-body translation of the u-curve in the $[u, \varepsilon_e]$-plane. In Fig. 4.4.1, the displaced *u*-curve is represented dashed. The *Ts* line is not significantly affected by plastic deformation and, therefore, will remain to a good approximation unaltered. Fig. 4.4.1 shows how the new elastic range and the new elastic limits resulting from the plastic deformation process can be obtained graphically from the intersections of the displaced *u*-curve with the *Ts*-line. This provides a graphical, thermodynamic explanation of a well-known phenomenon in the theory of metal plasticity, which is referred to as strain hardening with an associated Baushinger effect.

An analogous graphical analysis can be used to study the evolution of the elastic range of the steel bar under other processes, such as quenching and annealing. The rapid cooling that is associated with quenching freezes into the material a micro-crystalline structure that is stable at higher temperatures. This prevents the transition of the crystal structure of the material into a lower-entropy phase and thus increases the entropy of the material as it recovers the ordinary temperature at the end of quenching. The process leaves E and α almost unaltered. In view of eq. (4.4.2), this means that the increase in s is due to an increase in the inelastic term, $g(\boldsymbol{\xi}_p,T)$. As a result, the *Ts* line will move upward, as shown in Fig. 4.4.2(a). Because it does not essentially alter the values of E and α, the quenching process has little effect on the u curve, which is therefore left unchanged. The widening in the elastic range of the material is evident from Fig. 4.4.2(a), and it explains why the process increases the elastic limits of the bar and thus its strength.

Annealing has the opposite effect. It produces re-crystallization of the material to a more stable crystalline phase with lower entropy. In this case, the *Ts* line moves downward as the entropy of the material decreases in the process. Fig. 4.4.2(b) shows why annealing makes the elastic range shrink, thus reducing the elastic limits and the strength of the bar. The steel becomes mild and more machinable as a result.

The graphical representation described in this chapter is rather intuitive. It has the potential of explaining and unifying many disparate phenomena concerning the behaviour of a system, which would otherwise appear to be unrelated. It also suggests how some basic features of the system can be controlled and modified. Such representation applies to any system, whether inanimate or living. It does not require delving into the details of the microscopic structure of the system. In fact, the system may be a single cell, a homogeneous cell tissue, or a colony of cells; a chromosome or any polymeric chain of molecules; an inorganic liquid; an amorphous glass; or a microcrystalline solid. This representation only requires the knowledge of the state variables of the system and the expressions of its internal energy and entropy.

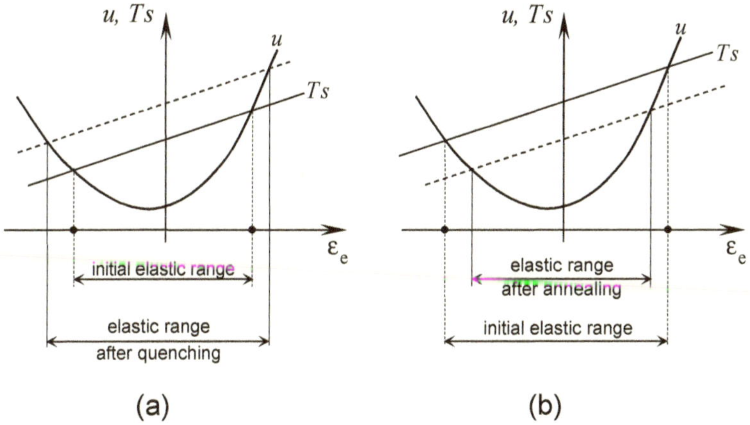

Fig. 4.4.2. (a) Quenching a steel bar makes the Ts line translate rigidly upward, thus increasing its elastic range and elastic limits. (b) Annealing displaces the Ts line downward and reduces elastic range and elastic limits.

5

System Formation

The capacity to supply and absorb energy to and from surroundings is one of the most fundamental properties of any system, whether inanimate or living. As discussed in Section 2.2, this property enables the system to retain its constitutive features in spite of all the energy exchanges that it has to negotiate for everyday existence. We often focus on these energy exchanges when the system operates in an environment at constant or near-constant temperature. In this case, the maximum gap between U and $T\,S$ represents the maximum amount of non-thermal energy that the system is capable of storing and releasing at the considered temperature (Section 4.3).

In the present chapter, we consider the processes that make the system acquire or modify that capacity. These processes are referred to collectively as *formation processes*. They must drive the system out of its original admissible range, because no change in the system's constitutive properties can be produced if the process belongs to that range.

Any increase in the system's capacity to store and supply non-thermal energy comes at a cost. This is the amount of non-thermal energy that a formation process invariably requires to be spent or dissipated to be executed. That energy is calculated in the next section, while subsequent sections consider the various ways in which a formation process can actually take place.

5.1 Primary Formation and Energy Cost of the Admissible Range

In Section 3.7, we observed that, without any loss in generality, we can choose the reference value of U in such a way that the constant Ψ_{max}, defined by eq. (3.6.4), is reduced to zero. This simplifies the expression of the admissible range of the system and, unless otherwise stated, is understood to apply. The quantity $\Delta\Psi|_{max}$, defined by eq. (4.3.1), then represents the maximum distance or gap between surface U and surface TS in the admissible range. It also represents the maximum amount of free energy that the system can store or spend in an isothermal process that is compatible with the system's constitutive equations.

$\Delta\Psi|_{max}$ is acquired at system formation. It cannot be modified by a process that belongs to the system's admissible range. To change $\Delta\Psi|_{max}$, the constitutive functions U and/or S must be modified. This requires another formation process.

Of course, systems in which $\Delta\Psi|_{max}$ is equal to zero are a possibility. This happens if $U \equiv TS$. A system with this property is, however, expected to be extremely volatile. Since it is unable to store any free energy, such a system would be at the mercy of even the smallest energy input. In the natural world, systems of this sort are expected to be the exception rather than the rule. For this reason, unless otherwise stated, we henceforth concentrate on systems that exhibit a non-vanishing gap $\Delta\Psi|_{max}$ and thus a non-vanishing admissible range. Finally, as stated at the beginning of Section 4.3, we exclude the unrealistic case of systems whose free energy is a constant.

Throughout this book, we confine our attention to formation processes that occur isothermally or almost so. Isothermal formation is quite common in the case of inanimate systems. Phase changes and the hardening of a metal due to cold working are two familiar instances of isothermal formation processes of inanimate systems. Also for living systems, isothermal formation is the rule rather than the exception. This is a consequence of the limited capacity of a living system to withstand even comparatively small temperature changes.

According to relation (3.7.1), U should be less than or equal to TS at any state of the system that belongs to the system's admissible range. For U to be less than TS, however, the system must have suffered a formation process where $\dot{U} < T\dot{S}$ for some finite time interval during its history, possibly far in the past. If $\dot{U}$ had been greater than $T\dot{S}$ for all of the past history of the system, the system could never have reached a state in which $U < TS$. Such a system could not possess an admissible range and therefore could not exist (see Fig. 4.3.1 and the discussion that precedes it). It follows that the formation process of any system must always contain at least one part in which $\dot{U} < T\dot{S}$ for a finite time interval. We refer to this part of the formation process as the *primary formation* (*process*) of the system.

The need for a primary formation process implies that the formation of every system requires the expenditure of some non-thermal energy to be executed. This can be proved as follows. From eqs (1.4.8) and (1.9.15), the energy dissipation during a primary formation process can be expressed as

$$D = \int_{pf} \left(T\dot{S} - \dot{U}\right) \mathrm{d}t + \int_{pf} \dot{W}_{\mathrm{in}} \, \mathrm{d}t \geq 0 \,, \tag{5.1.1}$$

where *pf* indicates primary formation. Now, a primary formation process that makes TS exceed U by, say, the amount $\Delta\Psi|_{\max}$, must be such that

$$\int_{pf} \left(T\dot{S} - \dot{U}\right) \mathrm{d}t = \Delta\Psi\big|_{\max} \,. \tag{5.1.2}$$

From this and from eq. $(5.1.1)_1$ we obtain

$$\begin{aligned} D &= \Delta\Psi\big|_{\max} + \int_{pf} \dot{W}_{\mathrm{in}} \, \mathrm{d}t \\ &= \Delta\Psi\big|_{\max} + W_{\mathrm{in}} \,, \end{aligned} \tag{5.1.3}$$

where

$$W_{\mathrm{in}} = \int_{pf} \dot{W}_{\mathrm{in}} \, \mathrm{d}t \,. \tag{5.1.4}$$

Equation $(5.1.3)_2$ and condition $\Delta\Psi|_{\max} > 0$ imply that

$$\Delta\Psi\big|_{\max} = D - W_{\mathrm{in}} > 0 \,. \tag{5.1.5}$$

This means that a primary formation process dissipates more energy than the non-thermal energy (W_{in}) that is absorbed by the forming system. The excess of dissipated energy over the non-thermal energy absorbed is $\Delta\Psi|_{max}$. This quantity is greater than zero in a primary formation process. Thus, a primary formation process always requires the expenditure of non-thermal energy to be performed. This completes the proof.

This conclusion does not necessarily mean that a primary formation process must be dissipative or that the forming system must absorb non-thermal energy from its surroundings. The following three alternatives help illustrate this point. The first alternative is when a primary formation process does not involve any exchange of non-thermal energy with the surroundings. In this case, $\dot{W}_{in} = 0$ throughout the process, and eq. (5.1.3) yields:

$$D = \Delta\Psi\big|_{max} . \tag{5.1.6}$$

This means that the process must be irreversible and dissipate as much energy as the width of the gap between TS and U that it generates. Note that irreversible formation is the only way in which a system can increase its $\Delta\Psi|_{max}$, i.e. its capacity of storing non-thermal energy, by a process in which $\dot{W}_{in} = 0$.

The second alternative is when the primary formation process occurs reversibly. In that case, $D = 0$, and, from eq. $(5.1.3)_1$, we obtain:

$$W_{in} = \int_{pf} \dot{W}_{in}\, \mathrm{d}t = -\Delta\Psi\big|_{max} . \tag{5.1.7}$$

This can also be written as

$$W_{out} = \int_{pf} \dot{W}_{out}\, \mathrm{d}t = -\int_{pf} \dot{W}_{in}\, \mathrm{d}t = \Delta\Psi\big|_{max} , \tag{5.1.8}$$

since $\dot{W}_{out} = -\dot{W}_{in}$, according to eq. (1.3.2). Equation (5.1.8) shows that, in a reversible formation process, the system must spend as much non-thermal energy as the width of the gap $\Delta\Psi|_{max}$ that it generates.

The third and last alternative is when a primary formation process is both dissipative and occurs at non-vanishing $\dot{W}_{out}$. For this process, we have

$$D = \Delta\Psi\big|_{max} - W_{out} , \tag{5.1.9}$$

according to relation (5.1.3) and eq. (1.3.2). In this case, the system dissipates as much energy as the difference between $\Delta\Psi|_{max}$ and W_{out}.

5.2 More General System-Forming Processes

Though essential to the formation of a system, primary formation processes are not the only processes that contribute to the formation of a system. More generally, a *system-forming process* is any process that modifies the constitutive equations of U and S of the forming system. In general, a system-forming process produces changes in how U and S depend on the system's state variables (i.e., the primary state variables, as defined in Section 2.4) and/or modifies the type and number of these variables. The primary formation processes considered in Section 5.1 are those particular system-forming processes in which $\dot{U} < T\dot{S}$. They are the processes that increase the gap $\Delta\Psi|_{max}$.

The definition of a system-forming process given above refers exclusively to functions U and S. This is due to the fact that U and S are the functions that control the system's admissible range and the gap $\Delta\Psi|_{max}$ (i.e., the maximum amount of non-thermal energy that the system can absorb or spend). According to that definition, a process that modifies a constitutive equation other than the equations of U and S is not a system-forming process. For instance, size and weight are constitutive properties of the system, which may be related to the amount of material reserves (such as fuel and fat) that the system stores. The amount of these reserves can be changed by appropriate refuelling processes. However, these reserves do not affect the system's admissible range, or the capacity of the system to spend its free energy apart from the energy of the above energy reserves. The power of a car engine, to use a common instance, is largely independent of the amount of fuel in the car's tank, as long as there is some fuel in it.

Because it modifies the constitutive equations of the system, a system-forming process is not an admissible process as defined in Section 2.6. Consequently, system formation cannot take place within the admissible range of the forming system. That range only contains the states that the system can reach that are compatible with its actual constitutive equations for U and S (see Section 2.5). To modify those equations, the formation process must necessarily bring the system out of its original admissible range. This can be done either by making the state variables of the system cross the limit surface or by activating some new state variable, if only temporarily.

A system-forming process that is not a primary formation process need not proceed at $\dot{U} < T\dot{S}$. For instance, a formation process that reduces the existing gap between U and TS must proceed at $\dot{U} > T\dot{S}$

(rather than at $\dot{U} < T\dot{S}$), for an appropriately long time interval. Neither must a process in which $\dot{U} < T\dot{S}$ be a system-forming process. In fact, processes in which $\dot{U} < T\dot{S}$ that are not system-forming processes are rather common. Any irreversible process in which $\dot{W}_{in} < 0$ will make $\dot{U} < T\dot{S}$, because

$$\dot{U} - T\dot{S} \leq \dot{W}_{in}, \tag{5.2.1}$$

as immediately follows from eqs (1.4.8) and (1.8.1). Clearly, if such a process occurs within the admissible range of the system, it will not produce any change in the system's constitutive properties. In that case, the process would not be a formation process.

5.3 Evolution of the Limit Surface

As previously observed, a system is defined by its state variables and constitutive equations. The latter relate the state variables to the system's properties that are considered in the model. The admissible range in the space of the state variables is the region of the states that the system can attain through the processes that comply with the system's constitutive equations and the laws of physics. Thus, the boundary of this region (i.e., the limit surface) cannot be crossed without violating the system's constitutive equations, the laws of physics, or both.

Because a formation process modifies the constitutive equations of the system, it must cross the system's limit surface. However, outside the admissible range, there is no process that is compatible with the constitutive equations of the system and the laws of physics. Therefore, crossing the limit surface must produce changes in the system's constitutive equations. The admissible range and limit surface of the system are modified as a result.

A change in the constitutive equations is a change in the system. Consequently, all subsequent admissible ranges that the system attains while performing a formation process are, strictly speaking, relevant to different systems. Once the limit surface is crossed and a new limit surface is produced, the system becomes a different system from that it was before. In particular, the system cannot recover the original state and properties without leaving the newly acquired admissible range. To revert back to the original condition, the system must undergo a further formation process, provided that such a process is possible (which, in general, cannot be taken for granted, especially under isothermal conditions).

The open curve ABCD of Fig. 5.3.1 represents a generic formation process in the system's state space. The closed curves are the limit surfaces that the system attains sequentially during the process. The area within them is the corresponding admissible range. The figure shows the limit surfaces and admissible ranges of the system as the process reaches states A, B, C,..., respectively. Points O, O′, O″, ... are the exhaustion states relevant to each range. The position of the exhaustion state is supposed to change during the process, because a formation process may actually affect other fundamental properties of the system besides the admissible range.

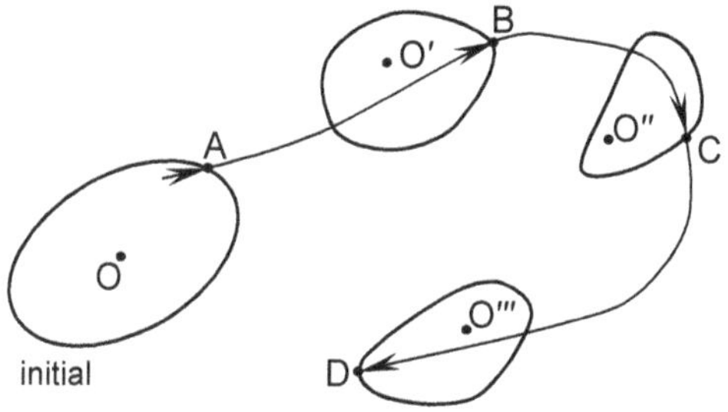

Fig. 5.3.1. Representation of the evolution of the admissible range of a system during a formation process (represented by curve ABCD).

An important property of the formation processes is the following:

Each point of a formation process must belong to the system's limit surface relevant to that point of the process.

In other words, a formation process always occurs at the boundary of the actual admissible range of the system. This is the direct result of the fact that the limit surface is the only place where a change in the system's constitutive equations can occur.

As discussed at length in Chapter 3, the limit surface must be equipotential for Ψ or, more generally, for the part of Ψ that does not include the contributions of non-thermodynamic or pure thermo-entropic components that the system may contain (see Sections 3.11-3.12). This justifies the following statement:

At any time during a formation process, the free energy of the forming system is consistently equal to the maximum amount of free energy that the system can store at that time.

The free energy referred to here is the free energy that the system stores in addition to that of the non-thermodynamic or pure thermo-entropic components that it may contain. For systems that include non-thermodynamic or pure thermo-entropic components, this is the free energy Ψ'' rather that the total free energy, Ψ, of the system.

Besides taking place at a point on the limit surface, a formation process must bring the system out of its admissible range. This requires that the process be outwardly directed with respect to the limit surface (Fig. 5.3.1). Otherwise, no change in the constitutive properties of the system could take place, because the process would occur entirely within the admissible range and, as such, would be compatible with the actual constitutive equations of the system.

5.4 Autonomous and Non-Autonomous Formation Processes

The realm of system-forming processes is wide and varied. Phase transitions, formation of new compounds, crystallization, and re-crystallization are just a few instances of system-forming processes that concern inanimate systems. Plastic yielding and rupture are additional instances of processes that modify the original response of a solid and must therefore be classified as system-forming processes. Even more varied, if possible, are the formation processes of biological systems, which range from the transformation of inert matter into a living system (*life-producing processes*) to the transformation of a living system into a lump of inanimate matter (*death-inducing processes*). Between these extremes, there is a whole spectrum of processes of growth and formation of a living system into a complex organism and processes in which complex organisms decay into a host of smaller and simpler living systems.

The relations that describe how the constitutive equations or some state variables change during a formation process are referred to as *evolution rules* or *evolution equations*. These relations are constitutive in character in that they describe a property of the system. Accordingly, their actual expression depends on the system. When an evolution equation concerns the growth of a variable that is related to displacement of matter, the evolution equation is also called *flow rule* or *flow equation*. For instance, in the theory of metal plasticity, the equation that governs the evolution of plastic strain is usually referred to as the plastic flow rule.

The systematic study of the physics that controls the evolution equations of a system is still in its infancy. The lack of a general theoretical framework in which the evolution rules can be classified and analyzed has undoubtedly hampered the development of this important subject. In many instances the evolution equations of particular systems have been known, or at least used, for a long time (cf., e.g., the evolution rules of the elastic range of a ductile metal during plastic deformation or those that describe the change in the state of aggregation of a material, say in the transition from solid to liquid). However, a general knowledge of the role of thermodynamics in shaping these rules is rather patchy, to say the least.

In what follows, we distinguish between two main classes of formation processes, depending on whether they involve absorption of non-thermal energy from the surroundings. The first class refers to the formation processes that do not absorb non-thermal energy from the surroundings. At any time during these processes, $\dot{W}_{in} \leq 0$. A process in this class is referred to as an *autonomous formation process*. The other class of formation processes refers to *non-autonomous formation processes*. These are the formation processes that require the expenditure of non-thermal energy on the forming system. During these processes, therefore, $\dot{W}_{in} > 0$.

In the following sections, the above two main classes of processes are divided further into more specific subclasses, depending on the changes in U and S they produce and whether they are adiabatic or diathermic. To keep the discussion simple, we assume that the formation process leaves the free energy of the exhaustion state, Ψ_{exh}, unaltered. The more general case can be treated in a similar way. However, any conclusion concerning the change in the gap $\Delta\Psi|_{max}$ will generally depend on the change in the free energy of the exhaustion state. We also assume that the system does not contain any non-thermodynamic component or any pure thermo-entropic component (Sections 3.11-3.12). This makes the limit surface equipotential for Ψ. However, the analysis that follows can be extended immediately to the cases in which the free energy includes non-thermodynamic and/or pure thermo-entropic components, provided that it is rephrased in terms of Ψ'' rather than Ψ.

Let $U(t)$ and $TS(t)$ be the values of U and TS at time t of a formation process. Because the system is consistently on the limit surface during a formation process (Section 5.3), the difference $U(t)-TS(t)$ represents the free energy of the limit surface at that time of the process. By recalling that the largest value of Ψ is attained on the limit surface (Section 3.6), we have

$$\Psi_{max}(t) = U(t) - T\,S(t), \tag{5.4.1}$$

at any time of the formation process. At time t, let $\overline{\delta\Psi}$ denote the increase in Ψ_{max} with respect to the value, $\Psi_{max}(0)$, of Ψ_{max} at the beginning of the formation process (assumed to start at time $t = 0$). In general, $\overline{\delta\Psi} = \overline{\delta\Psi}(t)$ or, more explicitly,

$$\overline{\delta\Psi} = \overline{\delta\Psi}(t) = \Psi_{max}(t) - \Psi_{max}(0) = U(t) - T\,S(t) - \Psi_{max}(0). \tag{5.4.2}$$

Because we are considering formation processes that leave Ψ_{exh} unchanged, from eqs $(5.4.2)_2$ and $(4.3.1)_4$, we calculate that at time t the value of the gap $\Delta\Psi|_{max}$ is given by

$$\begin{aligned} \Delta\Psi(t)|_{max} &= \Psi_{max}(t) - \Psi_{exh} \\ &= \Psi_{max}(0) - \Psi_{exh} + \Psi_{max}(t) - \Psi_{max}(0) \\ &= (\Delta\Psi|_{max})_0 + \overline{\delta\Psi}(t), \end{aligned} \tag{5.4.3}$$

where

$$(\Delta\Psi|_{max})_0 = \Psi_{max}(0) - \Psi_{exh} \tag{5.4.4}$$

is the initial value of the gap $\Delta\Psi|_{max}$, as obtained by applying eq. (4.3.1) at time $t = 0$.

5.5 Autonomous Formation

According to the definition given in the previous section, any formation process that absorbs no energy other than heat from its surroundings is classified as an autonomous formation process. In such a process, the quantity $\dot{W}_{in}$ will either vanish or be less than zero. When $\dot{W}_{in}$ vanishes, the system is also referred to as mechanically isolated. This terminology should be taken in its broadest sense here, meaning that any exchange of non-thermal energy (i.e., mechanical, magnetic, electric, etc.) is precluded.

In many practical cases, the value of $\dot{W}_{in}$ may be so small that it can be ignored altogether, whether it is positive or negative. An instance of an autonomous formation process that occurs at negligible $\dot{W}_{in}$ is the egg incubation process that leads to chick formation. The egg shell is almost rigid, which prevents any volume change during incubation. This makes $\dot{W}_{in} = 0$, because the forces acting on a rigid shell cannot perform any work. Gas exchange through the shell pores does take place between the egg and the surroundings during incubation. This involves negligible work, though, since it is mainly due to gas diffusion, occurs at small pressure differences, and the moles of produced gas equals to the moles of absorbed gas. Every energy exchange between the egg and its surroundings can then be regarded as thermal. This makes the chick formation during egg incubation essentially an autonomous formation process.

Other common instances of formation processes that occur at vanishing $\dot{W}_{in}$ are the processes of solidification and fusion of ordinary materials at atmospheric pressure. The small volume change that these processes produce makes $\dot{W}_{in}$ negligible with respect to the heat absorbed/lost by the system, a conclusion that rigorously applies when the process occurs in a vacuum.

An isothermal autonomous formation process can never produce a decrease in the gap $\Delta\Psi|_{max}$. This can be proved as follows. First, in such a process, we have

$$\dot{U} \leq \dot{Q}\,, \tag{5.5.1}$$

as follows from the first law (1.4.8), since $\dot{W}_{in} \leq 0$ because the process is autonomous. From the second law (1.8.1), we then infer that

$$\dot{U} - T\,\dot{S} \leq 0 \tag{5.5.2}$$

should also apply to the process. This means that the difference U–TS cannot increase as a result of autonomous formation. From eq. $(4.3.1)_2$, it can then be concluded that $\Delta\Psi|_{max}$ cannot decrease in the process.

In other words, a system cannot reduce its maximum capacity of storing/releasing non-thermal energy while operating on its own, i.e. in a process that does not take any non-thermal energy from the surroundings. That capacity, i.e. the gap $\Delta\Psi|_{max}$, can be reduced only by a non-autonomous formation process (a process in which $\dot{W}_{in} > 0$).

Incidentally, as observed in Section 5.1, in a primary formation process we have

$$\dot{U} < T\,\dot{S}\,, \tag{5.5.3}$$

which is equivalent to relation (5.5.2). Thus, a primary formation process can be an autonomous process, although by no means must this be the case.

In the two subsections that follow, we distinguish between adiabatic and diathermic autonomous formation. Generally speaking, these two kinds of autonomous formation are not isothermal. However, even under adiabatic conditions, the temperature changes may be small and for some purposes negligible, if the heat produced in the process is comparatively small.

5.5.1 Adiabatic autonomous formation

If the process takes place adiabatically, $\dot{Q}$ vanishes. From inequality (1.8.1), it then follows that

$$\dot{S} \geq 0. \tag{5.5.4}$$

That is, any autonomous formation process that occurs adiabatically cannot produce a decrease in the entropy of the system. The process makes $\dot{U} \leq 0$, as immediately follows from the first law (1.4.8), since, in an autonomous formation process, $\dot{W}_{in} \leq 0$. It follows that, in an adiabatic autonomous formation process,

$$\dot{\Psi} = \dot{U} - T\,\dot{S} \leq 0. \tag{5.5.5}$$

That is, the free energy of the system cannot increase.

In some cases, the process produces small temperature changes, so that T can be treated as constant when applying eqs (5.5.5), (5.4.1) and (5.4.2). From these equations we can then infer that Ψ_{max} does not increase during the process and that $\overline{\delta\Psi}$ remains consistently less than or equal to zero. From eq. (5.4.3)$_3$, we can then conclude that the considered formation process either produces a decrease in the gap $\Delta\Psi|_{max}$ or leaves that gap unaltered. Of course, this conclusion applies if, as assumed here, Ψ_{exh} is not altered by the process and the temperature changes that the process produces are appropriately small.

An important particular case is when $\dot{W}_{in} = 0$. In this case, from the first law (1.4.8), we deduce that $\dot{U} = 0$. The entropy increase that produces a decrease in the gap $\Delta\Psi|_{max}$ is then due entirely to some internal rearrangement of the system at constant internal energy. A typical time plot of the functions $U(t)$ and $TS(t)$ during an adiabatic autonomous formation process occurring at $\dot{W}_{in} = 0$ is shown in Fig. 5.5.1. Of course, the process must start from a point on the initial limit surface. The two insets in the same figure are two-dimensional representations of the system's admissible range at the beginning of the process ($t = 0$) and at time t. In each inset, a different internal energy constant is assumed to make U coincide with TS at the boundary of the admissible range. This can be done at any time of the process and does not affect the admissible range of the system or the gap $\Delta\Psi|_{max}$ at that time.

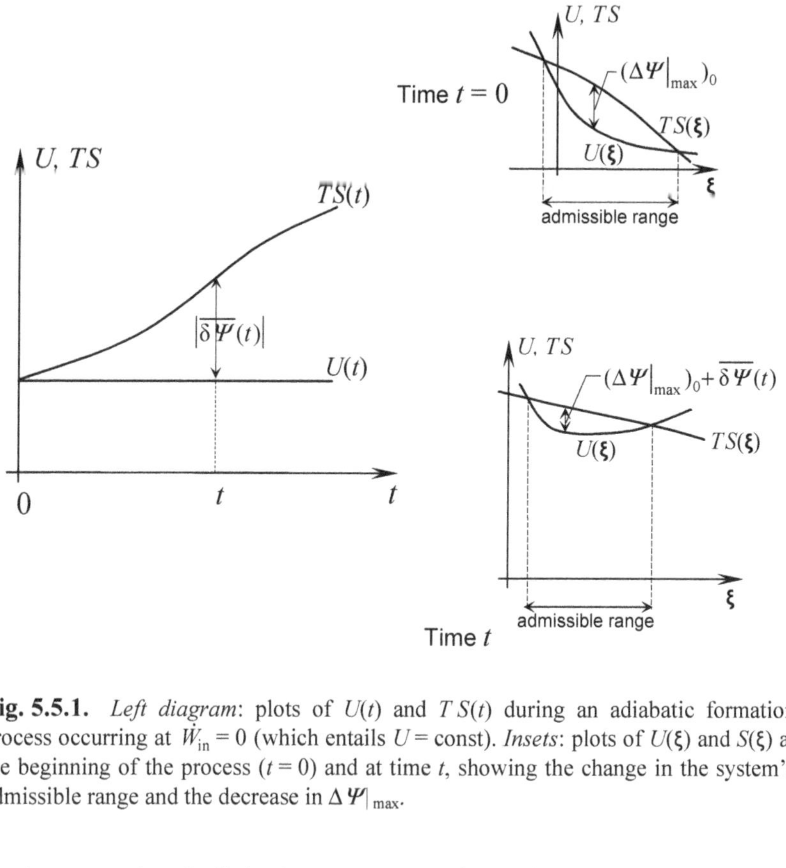

Fig. 5.5.1. *Left diagram*: plots of $U(t)$ and $TS(t)$ during an adiabatic formation process occurring at $\dot{W}_{in} = 0$ (which entails $U = $ const). *Insets*: plots of $U(\xi)$ and $S(\xi)$ at the beginning of the process ($t = 0$) and at time t, showing the change in the system's admissible range and the decrease in $\Delta\Psi|_{max}$.

An example of adiabatic autonomous formation process is provided by the splitting (mitosis) of a living cell into two sister cells, if temperature and volume changes are neglected and $\dot{Q}$ is taken to be equal to zero, which appear to be reasonable approximations. The constant volume hypothesis makes $\dot{W}_{in}$ vanish, thus making cell mitosis an autonomous process.

Autonomous adiabatic primary formation does also occur in the realm of inanimate systems. Any spontaneous chemical reaction that takes place adiabatically and at constant volume is an example of such a process. These processes must produce negligible amounts of heat, for the constant temperature hypothesis to be valid. Some of the most common instances of adiabatic autonomous formation occurring at near constant temperature are processes of atomic or molecular diffusion taking place in a rigid adiabatic container. In these conditions, the entropy of the system must increase, in agreement with relation (5.5.4).

In most cases, however, a spontaneous chemical reaction produces so much heat that the isothermal hypothesis is hardly applicable under adiabatic conditions. In these cases, diathermic rather than adiabatic conditions are the appropriate conditions under which a formation process can proceed isothermally.

5.5.2 Diathermic autonomous formation

In a diathermic formation process, the system can exchange heat with its surroundings. In particular, the process may take place isothermally or almost so, if adequate heat exchange with the surroundings is possible. As previously observed, isothermal or near-isothermal formation occurs frequently in the realm of biology, since living organisms avoid extreme temperature variations. Condition $\dot{W}_{in} \leq 0$ must apply to these processes, too, since they are autonomous. At variance with the adiabatic case, however, diathermic formation may increase the gap between U and TS and still reduce the entropy of the forming system. This situation is illustrated in Fig. 5.5.2, which shows typical plots of $U(t)$ and $TS(t)$ for a diathermic autonomous formation process in which $\dot{S} < 0$.

Diathermic autonomous formation may also result in an increase in the entropy of the system. However, if that is not the case, i.e., if $\dot{S} < 0$, then the process requires that

$$\dot{Q} < 0, \tag{5.5.6}$$

as immediately follows from the second law (1.8.1). Inequality (5.5.6) means that, in these processes,

$$\dot{U} < 0. \tag{5.5.7}$$

This is a consequence of the first law (1.4.8) and of the fact that the processes are autonomous ($\dot{W}_{in} \leq 0$) and reduce the system's entropy ($\dot{S} < 0$). From these observations, we can conclude that a diathermic autonomous formation that increases the gap between U and TS while decreasing the entropy of the system must occur at the expense of the system's internal energy. This is true if the process occurs at $\dot{W}_{in} = 0$ or $\dot{W}_{in} < 0$.

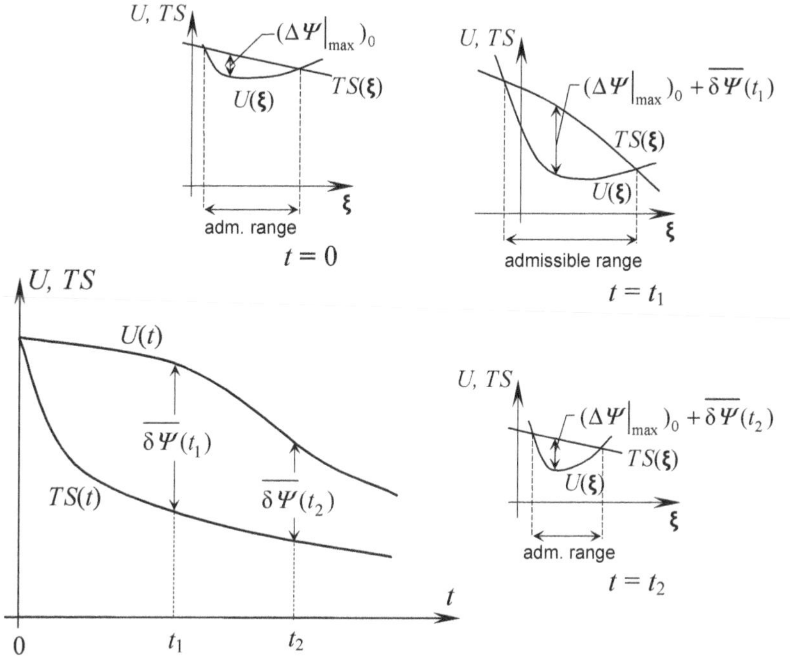

Fig. 5.5.2. Example of a diathermic autonomous formation process that results in a larger gap $\Delta\Psi|_{max}$ and decreases the system's entropy. The process modifies the constitutive equations for U and S and changes the system's admissible range (insets). The figure assumes $U = TS$ at the beginning of the process and at the boundary of the admissible ranges ($\Psi < 0$ within these ranges). The free energy of the exhaustion state is assumed to stay constant.

A familiar example of a diathermic autonomous formation process that takes place isothermally and results in an increase in the maximum gap between U and TS is the transformation of an egg into a chick. Because the chick is a much more organized system than the egg from which it comes, the prevailing wisdom is that the just-hatched chick should possess less entropy than the freshly-laid egg from which the chick developed. For this to be true, however, the egg should lose heat during incubation, as follows from inequality (1.8.1). In other words, to become a chick, an egg loses energy in the form of heat. Much experimental evidence supports this conclusion (e.g., [59] and [37]). This provides support to our general deduction (5.5.7), although it contrasts with the popular belief that the hen must supply heat for the egg to hatch.

Actually, the role of the brooding hen is to maintain the egg at the right incubation temperature for the delicate chemical reactions within the egg to occur. This does not mean that the hen must supply heat to the egg. In a cold environment, this may well be the case, as the heat produced by the reactions inside the egg may not be enough to keep it at the right temperature. The brooding hen will then reduce the heat losses with her own body and supply the extra heat that the egg may still need. On the other hand, if the nest becomes too hot, the hen could eliminate the excess heat through its breath or, and perhaps more effectively, leave the nest unattended for a while to increase ventilation. Better ventilation is the strategy used by bees to keep their eggs cool in the hive. In hot weather, bees gather near their eggs, keeping their wings flapping for hours to cool the eggs in spite of the heat that they produce.

An example of diathermic autonomous formation from the inanimate world is the liquid-to-solid transition of a material. The solidification process occurs at constant temperature (the melting point of the material) and usually involves heat loss ($\dot{Q} < 0$). The process can increase the gap between U and TS. For this to occur, however, the process must take place irreversibly, which is usually the case if the process is quick enough. At slow cooling rates, the solidification process tends to be reversible. This makes $T\dot{S} = \dot{Q} = \dot{U}$, provided that $\dot{W}_{\text{in}}$ can be neglected, as is usually the case when the process occurs at atmospheric pressure. Under these conditions, $T\dot{S} = \dot{U}$. This implies that the maximum gap between U and TS of the solidified material cannot be different from that of the liquid phase. We conclude that solidification, when taking place reversibly, changes the constitutive equations of the system, without changing its capacity for storing and releasing energy. On the contrary, rapid cooling produces $T\dot{S} > \dot{Q} = \dot{U}$, thus making the process irreversible. In that case, the cooling process is bound to increase the maximum gap between U and TS.

Any solidification process can be made more irreversible by increasing its cooling rate. Usually, the maximum gap between U and TS will increase accordingly. There are many consequences of this that have practical relevance. In metals, for instance, higher cooling rates result in a solid with a finer crystalline structure. As the size of the micro-crystals becomes smaller, the macroscopic elastic limit of the material increases along with its capacity for storing elastic energy without yielding. This phenomenon has long been exploited in the quenching of cast iron and steel.

If cooling from the liquid phase is quick enough, some materials may even vitrify and exhibit no crystal structure at all. However, at high cooling rates, other phenomena may take place (e.g., the formation of states of self-tension inside the material) that may adversely affect the admissible range of the material.

5.6 Non-Autonomous Formation

A formation process is non-autonomous if it makes the system absorb non-thermal energy from the surroundings. In this case,

$$\dot{W}_{\text{in}} > 0. \tag{5.6.1}$$

Again, it may be useful to distinguish between adiabatic and diathermic cases.

5.6.1 Adiabatic non-autonomous formation

If the process takes place adiabatically, then $\dot{Q} = 0$. From eq. (1.4.8) and inequality (5.6.1), it follows that the relation

$$\dot{U} = \dot{W}_{\text{in}} > 0 \tag{5.6.2}$$

must be met at all times during an adiabatic non-autonomous formation process. On the other hand, from eqs (5.4.1) and (5.4.3)$_1$ we infer that $\dot{U} < T\dot{S}$ for formation processes that produce a decrease in the gap between U and TS while leaving Ψ_{exh} unchanged. Therefore, from eq. (5.6.2) it follows that, in order to produce a decrease in the gap between U and TS, an adiabatic non-autonomous formation process should meet the following inequality:

$$T\dot{S} > \dot{W}_{\text{in}} > 0. \tag{5.6.3}$$

Under adiabatic conditions, however, eq. (1.9.14)$_1$ yields $\dot{D} = T\dot{S}$. According to relation (5.6.3), this means that such a process should dissipate more non-thermal energy than is supplied to the system. For processes that do not affect Ψ_{exh}, this could only occur if $\dot{U} < 0$. In this case, the system would lose some of its internal energy to dissipate it into heat. Such a possibility is, however, excluded by relation (5.6.2). It must be concluded that condition (5.6.3) cannot be met.

In other words, an adiabatic non-autonomous formation process that leaves the free energy of the exhaustion state unaltered can never result in a decrease of the maximum gap between U and TS. The same process can only increase the gap between U and TS and hence increase the system's capacity of storing and supplying energy isothermally. Figure 5.6.1 shows the plots of $U(t)$ and $TS(t)$ in these kinds of processes and their effect on the maximum amount of free energy that the system can store/supply.

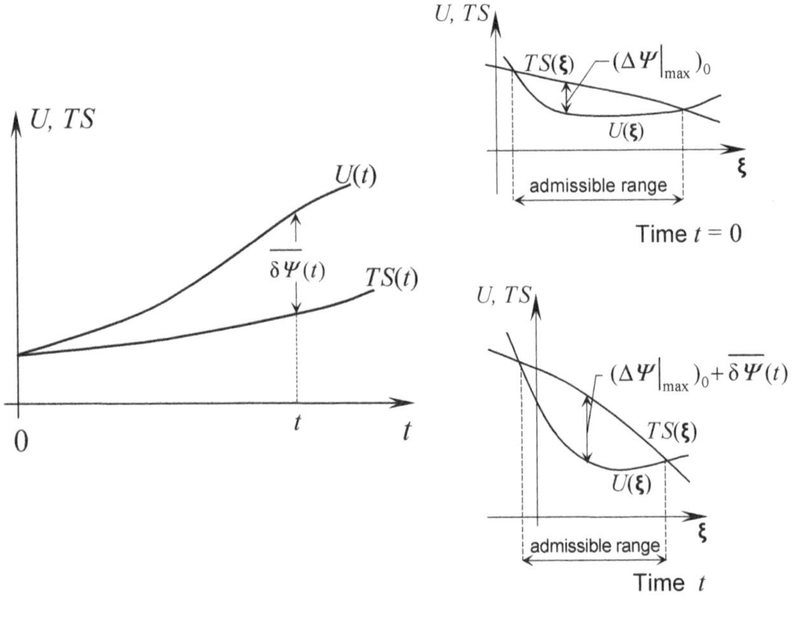

Fig. 5.6.1. Time plots of U and TS for an adiabatic non-autonomous formation process: if Ψ_{exh} remains unaffected, the process produces an increase in $\Delta\Psi|_{max}$, modifies the constitutive equations for U and S, and changes the system's admissible range (insets).

An example of such a process is cutting a lump of material into pieces while avoiding excessive dissipation so that temperature changes can be ignored. The process increases the surface of the system. Because the formation of a new surface requires energy, the process absorbs work from its surroundings. However, no substantial changes in the specific heat of the system, and thus in its entropy are produced. As observed above, this kind of formation process cannot decrease the maximum gap between U and TS. It follows that the maximum amount of non-thermal energy that the ground material can store and supply cannot be less than that of the solid piece of material it came from. Generally, it will be greater.

5.6.2 Diathermic non-autonomous formation

In a diathermic ($\dot{Q} \neq 0$), non-autonomous ($\dot{W}_{\text{in}} > 0$) formation process, $\dot{U} \neq \dot{W}_{\text{in}}$, as immediately follows from eq. (1.4.8). From the latter equation and from eq. (1.8.1), it then follows that

$$\dot{U} - T\dot{S} \leq \dot{W}_{\text{in}} . \tag{5.6.4}$$

Because $\dot{W}_{\text{in}} > 0$, relation (5.6.4) does not restrict the sign that the difference $\dot{U} - T\dot{S}$ can assume in the process. Therefore, that difference can be either positive or negative.

At variance with the adiabatic case, a diathermic non-autonomous formation process can result in an increase or a decrease in the maximum gap between U and TS. An increase in that gap occurs if $\dot{U} - T\dot{S} > 0$. The increase, however, cannot exceed the work done on the system, as can easily be verified by integrating relation (5.6.4) along the process.

An example of a diathermic non-autonomous formation is sintering. This is a process in which a powdered material is compressed ($\dot{W}_{\text{in}} > 0$) in a mould at high pressure and temperature. At the sintering temperature, the powder transforms into a solid body, greatly reducing the surface area of the material and thus the internal energy of the system. The transformation only produces minor changes in the specific heat of the system and hence in its entropy. The result is an increase in the maximum gap between U and TS and, thus, in the free energy of deformation that the system can store at the considered temperature.

5.7 Exchange of Matter with Surroundings

Exchange of matter during system formation is quite common, especially in the formation of living systems. By modifying the content of the system, a matter-exchanging process modifies the system itself. This process generally changes the system's constitutive equations, admissible range, and limit surface. The existence of a limit to the maximum amount of free energy that a finite amount of material can store isothermally is one of the main reasons for the existence of the admissible range and limit surface of a system. It may not be obvious whether the same notions are valid for open systems (i.e., systems that can exchange matter with the surroundings), given that the amount of free energy that a system can store depends on the type and amount of material that it contains.

To clarify the issue, we refer to a fixed portion of an open system, say, the part that corresponds to a given mass, a given volume, or a given number of moles of solvent. For a finite portion, it makes sense to speak of the maximum amount of free energy that the portion can store isothermally, even if the composition of the portion can vary as a result of exchange of matter with the surroundings. Thus, the analysis of Chapter 3 concerning admissible range and limit surface of any finite system can be applied to the considered portion of open system. The easiest way to do that is to base the analysis on the specific values of the quantities involved, e.g., per unit mass, per unit volume, or per unit mole number of solvent.

Of course, mass conservation generally does not apply to open systems. Consequently, the system's state variables related to mass and composition are likely to change in value, suddenly be activated, or disappear altogether, quite independently of any process or chemical transformation that takes place in the system. It all depends on the substances that are being exchanged between the system and its surroundings and on how the exchange takes place.

However, apart from the burden of dealing with more independent state variables to keep account of the exchanged matter, the notions of admissible range and limit surface for the considered portion of the system are the same as those considered for a closed system. Admissible range and limit surface are defined by the constitutive equations of the system and not by the system's state; they refer to isothermal conditions. Thus, the only restrictions needed for defining admissible range and limit surface are that temperature should be uniform throughout the system and that the system itself should be homogeneous as far as its constitutive properties are concerned. In any case, to determine the admissible range and limit surface of an open system at any time during its formation, we only need to know its constitutive equations at that time. It is irrelevant, therefore, whether or not the system is open or not.

6

The Elastic Limit in Solids

Elastic materials are characterized by the capacity to spring back to a privileged state of deformation once set free from external forces. That privileged state of deformation depends on temperature and is often referred to as the *stress-free state*. Under isothermal conditions, most solids behave this way provided that their deformation does not exceed certain limits, called the *elastic limits* of the material. These limits determine the boundary of a region in the space of the deformation variables, which is the *elastic range* of the material at the considered temperature. If the material is deformed beyond that range, it ceases to behave elastically. It may break (brittle materials), yield plastically (ductile materials), or exhibit any other sort of inelastic behaviour.

Knowledge of the elastic range of a solid is of a primary importance in both mechanical and civil engineering. It is vital in the design of buildings and structures capable of bearing the prescribed service loads safely. Different materials possess different elastic ranges and thus different elastic limits to deformation. To predict these limits, engineers use appropriate rupture or yield criteria, which are empirical rules that are based on *ad hoc* assumptions (cf. [54] for an in-depth account of the basics of this classical topic).

In the present chapter, however, we show that the elastic range and the elastic limits of a material are determined by the constitutive equations of the internal energy and entropy of the material once a point of the limit surface is known. The following analysis is an example of application of the notion of thermodynamically orthogonal state variables we presented in Section 3.10. It is assumed that the reader is familiar with the basic concepts of continuum mechanics, as can be found in many excellent textbooks on the subject (see, e.g., [28], [20], [41], [34], and [39]).

6.1 State Variables for Linear Thermoelastic Isotropic Materials

The elastic deformations of many materials of practical interest are very small. The same applies to the deformation produced by temperature variations within a wide range of temperatures. For these materials, it is legitimate to assume that stress depends linearly on strain and temperature change. Accordingly, in the present chapter, we consistently refer to the particular, but practically important, case of linear thermoelastic materials. A similar analysis, however, can be pursued for other classes of elastic materials once their constitutive equations have been specified.

The state of a linear thermoelastic material is defined by its temperature and deformation. Thus, any set of variables that describe the state of these materials must determine their temperature and deformation completely. In the case of materials that behave elastically, these variables are usually defined by referring to the stress-free configuration of the material at a reference temperature, say, the ambient temperature. We denote the reference temperature as T_0. As is well known from classical continuum mechanics, the deformation of a material from a given reference state is described by six scalars, the components of a symmetric second order tensor called the strain tensor. The latter, denoted by $\boldsymbol{\varepsilon}$, is referred to as *total strain*, in order to distinguish it from the analogous tensors, *elastic strain* and *thermal strain*, to be introduced below.

Tensor $\boldsymbol{\varepsilon}$ or, equivalently, its six independent scalar components, vanish in the reference configuration. That is, $\boldsymbol{\varepsilon} = \mathbf{0}$ at the reference state, $\mathbf{0}$ denoting the null tensor. On the other hand, the state variable that determines the temperature, T, of the material may conveniently be assumed as the difference between T and T_0, which is denoted as ΔT. That is:

$$\Delta T = T - T_0 . \tag{6.1.1}$$

Again, $\Delta T = 0$ in the reference state.

Without introducing any serious restrictions, we henceforth confine our attention to homogeneous bodies that undergo homogeneous processes. In that case, the variables $\boldsymbol{\varepsilon}$ and T do not depend on the point of the body to which they refer. The case of non-homogeneous bodies and non-homogeneous processes can be studied by applying the results of the homogeneous case to each infinitesimal material element of the body and by integrating the relevant equations over the entire body. For simplicity, we assume that the material is *isotropic*. This is true for many polycrystalline metals in virgin conditions, i.e., materials that have not suffered previous inelastic deformations.

When dealing with small deformations, it is possible to decompose the total strain tensor according to the relation:

$$\boldsymbol{\varepsilon} = \boldsymbol{\varepsilon}^{\mathrm{e}} + \boldsymbol{\varepsilon}^{\vartheta} , \tag{6.1.2}$$

where $\boldsymbol{\varepsilon}^{\mathrm{e}}$ and $\boldsymbol{\varepsilon}^{\vartheta}$ are the *elastic strain* and the *thermal strain* components of $\boldsymbol{\varepsilon}$, respectively. More precisely, $\boldsymbol{\varepsilon}^{\mathrm{e}}$ is the contribution to deformation that is due to stress in addition to the deformation produced by temperature. It is related to the stress tensor, $\boldsymbol{\sigma}$, acting on the material by the well-known stress-strain relation of a classical linear elastic isotropic material:

$$\boldsymbol{\varepsilon}^{\mathrm{e}} = \frac{1+\nu}{E} \boldsymbol{\sigma} - \frac{\nu}{E} \mathrm{tr}(\boldsymbol{\sigma}) \boldsymbol{1} , \tag{6.1.3}$$

or, in component form,

$$\varepsilon_{\mathrm{ij}}^{\mathrm{e}} = \frac{1+\nu}{E} \sigma_{\mathrm{ij}} - \frac{\nu}{E} (\sigma_{11} + \sigma_{22} + \sigma_{33}) \delta_{\mathrm{ij}} . \tag{6.1.4}$$

The symbol $\boldsymbol{1}$ that appears in eq. (6.1.3) is the unit tensor, while constants E and ν are the material's Young's modulus and Poisson's ratio, respectively. In the linear theory of thermoelasticity, these constants are assumed to be independent of temperature.

On the other hand, tensor $\boldsymbol{\varepsilon}^{\vartheta}$, introduced by eq. (6.1.2), represents the strain that results at zero stress from the temperature variation ΔT. For isotropic materials, this strain is given by

$$\boldsymbol{\varepsilon}^{\vartheta} = \alpha\, \Delta T\, \boldsymbol{1}, \tag{6.1.5}$$

where α denotes the thermal expansion coefficient of the material. In component form, eq. (6.1.5) becomes

$$\varepsilon_{\mathrm{ij}}^{\vartheta} = \alpha\, \Delta T\, \delta_{\mathrm{ij}}\,. \tag{6.1.6}$$

By inverting relation (6.1.3), we obtain

$$\boldsymbol{\sigma} = \lambda\, \mathrm{tr}(\boldsymbol{\varepsilon}^{\mathrm{e}})\boldsymbol{1} + 2\mu\, \boldsymbol{\varepsilon}^{\mathrm{e}}, \tag{6.1.7}$$

where the coefficients λ and μ are the so-called Lamé's constants, which are related to E and ν, respectively, as follows:

$$\lambda = \frac{\nu E}{(1+\nu)(1-2\nu)} \qquad \text{and} \qquad \mu = \frac{E}{2(1+\nu)} = G\,. \tag{6.1.8}$$

The quantity G introduced above is the shear modulus of the material. μ and G denote the same quantity. The symbol μ is preferred when expressing $\boldsymbol{\sigma}$ as a function of $\boldsymbol{\varepsilon}^{\mathrm{e}}$, as in eq. (6.1.7). This double notation is so diffused in elasticity theory, that its elimination may annoy the eyes of the practitioner. In linear elasticity (as distinguished from linear thermoelasticity), we have $\boldsymbol{\varepsilon}^{\vartheta}=\boldsymbol{0}$; therefore, $\boldsymbol{\varepsilon}^{\mathrm{e}}=\boldsymbol{\varepsilon}$, as immediately follows from eq. (6.1.2). In that case, eq. (6.1.7) is referred to as the *generalized Hooke's law*.

In the rest of this chapter, we adopt $\boldsymbol{\varepsilon}^{\mathrm{e}}$ and ΔT as the state variables of the thermoelastic material. In the literature, the pair of variables $\boldsymbol{\varepsilon}$ and ΔT are often used instead. In terms of $\boldsymbol{\varepsilon}$ and ΔT, the stress-strain relation of a thermoelastic material can be expressed as

$$\boldsymbol{\sigma} = \lambda\, \mathrm{tr}(\boldsymbol{\varepsilon})\boldsymbol{1} + 2\mu\, \boldsymbol{\varepsilon} - \beta \Delta T\, \boldsymbol{1}, \tag{6.1.9}$$

which is known as the Duhamel-Neumann stress-strain relation of linear thermoelasticity (cf., e.g., [28, p. 553] or [20, p. 422]). This relation is obtained by introducing eq. (6.1.2) into eq. (6.1.9) and by using eq. (6.1.5). The coefficient β introduced there is given by

$$\beta = \frac{\alpha E}{1-2\nu} = \alpha(3\lambda + 2\mu)\,. \tag{6.1.10}$$

Equation (6.1.7) shows that the stress tensor of a linear thermoelastic material does not depend explicitly on temperature if it is expressed as a function of $\boldsymbol{\varepsilon}^{\mathrm{e}}$. However, as shown by eq. (6.1.9), the same stress tensor explicitly depends on temperature if it is expressed as a function of $\boldsymbol{\varepsilon}$. This makes the variable $\boldsymbol{\varepsilon}$ less convenient to use than the variable $\boldsymbol{\varepsilon}^{e}$ as far as the analysis of the following sections is concerned. The fact that $\boldsymbol{\sigma}$ does not depend on ΔT when expressed as a function of $\boldsymbol{\varepsilon}^{e}$ does not mean that the stress of the material is unaffected by a change in temperature if the total deformation, $\boldsymbol{\varepsilon}$, is kept constant. Quite the opposite is true: a change in ΔT at constant $\boldsymbol{\varepsilon}$ produces a change in $\boldsymbol{\varepsilon}^{e}$ and thus in $\boldsymbol{\sigma}$, as is evident from eqs (6.1.2) and (6.1.7).

6.2 Deviatoric and Spherical Strain and Stress

The decomposition of the strain tensor and the stress tensor into their deviatoric and spherical parts leads to important simplifications in the analysis of the elastic limits of solids. For this reason, we recall here some of the most important formulae concerning this decomposition, which we use in the next sections.

Every symmetric second order tensor, $\mathbf{A}$, can be decomposed into a *spherical part*, $\bar{\mathbf{A}}$, and a *deviatoric part*, $\mathbf{A}'$, according to

$$\mathbf{A} = \bar{\mathbf{A}} + \mathbf{A}' . \tag{6.2.1}$$

The spherical part of $\mathbf{A}$ is defined as

$$\bar{\mathbf{A}} = \frac{1}{3}\,\mathrm{tr}(\mathbf{A})\,\boldsymbol{I} = \frac{1}{3}\,I_1(\mathbf{A})\,\boldsymbol{I} , \tag{6.2.2}$$

where $I_1(\mathbf{A}) = \mathrm{tr}(\mathbf{A}) = \mathrm{A_{kk}}$ is the first principal invariant of $\mathbf{A}$. In component form, eq. (6.2.2) reads

$$\bar{\mathrm{A}}_{\mathrm{ij}} = \frac{1}{3}\,\mathrm{A_{kk}}\,\delta_{\mathrm{ij}} = \frac{1}{3}\,(\mathrm{A_{11}} + \mathrm{A_{22}} + \mathrm{A_{33}})\,\delta_{\mathrm{ij}} . \tag{6.2.3}$$

As apparent from this definition, $\bar{\mathbf{A}}$ is a diagonal tensor. Its non-vanishing components are all equal to the mean value of the diagonal components of the original tensor $\mathbf{A}$. Since $\mathrm{tr}(\boldsymbol{I}) = \delta_{\mathrm{kk}} = 3$, from eq. (6.2.2) or (6.2.3) it follows that

$$I_1(\bar{\mathbf{A}}) = I_1(\mathbf{A}) . \tag{6.2.4}$$

On the other hand, the deviatoric part of $\mathbf{A}$ is given trivially by the difference between $\mathbf{A}$ and its spherical part:

$$\mathbf{A}' = \mathbf{A} - \bar{\mathbf{A}} \qquad or \qquad \mathrm{A}'_{\mathrm{ij}} = \mathrm{A_{ij}} - \frac{1}{3}\,\bar{\mathrm{A}}_{\mathrm{kk}}\,\delta_{\mathrm{ij}} . \tag{6.2.5}$$

Tensor $\mathbf{A}'$ is traceless, as evident from eqs (6.2.5) and (6.2.4).

By applying decomposition (6.2.1) to strain tensor $\boldsymbol{\varepsilon}^{\mathrm{e}}$, we obtain

$$\boldsymbol{\varepsilon}^{\mathrm{e}} = \bar{\boldsymbol{\varepsilon}}^{\mathrm{e}} + \boldsymbol{e}^{\mathrm{e}} . \tag{6.2.6}$$

Tensor $\bar{\boldsymbol{\varepsilon}}^{\mathrm{e}}$ introduced in this relation is the spherical part of $\boldsymbol{\varepsilon}^{\mathrm{e}}$:

$$\bar{\boldsymbol{\varepsilon}}^{\mathrm{e}} = \frac{1}{3}\,\mathrm{tr}(\boldsymbol{\varepsilon}^{\mathrm{e}})\,\boldsymbol{1} = \frac{1}{3}\,I_1(\boldsymbol{\varepsilon}^{\mathrm{e}})\,\boldsymbol{1} . \tag{6.2.7}$$

Its components can be expressed variously as

$$\bar{\varepsilon}^{\mathrm{e}}_{\mathrm{ij}} = \frac{1}{3} I_1(\boldsymbol{\varepsilon}^{\mathrm{e}})\,\delta_{\mathrm{ij}} = \frac{1}{3}\mathrm{tr}(\boldsymbol{\varepsilon}^{\mathrm{e}})\,\delta_{\mathrm{ij}} = \frac{1}{3}(\varepsilon^{\mathrm{e}}_{11} + \varepsilon^{\mathrm{e}}_{22} + \varepsilon^{\mathrm{e}}_{33})\,\delta_{\mathrm{ij}} . \tag{6.2.8}$$

$\bar{\boldsymbol{\varepsilon}}^{\mathrm{e}}$ is also called the *volumetric* part of $\boldsymbol{\varepsilon}^{\mathrm{e}}$, since in linear elasticity the strain invariant $I_1(\boldsymbol{\varepsilon}^{\mathrm{e}})$, the so-called *dilatation*, represents the volume increase per unit volume of material. The other tensor, $\boldsymbol{e}^{\mathrm{e}}$, that appears in the decomposition (6.2.6) is the *strain deviator*. It is defined as

$$\boldsymbol{e}^{\mathrm{e}} = \boldsymbol{\varepsilon}^{\mathrm{e}} - \bar{\boldsymbol{\varepsilon}}^{\mathrm{e}} , \tag{6.2.9}$$

or, in component form,

$$e^{\mathrm{e}}_{\mathrm{ij}} = \varepsilon^{\mathrm{e}}_{\mathrm{ij}} - \frac{1}{3}\,\varepsilon^{\mathrm{e}}_{\mathrm{kk}}\,\delta_{\mathrm{ij}} . \tag{6.2.10}$$

Being traceless [$I_1(\boldsymbol{e}^{\mathrm{e}})=0$], this part of $\boldsymbol{\varepsilon}^{\mathrm{e}}$ does not produce any change in volume as long as we remain within the realm of linear elasticity. For this reason, $\boldsymbol{e}^{\mathrm{e}}$ is also referred to as the distortional part of $\boldsymbol{\varepsilon}^{\mathrm{e}}$.

Likewise, stress tensor $\boldsymbol{\sigma}$ can be decomposed into a spherical and a deviatoric part, according to the following relation:

$$\boldsymbol{\sigma} = \bar{\boldsymbol{\sigma}} + \boldsymbol{\sigma}' . \tag{6.2.11}$$

Here, $\bar{\boldsymbol{\sigma}}$ is the spherical part of $\boldsymbol{\sigma}$, given by

$$\bar{\boldsymbol{\sigma}} = \frac{1}{3}\,\mathrm{tr}(\boldsymbol{\sigma})\,\boldsymbol{1} = \frac{1}{3}\,I_1(\boldsymbol{\sigma})\,\boldsymbol{1} , \tag{6.2.12}$$

where $I_1(\boldsymbol{\sigma})$ is the first principal invariant of $\boldsymbol{\sigma}$. In component form, eq. (6.2.12) reads:

$$\bar{\sigma}_{\mathrm{ij}} = \frac{1}{3} I_1(\boldsymbol{\sigma})\,\delta_{\mathrm{ij}} = \frac{1}{3}\mathrm{tr}(\boldsymbol{\sigma})\,\delta_{\mathrm{ij}} = \frac{1}{3}(\sigma_{11} + \sigma_{22} + \sigma_{33})\,\delta_{\mathrm{ij}} . \tag{6.2.13}$$

The mean normal stress exerted by $\boldsymbol{\sigma}$ is given by

$$p = -\frac{1}{3}(\sigma_{11} + \sigma_{22} + \sigma_{33}) = -\frac{1}{3}\mathrm{tr}(\boldsymbol{\sigma}) = -\frac{1}{3}I_1(\boldsymbol{\sigma}) \qquad (6.2.14)$$

and is referred to as *mean pressure*. In terms of p, the components of $\bar{\boldsymbol{\sigma}}$ are

$$\bar{\sigma}_{\mathrm{ij}} = -p\,\delta_{\mathrm{ij}}\,. \qquad (6.2.15)$$

This justifies calling $\bar{\boldsymbol{\sigma}}$ the *hydrostatic part* of $\boldsymbol{\sigma}$. The negative sign introduced in definition (6.2.14) makes p positive when representing a compressive force per unit area.

In contrast, the *stress deviator*, $\boldsymbol{\sigma}'$, that appears in eq. (6.2.11) is defined by

$$\boldsymbol{\sigma}' = \boldsymbol{\sigma} - \frac{1}{3}I_1(\boldsymbol{\sigma})\boldsymbol{1} \qquad (6.2.16)$$

and is traceless. As such, it does not produce any hydrostatic stress on the material. In component form, eqs (6.2.11) and (6.2.16) become

$$\sigma_{\mathrm{ij}} = -\,p\,\delta_{\mathrm{ij}} + \sigma'_{\mathrm{ij}} \qquad (6.2.17)$$

and

$$\sigma'_{\mathrm{ij}} = \sigma_{\mathrm{ij}} + p\,\delta_{\mathrm{ij}}\,, \qquad (6.2.18)$$

respectively.

From eqs (6.2.6) and (6.2.11), it is not difficult to decompose eq. (6.1.7) into two separate expressions, one for the spherical components of stress and strain and one for the deviatoric components of stress and strain. Thus, the generalized Hooke's law becomes:

$$\begin{aligned} \bar{\boldsymbol{\sigma}} &= 3\,K\,\bar{\boldsymbol{\varepsilon}}^{\mathrm{e}}, \\ \boldsymbol{\sigma}' &= 2\mu\,\boldsymbol{e}^{\mathrm{e}} \end{aligned} \qquad (6.2.19)$$

or, in component form,

$$\begin{aligned} \sigma_{\mathrm{kk}} &= 3\,K\,\varepsilon^{\mathrm{e}}_{\mathrm{kk}}, \\ \sigma'_{\mathrm{ij}} &= 2\mu\,e^{\mathrm{e}}_{\mathrm{ij}}\,. \end{aligned} \qquad (6.2.20)$$

The scalar K introduced here is the *bulk modulus* of the material:

$$K = \lambda + \frac{2}{3}\mu = \frac{E}{3(1-2\nu)} = \frac{\beta}{3\alpha}. \tag{6.2.21}$$

6.3 Internal Energy of a Linear Thermoelastic Material

Let u denote the specific internal energy (per unit mass) of a linear thermoelastic material. It is a function that depends on the values of $\boldsymbol{\varepsilon}^{\mathrm{e}}$ and ΔT at the particular point of the body at which it is calculated. Thus, $u = u(\boldsymbol{\varepsilon}^{\mathrm{e}}, \Delta T)$. Therefore, if U is the internal energy of the entire body, $\mathscr{B}$, we have

$$U = \int_{\mathscr{B}} u(\boldsymbol{\varepsilon}^{\mathrm{e}}, \Delta T)\mathrm{d}m = \int_{\mathscr{B}} \rho\, u(\boldsymbol{\varepsilon}^{\mathrm{e}}, \Delta T)\mathrm{d}V, \tag{6.3.1}$$

where ρ is the mass density of the considered material.

Because internal energy is a state function, the change in U and u in any process between any two states A and C, does not depend on the process that joins them. Thus, to calculate the change that u undergoes in any process, $\mathscr{P}$, from state $\mathsf{A} \equiv \{\boldsymbol{\varepsilon}^{\mathrm{e}} = \mathbf{0},\ \Delta T = 0\}$ to state $\mathsf{C} \equiv \{\boldsymbol{\varepsilon}^{\mathrm{e}},\ \Delta T\}$, we may refer to the sequence of reversible processes from A to B to C, as illustrated in Fig. 6.3.1, whether or not $\mathscr{P}$ is reversible.

This sequence begins with a simple heating/cooling processat zero stress, starting from the reference state, A. This process produces the temperature change ΔT and it leaves $\boldsymbol{\varepsilon}^{\mathrm{e}} = \mathbf{0}$ since $\boldsymbol{\sigma} = \mathbf{0}$, cf. eq. (6.1.3). The process can easily be performed reversibly and it brings the material to the intermediate state $\mathsf{B} \equiv \{\boldsymbol{\varepsilon}^{\mathrm{e}} = \mathbf{0},\ \Delta T\}$. As a result, the material absorbs the amount of heat per unit mass given by $c\,\Delta T$, where c is the appropriate value of the specific heat of the material for the process under consideration. However, within the limits of validity of linear thermoelasticity, the specific heat of any solid can, for most purposes, be treated as a constant, independent of the process. Denoting the change in u relevant to the process from A to B by $\Delta_{\mathrm{AB}}\, u$, we have

$$\Delta_{\mathrm{AB}} u = u(\mathbf{0}, \Delta T) - u(\mathbf{0}, 0) = c\,\Delta T, \tag{6.3.2}$$

as immediately follows from eq. (1.4.6), since no work is done in the process.

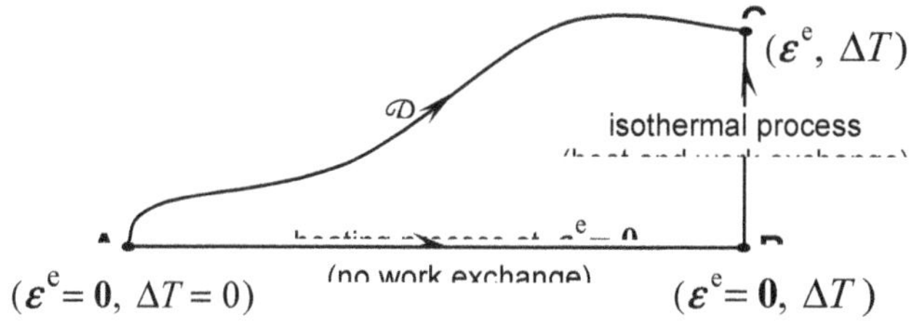

Fig. 6.3.1. In any process, $\mathcal{P}$, from A to C, the internal energy change is the same as that produced by the sequence of reversible processes from A to B and from B to C.

In the process from B to C the stress tensor does not vanish identically. In this part of the sequence, therefore, some external work is done on the material. However, during the process, $\boldsymbol{\varepsilon}^{\vartheta}$ remains constant since the process is isothermal. Consequently, if the process from B to C is performed reversibly, the material behaves as a classical linear elastic material with a stress-free state at B. Accordingly, if ψ denotes the material's specific strain energy per unit volume, we have that, at constant temperature $T_0+\Delta T$:

$$(w_{\text{in}})_{\text{BC}} = \Delta_{\text{BC}}\psi\,, \tag{6.3.3}$$

where $(w_{\text{in}})_{\text{BC}}$ indicates the work per unit volume done on the material in the reversible process from B to C.

To determine the function ψ explicitly, recall from classical elasticity that

$$\psi = \frac{1}{2}\operatorname{tr}(\boldsymbol{\sigma}^{\mathsf{T}}\boldsymbol{\varepsilon}^{\text{e}}) = \frac{1}{2}\sigma_{\text{ij}}\,\varepsilon^{\text{e}}_{\text{ij}}\,, \tag{6.3.4}$$

where $\boldsymbol{\varepsilon}^{\text{e}}$ is the elastic strain of the material from its stress-free state. Using eq. (6.1.7), we can express ψ in a number of equivalent ways: as a function of $\boldsymbol{\varepsilon}^{\text{e}}$ alone, as a function of $\boldsymbol{\sigma}$ alone, or as a function of both of $\boldsymbol{\varepsilon}^{\text{e}}$ and $\boldsymbol{\sigma}$. In what follows we shall refer to the following expression of ψ (cf., e.g., [20, p. 420]):

$$\psi = \psi(\boldsymbol{\varepsilon}^{\text{e}}) = \frac{1}{2}(\lambda + \frac{2}{3}\mu)\, I_1(\boldsymbol{\varepsilon}^{\text{e}})^2 + 2\mu\, J_2(\boldsymbol{\varepsilon}^{\text{e}})\,. \tag{6.3.5}$$

The function $J_2(\boldsymbol{\varepsilon}^{\mathrm{e}})$ introduced here is the second principal invariant of the deviatoric part of $\boldsymbol{\varepsilon}^{\mathrm{e}}$. That is,

$$J_2(\boldsymbol{\varepsilon}^{\mathrm{e}})=\frac{1}{2}\left[\varepsilon_{\mathrm{ij}}^{\mathrm{e}}-\frac{1}{3}\varepsilon_{\mathrm{kk}}^{\mathrm{e}}\,\delta_{\mathrm{ij}}\right]\left[\varepsilon_{\mathrm{ij}}^{\mathrm{e}}-\frac{1}{3}\varepsilon_{\mathrm{kk}}^{\mathrm{e}}\,\delta_{\mathrm{ij}}\right]=\frac{1}{2}e_{\mathrm{ij}}^{\mathrm{e}}\,e_{\mathrm{ij}}^{\mathrm{e}}, \qquad (6.3.6)$$

where $\boldsymbol{e}^{\mathrm{e}}$ is the strain deviator, defined by eq. (6.2.9). As apparent from definitions (6.3.6) and (6.2.10), $J_2(\boldsymbol{\varepsilon}^{\mathrm{e}})$ is a positive-definite quadratic function of $\boldsymbol{e}^{\mathrm{e}}$. From eqs (6.3.3) and (6.3.5), it follows that

$$(w_{\mathrm{in}})_{\mathrm{BC}}=\psi(\boldsymbol{\varepsilon}^{\mathrm{e}})-\psi(\mathbf{0})=\frac{1}{2}(\lambda+\frac{2}{3}\mu)\,I_1(\boldsymbol{\varepsilon}^{\mathrm{e}})^2+2\mu\,J_2(\boldsymbol{\varepsilon}^{\mathrm{e}}), \qquad (6.3.7)$$

since $\psi(\mathbf{0})=0$.

In the process from B to C, the material absorbs heat from its surroundings. This heat can be calculated from the so-called *Kelvin's formula*, which gives the temperature change of a thermoelastic material subjected to an adiabatic change in strain (cf. [20, Sect. 14.2]). For isotropic materials, that formula reads

$$(\dot{T})_{\mathrm{ad}}=-\frac{T}{\rho c_{\mathrm{v}}}\beta\dot{\varepsilon}_{\mathrm{ij}}^{\mathrm{e}}\delta_{\mathrm{ij}}=-\frac{T}{\rho c_{\mathrm{v}}}\beta\,\mathrm{tr}(\dot{\boldsymbol{\varepsilon}}^{\mathrm{e}}), \qquad (6.3.8)$$

where β is given by eq. (6.1.10), c_{v} denotes the specific heat at constant volume, and the index "ad" appended to $\dot{T}$ on the far left-hand side of the equation stands for adiabatic. For a solid material, $c_{\mathrm{v}}\cong c$, to a good approximation.

If, rather than being adiabatic, the deformation process occurs isothermally, then the material must exchange heat in order to maintain its temperature constant. In general, the amount of heat per unit mass that must be supplied to the material in order to compensate for the adiabatic temperature change $\mathrm{d}T=(\dot{T})_{\mathrm{ad}}\,\mathrm{d}t$ is given by

$$\mathrm{d}q=-c_{\mathrm{v}}\,(\dot{T})_{\mathrm{ad}}\,\mathrm{d}t\,. \qquad (6.3.9)$$

The negative sign on the right-hand side of this equation is due to the fact that the material absorbs heat if $(\dot{T})_{\mathrm{ad}}<0$, and it releases heat if $(\dot{T})_{\mathrm{ad}}>0$. From eqs (6.3.9) and (6.3.8), we then obtain:

$$\mathrm{d}q=\frac{T}{\rho}\,\beta\,\mathrm{d}\varepsilon_{\mathrm{ij}}^{\mathrm{e}}\,\delta_{\mathrm{ij}}\,. \qquad (6.3.10)$$

Therefore the heat, q_{BC}, absorbed per unit mass by the material to keep the deformation process from B to C isothermal is

$$q_{\mathrm{BC}} = \frac{T}{\rho} \beta I_1(\boldsymbol{\varepsilon}^{\mathrm{e}}), \tag{6.3.11}$$

as can be obtained immediately by integrating eq. (6.3.10) along the process and by recalling eq. (6.2.7).

The internal energy change $\Delta_{\mathrm{BC}}\, u$ can be calculated from eq. (1.4.6):

$$\Delta_{\mathrm{BC}}\, u = \frac{1}{\rho} (w_{\mathrm{in}})_{\mathrm{BC}} + q_{\mathrm{BC}} \,. \tag{6.3.12}$$

In view of eqs (6.3.7) and (6.3.11), this can be written as

$$\Delta_{\mathrm{BC}} u = \frac{1}{2\rho} (\lambda + \frac{2}{3}\mu)\, I_1(\boldsymbol{\varepsilon}^{\mathrm{e}})^2 + \frac{2}{\rho} \mu\, J_2(\boldsymbol{\varepsilon}^{\mathrm{e}}) + \frac{T}{\rho} \beta I_1(\boldsymbol{\varepsilon}^{\mathrm{e}}) \,. \tag{6.3.13}$$

If we observe that

$$\Delta_{\mathrm{AC}} u = u(\boldsymbol{\varepsilon}^{\mathrm{e}}, \Delta T) - u(\mathbf{0}, 0) \tag{6.3.14}$$

and

$$\Delta_{\mathrm{AC}} u = \Delta_{\mathrm{AB}} u + \Delta_{\mathrm{BC}} u \,, \tag{6.3.15}$$

then from eqs (6.3.2), (6.3.13), and (6.1.1), we obtain:

$$\begin{aligned} u(\boldsymbol{\varepsilon}^{\mathrm{e}}, \Delta T) = {} & \frac{1}{2\rho} \left[(\lambda + \frac{2}{3}\mu)\, I_1(\boldsymbol{\varepsilon}^{\mathrm{e}})^2 + 4\mu\, J_2(\boldsymbol{\varepsilon}^{\mathrm{e}}) \right] \\ & + \frac{T_0 + \Delta T}{\rho} \beta I_1(\boldsymbol{\varepsilon}^{\mathrm{e}}) + c\, \Delta T + u(\mathbf{0}, 0) \,. \end{aligned} \tag{6.3.16}$$

A more compact form of this equation can be obtained by expressing u as a function of the scalars ξ_1 and ξ_2 defined by

$$\xi_1 = I_1(\boldsymbol{\varepsilon}^{\mathrm{e}}) \tag{6.3.17}$$

and

$$\xi_2 = J_2(\boldsymbol{\varepsilon}^{\mathrm{e}}) \,. \tag{6.3.18}$$

Because the first principal invariant of a symmetric tensor is independent of the deviatoric part of the tensor, the variable ξ_1 is independent of the deviatoric part of $\boldsymbol{\varepsilon}^{\mathrm{e}}$. This means that $\xi_1 = 0$ for purely deviatoric (i.e., *distortional*) deformations. On the other hand, the variable ξ_2 is independent of the spherical part of $\boldsymbol{\varepsilon}^{\mathrm{e}}$, because so is

$J_2(\boldsymbol{\varepsilon}^{\mathrm{e}})$. It follows that $\xi_2 = 0$ for purely spherical (i.e., *volumetric*) deformations. Note that ξ_2 is a non-negative variable, because $J_2(\boldsymbol{\varepsilon}^{\mathrm{e}})$ is a positive definite function.

By taking ξ_1, ξ_2 and ΔT as independent variables and by recalling eq. $(6.2.21)_1$, we can rewrite eq. (6.3.16) as

$$\begin{aligned} u &= u(\xi_1, \xi_2, \Delta T) \\ &= \frac{1}{\rho}\left[\frac{1}{2}K\xi_1^{\,2} + 3\alpha K(T_0 + \Delta T)\xi_1 + 2\mu\xi_2\right] + c\,\Delta T + u_0\,, \end{aligned} \tag{6.3.19}$$

where $u_0 = u(\mathbf{0},0)$ is a constant. This shows that the specific internal energy of an isotropic linear thermoelastic material is a function of just three scalar variables: ξ_1, ξ_2, and ΔT.

6.4 Entropy and Free Energy of a Linear Thermoelastic Material

Similar to what we did for U, we shall refer to the material's specific entropy per unit mass, $s = s(\boldsymbol{\varepsilon}^{\mathrm{e}}, \Delta T)$, rather than the total entropy, S, of the entire body. The two quantities are related by the following equation:

$$S = \int_{\mathscr{B}} s(\boldsymbol{\varepsilon}^{\mathrm{e}}, \Delta T)\,\mathrm{d}m = \int_{\mathscr{B}} \rho\, s(\boldsymbol{\varepsilon}^{\mathrm{e}}, \Delta T)\,\mathrm{d}V\,, \tag{6.4.1}$$

which is analogous to eq. (6.3.1). Since entropy is a state function, its change in any process depends only on the initial and final state of the material. In particular, by referring to the processes of Fig. 6.3.1, we have

$$\Delta_{\mathrm{AC}}s = s(\boldsymbol{\varepsilon}^{\mathrm{e}}, \Delta T) - s(\mathbf{0}, 0) = \Delta_{\mathrm{AB}}s + \Delta_{\mathrm{BC}}s\,, \tag{6.4.2}$$

with obvious meaning of the notation used.

Because the processes from A to B and from B to C are assumed to be reversible, eq. (1.6.4) applies. Thus, the entropy change in these processes equals the amount of absorbed heat divided by the absolute temperature at which the heat is absorbed. Now, the quantity $c\,\mathrm{d}T$ is the heat that the material absorbs per unit mass in a simple heating/cooling process as the temperature of the material increases by the amount $\mathrm{d}T$. Therefore, if T denotes the instantaneous value of the temperature of the material in the process from A to B, then the corresponding increase in entropy is $\mathrm{d}s = c\,\mathrm{d}T/T$. By integrating this entropy change over the interval $[T_0, T_0+\Delta T]$, we obtain

$$\Delta_{AB}s = \int_{T_0}^{T_0+\Delta T} \frac{c}{T}\mathrm{d}T = c\,\ln\left(1+\frac{\Delta T}{T_0}\right). \tag{6.4.3}$$

On the other hand, the process from B to C occurs at the constant temperature $T = T_0 + \Delta T$ and absorbs the amount of heat (6.3.11). The relevant entropy change is

$$\Delta_{BC}s = \frac{q_{BC}}{T} = \frac{1}{\rho}\beta I_1(\boldsymbol{\varepsilon}^e). \tag{6.4.4}$$

From eqs (6.4.2) through (6.4.4), it follows that

$$s = s(\boldsymbol{\varepsilon}^e, \Delta T) = \frac{1}{\rho}\beta I_1(\boldsymbol{\varepsilon}^e) + c\,\ln\left(1+\frac{\Delta T}{T_0}\right) + s(\mathbf{0},0). \tag{6.4.5}$$

From eqs (6.3.17) and $(6.2.21)_3$, we can also express this equation in the following form:

$$s = s(\xi_1, \Delta T) = \frac{1}{\rho}3\alpha K\,\xi_1 + c\,\ln\left(1+\frac{\Delta T}{T_0}\right) + s_0, \tag{6.4.6}$$

where $s_0 = s(\mathbf{0},0)$. This shows that the specific entropy of an isotropic thermoelastic material is only a function of the two scalar variables, ξ_1 and ΔT.

The expression of the Helmholtz free energy of the material follows directly from definition (1.11.3) once the expressions of internal energy and entropy are inserted in it. By applying that definition to the specific values per unit mass of the quantities involved and by inserting eqs (6.3.19) and (6.4.6), the following expression of the Helmholtz free energy per unit mass is finally obtained:

$$\psi = \psi(\xi_1, \xi_2, \Delta T) = \frac{1}{2\rho}K\,\xi_1^{\;2} + \frac{2}{\rho}\mu\,\xi_2 + \overline{\psi}(\Delta T) + \psi_0. \tag{6.4.7}$$

Here, $\psi_0 = (u_0 - T_0 s_0)$ is a constant, while the function $\overline{\psi}(\Delta T)$ is given by

$$\overline{\psi}(\Delta T) = c\,\Delta T\left[1 - \left(\frac{T_0}{\Delta T}+1\right)\,\ln\left(1+\frac{\Delta T}{T_0}\right)\right] - \Delta T\,s_0. \tag{6.4.8}$$

6.5 Admissible Range and Limit Surface

Equations (6.3.19) and (6.4.6) have the following forms:

$$u = u(\xi_1, \xi_2, \Delta T) = u_1(\xi_1, \Delta T) + u_2(\xi_2) + \bar{u}(\Delta T) + u_0 \tag{6.5.1}$$

and

$$s = s(\xi_1) = s_1(\xi_1, \Delta T) + s_2 + \bar{s}(\Delta T) + s_0, \tag{6.5.2}$$

respectively. In these equations, we set

$$\begin{aligned} u_1(\xi_1, \Delta T) &= \frac{1}{\rho}\left[\frac{1}{2}K\xi_1^{\,2} + 3\alpha K(T_0 + \Delta T)\xi_1\right] \\ u_2(\xi_2) &= \frac{2\mu}{\rho}\xi_2 \\ \bar{u}(\Delta T) &= c\,\Delta T \end{aligned} \tag{6.5.3}$$

and

$$\begin{aligned} s_1(\xi_1) &= \frac{1}{\rho}3\alpha K\xi_1 \\ s_2 &\equiv 0 \\ \bar{s}(\Delta T) &= c\ln\left(1 + \frac{\Delta T}{T_0}\right). \end{aligned} \tag{6.5.4}$$

When expressed in terms of the variables ξ_1, ξ_2, and ΔT, specific internal energy and specific entropy decompose into the sum of independent functions of the form (3.10.1) and (3.10.2), respectively. This means that ξ_1 and ξ_2 are thermodynamically orthogonal variables. As discussed in Section 3.10, under isothermal conditions (i.e., for constant ΔT), the material possesses two separate and independent admissible ranges, one for the variable ξ_1 and one for the variable ξ_2.

In the present case, each of these ranges reduces to a one-dimensional segment, because it is relevant to just one variable, ξ_1 or ξ_2. Accordingly, the limit surface reduces to one or both of the endpoints of that segment. In the last case, the two-point limit surface is discontinuous; hence, there cannot be a continuous neutral process that links its two points. This means that the result of Section 3.4 does not apply to these points; therefore, it is not required that the two endpoints be equipotential for free energy.

Of course, the specific free energy (6.4.7) can be decomposed in a similar way:

$$\psi = \psi(\xi_1, \xi_2, \Delta T) = \psi_1(\xi_1) + \psi_2(\xi_2) + \bar{\psi}(\Delta T) + \psi_0, \qquad (6.5.5)$$

where

$$\psi_1(\xi_1) = \frac{1}{2\rho} K \xi_1^2, \qquad \psi_2(\xi_2) = \frac{2}{\rho} \mu \xi_2, \qquad (6.5.6)$$

while $\bar{\psi}(\Delta T)$ is defined by eq. (6.4.8). From the definitions of ξ_1 and ξ_2, it is clear that $\psi_1(\xi_1)$, $\psi_2(\xi_2)$, and $\bar{\psi}(\Delta T)$ are the volumetric, distortional, and thermal parts of ψ, respectively.

Admissible range for ξ_1

The expression of the admissible range for the variable ξ_1 can be found by applying relation (3.10.10) to u_1 and s_1. In general, the maximum amount of free energy that the material can store will be different for $\xi_1 > 0$ and $\xi_1 < 0$, because, as observed, the free energy of these points may not be the same. Thus, for a given value of ΔT, the admissible range for ξ_1 can be expressed quite generally as

$$\psi_1(\xi_1) = u_1(\xi_1, \Delta T) - (T_0 + \Delta T)\, s_1(\xi_1, \Delta T) \le C_1' \quad \text{for } \xi_1 > 0 \qquad (6.5.7)$$

and

$$\psi_1(\xi_1) = u_1(\xi_1, \Delta T) - (T_0 + \Delta T)\, s_1(\xi_1, \Delta T) \le C_1'' \quad \text{for } \xi_1 < 0, \qquad (6.5.8)$$

where C_1' and C_1'' denote appropriate positive constants representing the maximum amount of free energy $\psi_1(\xi_1)$ that, at constant ΔT, the material can store for $\xi_1 > 0$ and $\xi_1 < 0$, respectively. That is,

$$C_1' = \Delta_{\max} \psi_1 \big|_{\xi_1 > 0} \qquad (6.5.9)$$

and

$$C_1'' = \Delta_{\max} \psi_1 \big|_{\xi_1 < 0} \, . \qquad (6.5.10)$$

By using eqs $(6.5.3)_1$ and $(6.5.4)_1$, from relations (6.5.7) and (6.5.8) we obtain

$$\frac{1}{2\rho} K \xi_1^2 \le C_1' \quad \text{and} \quad \frac{1}{2\rho} K \xi_1^2 \le C_1'', \qquad (6.5.11)$$

which hold true for $\xi_1 > 0$ and for $\xi_1 < 0$, respectively. Taken together, these relations express the following limitation on the values of ξ_1:

$$-\sqrt{\frac{2\rho}{K}C_1''} \leq \xi_1 \leq \sqrt{\frac{2\rho}{K}C_1'}, \tag{6.5.12}$$

which represents an explicit expression of the admissible range for ξ_1. The limits to this range are

$$\xi_1 = +\sqrt{\frac{2\,\rho\,C_1'}{K}} \quad \text{and} \quad \xi_1 = -\sqrt{\frac{2\,\rho\,C_1''}{K}}. \tag{6.5.13}$$

These are the coordinates of the endpoints of the segment that represents the said range on the ξ_1-axis.

Admissible range for ξ_2

By a similar procedure, if we apply relation (3.10.10) to u_2 and s_2, we obtain:

$$u_2(\xi_2) - (T_0 + \Delta T)\, s_2 \leq C_2. \tag{6.5.14}$$

The constant C_2 introduced here represents the maximum amount of free energy ψ_2 that the material can store. That is,

$$C_2 = \Delta_{\max} \psi_2. \tag{6.5.15}$$

The explicit expression of admissible range for ξ_2 is obtained by introducing eqs $(6.5.3)_2$ and $(6.5.4)_2$ into relation (6.5.14):

$$\frac{2\mu}{\rho}\xi_2 \leq C_2. \tag{6.5.16}$$

In this case, thermodynamics sets a limit to the maximum value of ξ_2, while leaving its minimum value unrestricted. Thus, when expressed in terms of the state variables ξ_1, ξ_2, and ΔT, the limit surface of the variable ξ_2 reduces to the following single point:

$$\xi_2 = \frac{\rho\, C_2}{2\mu} = \frac{\rho}{2\mu}\Delta_{\max} \psi_2, \tag{6.5.17}$$

as immediately follows by considering the equality case of inequality (6.5.16) and by recalling eq. (6.5.15). We have here an instance of a finite, one-dimensional range, only one end of which is a limit point.

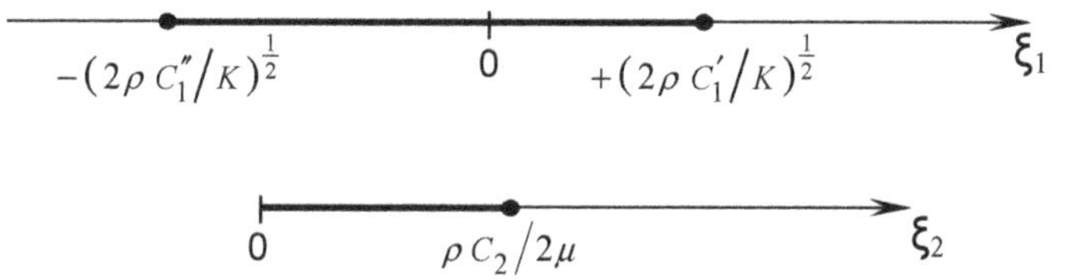

Fig. 6.5.1. Admissible range of a linear thermoelastic material in terms of the state variables ξ_1 and ξ_2: the admissible range is represented by the two bold segments of axis ξ_1 and axis ξ_2, respectively. The limit surface reduces to the three endpoints which are represented in bold in the figure.

To summarize, relations (6.5.11) and (6.5.16) show that, when represented in terms of state variables ξ_1, ξ_2, and ΔT, the admissible range of a linear thermoelastic material is given by two segments that belong to different axes, namely, the ξ_1- and ξ_2-axis. In terms of the same state variables, the limit surface reduces to the two endpoints of the admissible range for ξ_1 on the ξ_1-axis and to just the endpoint at $\xi_2 \neq 0$ of the segment that represents the admissible range for ξ_2 on the ξ_2-axis. The two ranges are represented in Fig. 6.5.1.

6.6 Elastic Limits and Elastic Range in Terms of Stress or Elastic Strain

The state variables ξ_1 and ξ_2 are perfectly adequate to determine the energy and the entropy of a linear thermoelastic material and thus its admissible range. In fact, together with ΔT, they are the primary state variables of that material (Section 2.4). However, the main variables of interest to mechanical engineers are not energy and entropy. By far, stress is the most important variable when designing machine components and structures. State variables ξ_1 and ξ_2 do not determine the state of stress of the material completely.

Stress is a symmetric, second-order tensor. It has six independent scalar components, σ_{ij}, which are related to the components of $\boldsymbol{\varepsilon}^e$ by eq. (6.1.7). The two scalars ξ_1 and ξ_2 cannot define all six components of $\boldsymbol{\varepsilon}^e$ and thus of $\boldsymbol{\sigma}$. The same is true if we refer to eq. (6.1.9) rather than eq. (6.1.7), because $\boldsymbol{\varepsilon}$ is not completely determined by ξ_1, ξ_2, and

ΔT. There is an infinite number of different states of stress and strain for any pair of values of ξ_1 and ξ_2. If we want to use stress to evaluate the position of the state of the material within its admissible range, we must express that range in terms of σ_{ij} rather than ξ_1 and ξ_2.

This can be done by first expressing ξ_1 and ξ_2 in terms of the components of $\boldsymbol{\varepsilon}^e$ and $\boldsymbol{e}^e$ and then using Hooke's law in the form (6.2.20) to convert strain components into stress components. Thus, the admissible range of ξ_1 will be transformed into the admissible range for the hydrostatic part of $\boldsymbol{\sigma}$. Likewise, the admissible range of ξ_2 will transform into the admissible range for the deviatoric part of $\boldsymbol{\sigma}$. For an isotropic material, both ranges can be represented in the space of principal stresses σ_1, σ_2, and σ_3. Each of the two ranges defines a different region of that space. Their intersection contains all states of stress that are compatible with both regions, i.e., the admissible range of the material in terms of stress. These ranges are considered in detail below.

Admissible range for the hydrostatic part of stress

From eq. $(6.2.19)_1$ and (6.3.17), we obtain

$$\xi_1 = \frac{I_1(\sigma)}{3K} = \frac{\sigma_{11} + \sigma_{22} + \sigma_{33}}{3K} = \frac{\sigma_1 + \sigma_2 + \sigma_3}{3K}, \tag{6.6.1}$$

where σ_1, σ_2, and σ_3 are the principal values of $\boldsymbol{\sigma}$. The last of the above equalities follows from the fact that the sum σ_{kk} is an invariant (actually, the first principal invariant of $\boldsymbol{\sigma}$). This means that the value of σ_{kk} is the same irrespective of the orientation of the reference axes in space (x_1, x_2, x_3). In particular, if the reference axes are directed as the principal axes of $\boldsymbol{\sigma}$, we have $\sigma_{11} = \sigma_1$, $\sigma_{22} = \sigma_2$, and $\sigma_{33} = \sigma_3$, and thus $\sigma_{kk} = \sigma_1 + \sigma_2 + \sigma_3$. By introducing eq. (6.6.1) into eq. (6.5.12), we obtain:

$$-p'' \leq \frac{\sigma_1 + \sigma_2 + \sigma_3}{3} \leq -p', \tag{6.6.2}$$

where the constants p' and p'' are defined as

$$p' = -\sqrt{(2\rho K C_1')} \qquad \text{and} \qquad p'' = +\sqrt{(2\rho K C'')}, \tag{6.6.3}$$

respectively. These two constants represent the limits to the hydrostatic pressure that the material can withstand while remaining elastic. The minus signs that precede p' and p'' in relation (6.6.2) is because pressure,

p, is assumed to be positive when representing compression, in agreement with definition (6.2.14). Thus, a negative value for p' denotes hydrostatic tension, which makes the far right-hand side of (6.6.2) positive.

Relation (6.6.2) defines the admissible range of the hydrostatic part of the stress tensor in the space of the variables σ_1, σ_2, and σ_3. The same relation makes it obvious that the direction of the principal axes of stress has no influence on that range. This is expected, since the material is assumed to be isotropic. It implies that all states of stress that have the same principal values fall within that range if one of these states does.

In space $(\sigma_1, \sigma_2, \sigma_3)$, the limit surfaces of this range are the two planes of equations:

$$\sigma_1 + \sigma_2 + \sigma_3 = -3\, p' \tag{6.6.4}$$

and

$$\sigma_1 + \sigma_2 + \sigma_3 = -3\, p'' . \tag{6.6.5}$$

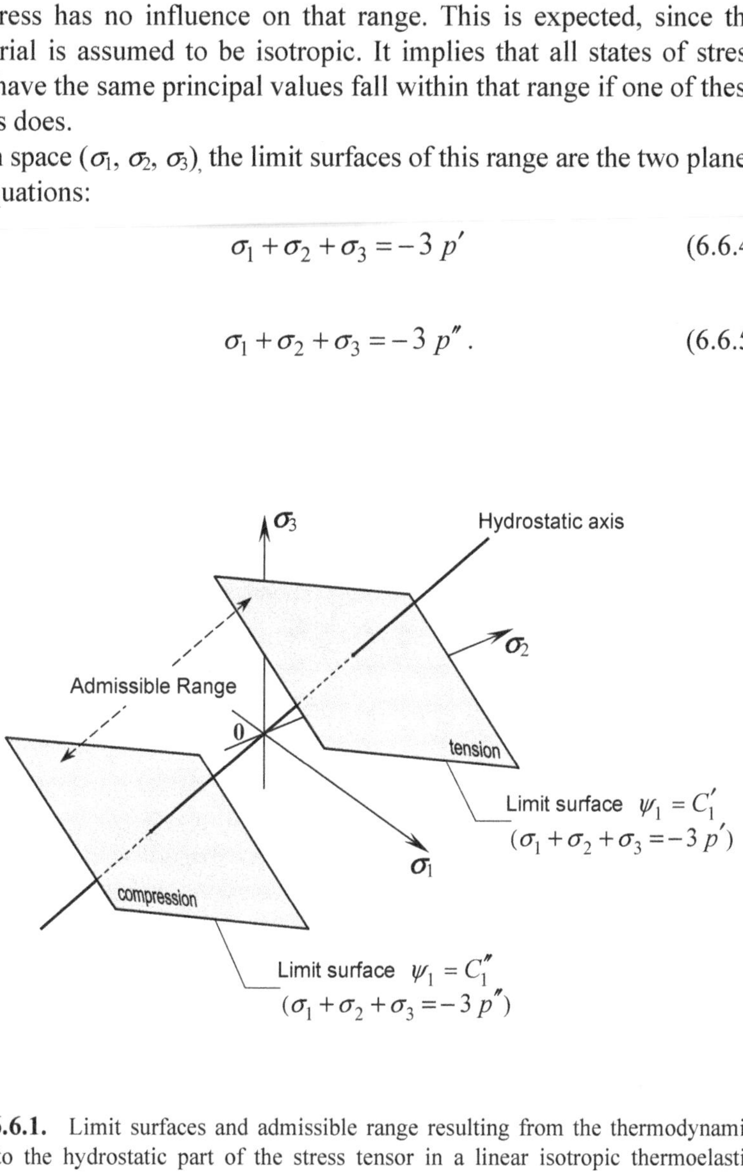

Fig. 6.6.1. Limit surfaces and admissible range resulting from the thermodynamic limit to the hydrostatic part of the stress tensor in a linear isotropic thermoelastic material.

These planes are normal to the hydrostatic axis $\sigma_1 = \sigma_2 = \sigma_3$ (i.e., the axis through the origin, making equal angles with each of the coordinate axes). The distances of the limit planes from the origin of the reference system are $-3p'$ and $-3p''$, respectively. Fig. 6.6.1 shows the two planes for the case in which $p' < 0$ and $p'' > 0$. As should be evident from eqs (6.6.4), (6.6.5), and (6.6.3), each plane can be determined when one of its points is known, since this suffices to determine p' or p'' (and, hence, C_1' or C_1'').

Admissible range for the deviatoric part of stress

By introducing eq. $(6.2.20)_2$ into eq. $(6.3.6)_2$, we obtain:

$$J_2(\boldsymbol{\varepsilon}^{\mathrm{e}}) = \frac{1}{4\mu^2} J_2(\boldsymbol{\sigma}) . \tag{6.6.6}$$

Here, $J_2(\boldsymbol{\sigma})$ is the second principal invariant of the stress deviator (6.2.18):

$$J_2(\boldsymbol{\sigma}) = \frac{1}{2}\sigma'_{\mathrm{ij}}\,\sigma'_{\mathrm{ij}} = \frac{1}{2}(\sigma_{\mathrm{ij}} + p\,\delta_{\mathrm{ij}})(\sigma_{\mathrm{ij}} + p\,\delta_{\mathrm{ij}}) , \tag{6.6.7}$$

where $p = -\sigma_{\mathrm{kk}}/3$, according to eq. (6.2.14). $J_2(\boldsymbol{\sigma})$ is independent of the hydrostatic part of $\boldsymbol{\sigma}$. This makes the analysis that follows depend only on the deviatoric part of $\boldsymbol{\sigma}$. By substituting eq. $(6.2.14)_1$ into eq. (6.6.7), we obtain, after some algebra:

$$J_2(\boldsymbol{\sigma}) = \frac{1}{6}\Big[(\sigma_{11} - \sigma_{22})^2 + (\sigma_{22} - \sigma_{33})^2 + (\sigma_{33} - \sigma_{11})^2 + \sigma_{23}{}^2 + \sigma_{31}{}^2 + \sigma_{12}{}^2\Big], \tag{6.6.8}$$

cf., e.g., [54, p. 476].

Again, being a tensor invariant, the quantity $J_2(\boldsymbol{\sigma})$ assumes the same value irrespective of how the reference axes of space (x_1, x_2, x_3) are rotated. Thus, by taking the latter axes as coinciding with the principal directions of $\boldsymbol{\sigma}$, we can also write eq. (6.6.8) as

$$J_2(\boldsymbol{\sigma}) = \frac{1}{6}\Big[(\sigma_1 - \sigma_2)^2 + (\sigma_2 - \sigma_3)^2 + (\sigma_3 - \sigma_1)^2\Big], \tag{6.6.9}$$

because, in such a reference system, all non-diagonal components of $\boldsymbol{\sigma}$ vanish, while components σ_{11}, σ_{22}, and σ_{33} coincide with the principal values σ_1, σ_2, and σ_3. Therefore, from eqs (6.6.6), (6.6.9), and (6.3.18), we can express ξ_2 as a function of σ_1, σ_2, and σ_3 as follows:

$$\xi_2 = \frac{1}{4\mu^2} J_2(\sigma)$$
$$= \frac{1}{24\mu^2}\left[(\sigma_1-\sigma_2)^2+(\sigma_2-\sigma_3)^2+(\sigma_3-\sigma_1)^2\right]. \qquad (6.6.10)$$

By inserting eq. $(6.6.10)_1$ or $(6.6.10)_2$ into eq. (6.5.16) and by taking eq. (6.5.15) into account, the admissible range for the deviatoric part of stress can be expressed by

$$\frac{1}{2\rho\mu} J_2(\sigma) \le \Delta_{max}\psi_2 , \qquad (6.6.11)$$

or

$$\frac{1}{12\rho\mu}\left[(\sigma_1-\sigma_2)^2+(\sigma_2-\sigma_3)^2+(\sigma_3-\sigma_1)^2\right] \le \Delta_{max}\psi_2 . \qquad (6.6.12)$$

This can also be put in the following form:

$$(\sigma_1-\sigma_2)^2+(\sigma_2-\sigma_3)^2+(\sigma_3-\sigma_1)^2 \le 6k^2 , \qquad (6.6.13)$$

where the constant k is defined as

$$k = \sqrt{2\rho\mu\Delta_{max}\psi_2} = \sqrt{2\rho\mu C_2} . \qquad (6.6.14)$$

The last equality in this expression follows from eq. (6.5.15) and shows that k depends on the particular material at hand, since C_2 does.

Constant k coincides with the value τ_{max} of the shear stress, τ, at the elastic limit of the material in pure shear. This can be proven easily by using relation (6.6.13) to calculate the maximum value of τ that is compatible with the same relation in the case of pure shear stress state ($\sigma_1 = -\sigma_2 = \tau$, $\sigma_3 = 0$). A similar analysis for the uniaxial stress case ($\sigma_1 = \sigma$, $\sigma_2 = \sigma_3 = 0$) reveals that the maximum uniaxial stress, say σ_{max}, is tied to τ_{max} and hence to k by the well-known relation $\sigma_{max} = \sqrt{3}\,\tau_{max} = \sqrt{3}\,k$.

In space (σ_1, σ_2, σ_3), the limit surface of the considered admissible range is obtained immediately by taking the equality sign case of relation (6.6.13). That is:

$$(\sigma_1-\sigma_2)^2+(\sigma_2-\sigma_3)^2+(\sigma_3-\sigma_1)^2 = 6k^2 . \qquad (6.6.15)$$

We thus recover the classical limit condition proposed a century ago by von Mises [46] as a convenient empirical formula to predict the limit stress beyond which a virgin ductile metal ceases to respond elastically and yields plastically. The above analysis shows that von Mises' limit is imposed by the laws of thermodynamics and the constitutive equations of internal energy and entropy of the material if it behaves as a linear isotropic thermoelastic material while in the elastic range. Ductile metals in virgin conditions belong to this class of materials.

In space $(\sigma_1,\sigma_2,\sigma_3)$, relation (6.6.13) defines a circular cylindrical region, the axis of which coincides with the deviatoric axis (Fig. 6.6.2). The boundary to this region is the cylindrical surface that is defined by eq. (6.6.15). It is fully determined once a point of it, and thus k, is known. That surface is equipotential for ψ_2 (the deviatoric part of the free energy of the material) and is often referred to as the von Mises limit surface.

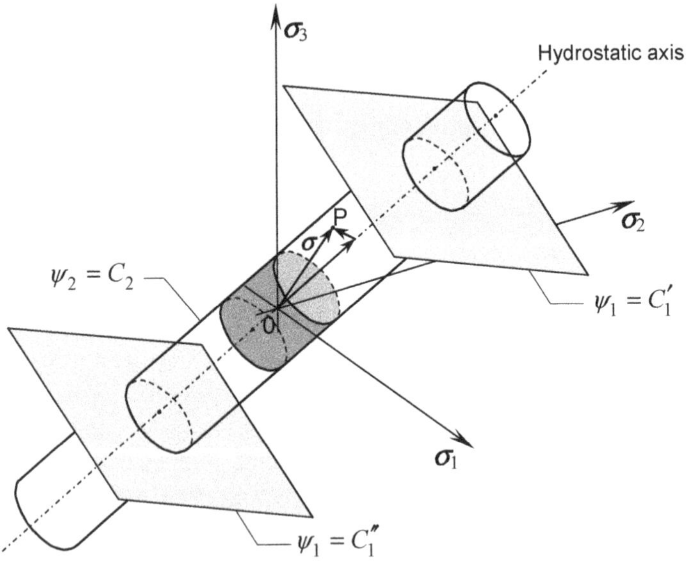

Fig. 6.6.2. *Linear isotropic thermoelastic materials*: The indefinite circular cylinder represented in the figure is the admissible range of the deviatoric part of stress. Its surface is equipotential for ψ_2. The elastic range is the part of that cylinder between planes $\psi_1 = C_1'$ and $\psi_1 = C_1''$. The darker part is the portion of the elastic range that is of interest in structural engineering.

Any point, P, of space $(\sigma_1,\sigma_2,\sigma_3)$ represents a stress tensor, $\boldsymbol{\sigma}$, apart from its principal directions. The projection of the position vector OP on the hydrostatic axis is the hydrostatic part of $\boldsymbol{\sigma}$. In contrast, the projection of OP on a plane normal to that axis has no hydrostatic component. Therefore, it represents the deviatoric part of $\boldsymbol{\sigma}$. Because limit surface (6.6.15) parallels the hydrostatic axis, it does not set any limit on the magnitude of the hydrostatic projection of OP; thus, there is no limit on the magnitude of the hydrostatic stress component. The same surface, however, requires the deviatoric component of stress to be confined within the circular cross-section of the cylinder.

The elastic range

Taken together, limitations (6.6.13) and (6.6.2) define a finite circular cylinder in space $(\sigma_1,\sigma_2,\sigma_3)$, the ends of which are the intersection of von Mises' indefinite cylinder (6.6.13) with planes (6.6.4) and (6.6.5), respectively (Fig. 6.6.2). This finite cylinder is the admissible range of the material and can be referred to as the material's *elastic range*. Any state of stress $\mathsf{P} \equiv [\sigma_1,\sigma_2,\sigma_3]$ that falls outside that range is not physically admissible in that it would require the material to store more free energy than it actually can.

A large body of evidence concerning the elastic limits of ductile metals indicates that the elastic limits of hydrostatic stress are, by far, greater than those of deviatoric stress. This makes the elastic range of these materials a highly elongated cylinder the height of which exceeds its diameter by a significant amount. However, states of stress at high hydrostatic compression are seldom encountered in practice, and states of stress at high hydrostatic tension are extremely difficult to produce. Therefore, in almost all practical applications, the elastic limit of a ductile metal is due to the deviatoric component of stress, the limits of which are far easier to reach. This means that the only part of the elastic range of ductile metals that is of practical interest is a small portion of the entire elastic range around the stress-free state. This portion is shown as the dark, cylindrical segment in Fig. 6.6.2.

Obviously, if that cylindrical segment is the only part of the elastic range that is of interest, it makes little difference whether the complete elastic range of the material is the entire indefinite von Mises' cylinder or just the finite part delimited by the two planes shown in Fig. 6.6.2. For this reason, when dealing with ductile metals, the hydrostatic limits of the material are ignored, and the indefinite von Mises' cylindrical surface is considered as the only limit to the elastic behaviour of the material. This amounts to assuming an infinite elastic limit for the hydrostatic stress component. Though unphysical, this assumption is acceptable in view of the small values of hydrostatic stress that occur in practical applications.

6.7 Elastic Range of More General Materials

Being symmetric, second-order tensors, stress and strain can always be decomposed into spherical and deviatoric components, as discussed in Section 6.2. However, such decomposition does not always lead to the definition of groups of thermodynamically orthogonal state variables, as happens in the case of the linear thermoelastic materials considered in the previous section. For more general thermoelastic materials, the elastic limit to the deviatoric part of stress may well depend on the hydrostatic part; reciprocally, the elastic limit to hydrostatic stress component may depend on the deviatoric stress component. Under these conditions, the free energy that the material accumulates due to pressure work may affect the elastic limit of the deviatoric stress component—an interesting possibility that should be of consequence in some applications, such as geophysics, where large hydrostatic pressures are commonplace.

The coupling between the spherical and deviatoric parts of the material's free energy may take place even in the presence of constitutive stress-strain equations that are quite simple. For instance, consider the case of a material that possesses a Hooke's type stress-strain relation, i.e., (6.2.19) or (6.2.20), except that the coefficients in these relations depend on stress or strain, rather than being constant. This entails a free energy function similar to the one given by eq. (6.4.7); however, the constants K and μ are replaced by appropriate functions of ξ_1 and ξ_2. That is

$$\begin{aligned}\psi &= \psi(\xi_1,\xi_2,\Delta T)\\ &= \frac{1}{2\rho}K(\xi_1,\xi_2)\,\xi_1^2 + \frac{2}{\rho}\mu(\xi_1,\xi_2)\xi_2 + \overline{\psi}(\Delta T) + \psi_0 .\end{aligned} \tag{6.7.1}$$

If K depends on ξ_2 and μ depends on ξ_1, then ξ_1 and ξ_2 cannot be thermodynamically orthogonal in the sense specified in Section 3.10.

When variables ξ_1 and ξ_2 are not thermodynamically orthogonal, there is not a separate admissible range for each of these variables, as in the case considered in Section 6.5. The admissible range and limit surface are then given by the general relations (3.6.5) and (3.6.6), respectively. If ψ is regular enough, the limit surface of the material will reduce to just one continuous equipotential surface, namely $\psi = \psi_{max}$.

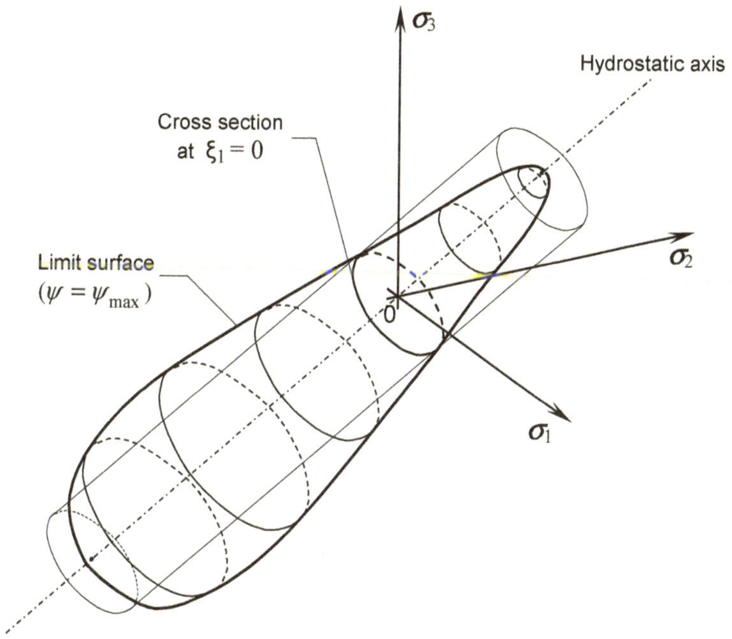

Fig. 6.7.1. Example of elastic range and limit surface of a thermoelastic material in which $K = K(\xi_2)$ and $\mu = \mu(\xi_1)$. (The circular cylinder represents the von Mises' surface of a linear thermoelastic material with $K = K(0) = \text{const}$ and $\mu = \mu(0) = \text{const}$.)

Fig. 6.7.1 provides a qualitative example of how the elastic domain may differ from the cylindrical domain discussed in the previous section, if

$$K = K(\xi_2) \tag{6.7.2}$$

and

$$\mu = \mu(\xi_1)\,. \tag{6.7.3}$$

In this case, the elastic limits to purely hydrostatic stress ($\xi_2 = 0$) or purely deviatoric stress ($\xi_1 = 0$) coincide with the analogous limits of a linear thermoelastic material, where K and μ are constant and equal to

$K(0)$ and $\mu(0)$, respectively. The cylinder that represents the elastic range of such linear thermoelastic material is shown by the lighter lines in Fig. 6.7.1. Its intersections with the hydrostatic axis correspond to purely hydrostatic limits ($\xi_2 = 0$) and therefore coincide with the analogous points of the admissible range of the more general material considered in this example. Observe also that the above cylinder and the limit surface of the considered material share the same cross-section with plane $\sigma_1+\sigma_2+\sigma_3 = 0$, because the points of that plane are relevant to purely deviatoric states of stress ($\xi_1 = 0$).

7

Electrochemical Cell's Admissible Range: The Transformation Cells

Electrochemistry is concerned with the transformation of the energy liberated by a chemical reaction into exploitable electric energy. It also deals with the opposite process, that is, using electric energy to drive a chemical reaction. Electrochemical cells are manmade systems that have been designed to perform these processes. They are referred to as galvanic (or voltaic) cells when they produce energy and as electrolytic cells when they drive a non-spontaneous chemical reaction. The ubiquitous rechargeable battery, which we depend upon so heavily nowadays, is perhaps the most common example of both kinds of cells.

Direct transformation of chemical energy into electric energy is possible, because the energy liberated by a reaction comes from the change in the electric potential of the valence electrons of the reacting substances as they move from one reactant to another to produce the reaction. In a galvanic cell, the electrons from the electron-losing species are collected at an electrode and driven through an external electric circuit before being released to their final, lower potential position at the other electrode. While flowing through the external circuit, the electrons can produce useful work. In contrast, by putting the reacting substances in direct contact with each other, the reaction occurs immediately. In this case, the valence electrons move directly to their final location, and their surplus energy is dissipated into heat of reaction in the material where the reaction takes place.

We distinguish between two types of electrochemical cells: transformation cells and concentration cells. The difference between the two types is in the kind of reaction that drives the cell. In transformation cells, the reaction is an ordinary redox reaction that transforms reactant species into different species. The present chapter focuses on this type of cell. Unlike transformation cells, concentration cells do not result in chemical changes in the species involved; only their concentrations are changed. We address this type of cell in the next chapter. The aim of both chapters is to determine the admissible range of the cell and how it relates to the cell's capacity to store energy.

Electrochemistry is at the heart of the energy exchange capacity of every living system. A proper understanding of the common features of electrochemical cells and living systems helps to discriminate between the subjective behaviours of living systems and the physics that makes their responses possible. For a complete treatment of classical electrochemistry, the reader may refer to general textbooks, such as [23] and [2], or to one of the many specialized treatises on the subject (e.g., [10] and [55]).

7.1 Basics of Electrochemical Cells

For simplicity, we consider electrochemical cells at the standard temperature (298.15 K) and pressure (1 bar = 10^5 Pa). Also, if not otherwise stated, the concentration of each solute that undergoes chemical reactions in the solution within the cell is assumed to be standard concentration, i.e., the concentration at which the activity of the solute is equal to unity (cf. Appendix C.5). For ideal solutions in an aqueous solvent, this concentration is usually taken to be one molality.

Of course, the concentration of a solute will generally change due to the chemical reactions that take place in the cell. Reference to a cell that operates at a standard concentration implies, therefore, that the change in the extent of the reactions should be small enough to produce a negligible change in the concentration of the solutes. Of course, non-standard concentrations must be considered when dealing with the processes of cell exhaustion and recharging. In that case, pressure and temperature will be the only variables that can be kept at their standard values.

7.1.1 Oxidation and reduction

An electrically neutral atom or molecule carries as many electrons as protons. The addition or release of one or more electrons transforms it into a negative or positive ion, respectively. The electron removal process is called *oxidation*. It may operate on atoms, molecules, and ions. In the last case, it makes a positive ion more positive and a negative ion less negative. If A denotes any neutral atom or molecule, then the oxidation process is symbolically described by the following relation:

$$\mathsf{A} \rightarrow \mathsf{A}^{\alpha+} + \alpha\, \mathrm{e}^{-}, \tag{7.1.1}$$

where α is an integer that equals the number of electrons that are removed from A, and e^{-} indicates the electron itself (α is omitted if equal to 1). The positive ion $\mathsf{A}^{\alpha+}$ (*cation*) that results from the process has the charge of α protons, i.e., αe, where e denotes the charge of a proton. This charge is also referred to as *elementary positive charge* and is opposite to the charge of an electron.

A relation similar to (7.1.1) applies to *reduction* processes, i.e., processes that reduce the charge of an atom, a molecule, or an ion. A reduction process adds electrons, thus reducing the positive charge of the particle to which the electrons are added and therefore increasing its negative charge. The reduction process of a neutral atom or molecule is represented by:

$$\mathsf{A} + \alpha\, \mathrm{e}^{-} \rightarrow \mathsf{A}^{\alpha-}, \tag{7.1.2}$$

where $\mathsf{A}^{\alpha-}$ is the negative ion (*anion*) resulting from A, carrying the negative charge $-\alpha e$.

Relations (7.1.1) and (7.1.2) also apply when A is an ion. In that case, the superscripts $\alpha+$ and $\alpha-$ will refer to the number of elementary positive or negative charges, respectively, that the ion acquires in the process.

7.1.2 Half-cells

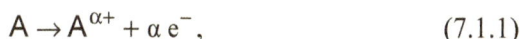

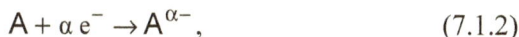

Oxidation and reduction processes may take place separately. A typical case is when they occur in a so-called *half-cell*. The latter consists of one single electrode and an electrolyte in electric contact with it. The electrode may be a metal or any other solid, liquid, or gaseous material. If the electrode is not a metal, then it must be connected to a conductor that supplies or collects the electrons that the electrode exchanges with the electrolyte.

When in contact with the electrolyte, some atoms from the electrode may leave their electrons at the electrode and go into solution as ions. The electrode will acquire a negative charge as a result. As far as the atoms that leave the electrode are concerned, this is an oxidation process. Conversely, some ions from the electrolyte may reach the electrode and become neutral upon transferring their charges to it. The particles thus neutralized may stick to the electrode, precipitate from the solution, or develop as gas. This process makes the electrode acquire a positive or negative charge, depending on the charge of the ions that are being neutralized. If the ions are positive, they will take electrons from the electrode. The process will reduce the charge of the ions and leave the electrode positively charged. In contrast, if the ions are negative, they will give up electrons to the electrode. The ions will be oxidized, whereas the electrode will end up negatively charged.

When a half-cell is electrically isolated from the surroundings, the above processes will quickly build up a potential difference between the electrode and the electrolyte. This will oppose further electron exchange and bring the process to a halt.

To measure the potential difference between electrode and electrolyte, we must introduce another electrode into the solution. Doing so, however, will introduce a new half-cell into the original half-cell, invariably affecting the very potential to be measured. For this reason, the equilibrium electric potential of a half-cell in electrically-isolated conditions is conventionally measured by the difference in the electric potential between its electrode and the electrode of another specific half-cell, i.e., a reference half-cell. By universal convention, the reference half cell is taken to be the so-called *standard hydrogen half-cell.*

The *standard reduction potential tables* comprise a collection of electric potential differences at standard conditions of a number of different half-cells with respect to the standard hydrogen cell. A positive value of this potential indicates that the electrode of the considered half-cell has the tendency to take electrons from the electrode of the hydrogen half-cell. In this case, the half-cell in question is the seat of a reduction process of the kind described by relation (7.1.2). The potential difference between any two half-cells at standard conditions can be calculated from the same tables as the difference between their standard reduction potentials.

7.1.3 Redox processes

In the vast majority of cases, the oxidation of a chemical species and the reduction of another species take place simultaneously. The combination of the two reactions is known as a *redox* process. A typical example is the reaction between solid zinc [Zn(s)] and cupric sulphate in aqueous solution [Cu_2SO_4(aq)]. The solution contains the cupric sulphate as dissociated ions, Cu^{2+} and SO_4^{2-}, in addition to the H^+ and OH^- ions from the solvent. If the solution is put in contact with solid zinc, the following reaction takes place:

$$\mathrm{Cu^{2+}(aq) + Zn(s) \rightarrow Zn^{2+}(aq) + Cu(s)}\,. \tag{7.1.3}$$

Under ordinary conditions of temperature and pressure, this is a spontaneous reaction, because its Gibbs energy of reaction, $\Delta_r G°$, is negative (Appendix C.4). This is a redox process. It reduces the cupric ions according to the relation

$$\mathrm{Cu^{2+}(aq) + 2\,e^- \rightarrow Cu(s)}\,, \tag{7.1.4}$$

and oxidizes the zinc atoms according to the relation

$$\mathrm{Zn(s) \rightarrow Zn^{2+}(aq) + 2\,e^-}\,. \tag{7.1.5}$$

As in any chemical reaction, the quantity $\Delta_r G°$ relevant to reaction (7.1.3) is readily obtained from eq. (C.22) as the difference between the Gibbs energy of formation of the reaction products and that of the reactants. The energy that is liberated per mole of each reactant is $-\Delta_r G°$. Reaction (7.1.3) transfers two electrons from each zinc atom to each cupric ion. This means that the reaction of one mole of Zn with one mole of Cu^{2+} produces a flow of two moles of electrons.

7.1.4 The electrochemical cells

An electrochemical cell exploits a spontaneous redox reaction to produce electric energy, or it absorbs electric energy to drive the reaction in the opposite direction. A full cell consists of two half-cells coupled together. One half-cell is the cell's negative electrode (*anode*). This is where the oxidation part of the process takes place. In normal operating conditions, this electrode supplies electrons to the external circuit by taking them from oxidizing species. The other half-cell is the positive electrode (*cathode*). It takes the electrons from the external circuit to supply them to the reducing species in the solution, thus performing the reduction part of the complete redox process.

The two half-cells may share the same electrolyte, as in an ordinary car battery, or they may not. If the two half-cells operate on separate electrolytes, then the two electrolytes must be allowed to exchange ions with each other. This may be achieved by separating them through a semi-permeable membrane that prevents their mixing while allowing the passage of certain kinds of ions. In the cell's most essential setup, a felt barrier can be used as a semi-permeable membrane.

It should be noted that, due to the different mobility of the different species of ions involved in the reaction, the membrane separating the two half-cells may develop a difference in the electric potentials of its two surfaces (*liquid junction potential*). As the cell operates, this potential difference will dissipate some of the energy available from the reaction and make the process irreversible. Also, it will reduce the potential difference between the electrodes of the cell, thus impairing its efficiency.

The usual way to eliminate the formation of the liquid junction potential is to keep the two half-cells separate while joining their electrolytes by a *salt bridge*. In its most basic setup, the latter is made of a tube filed with a solution that exchanges ions of equal mobility with the two half-cells connected to its semi-permeable ends. This device, or an equivalent device, substitutes for the semi-permeable membrane and eliminates the irreversibility that the membrane may produce as a result of the associated liquid junction potential.

In the following sections, we shall assume that the semi-permeable membrane that separates the two half-cells operates reversibly. In practice, this may imply the use of an appropriate salt bridge instead of the semi-permeable membrane.

The cell's emf

The electric potential difference that is established between the cell's electrodes in open-circuit conditions is denoted by $\boldsymbol{E}$ and referred to as the cell's *electromotive force* (emf). It is defined as the electric potential difference between the positive and the negative electrodes of the cell. The positive electrode is defined as the one at which reduction takes place under normal operating conditions of the cell, while the negative electrode is where the oxidation part of the cell reaction takes place.

Thus, under normal operating conditions, $\boldsymbol{E}$ is a positive quantity for a galvanic cell. Under general operating conditions, however, the positive electrode may assume a lower potential than that of the other electrode, depending on the concentration of the electrolytes. This means that $\boldsymbol{E}$ may have either positive or negative values, depending on the state of the cell. We shall see that $\boldsymbol{E}$ decreases to zero as the concentration of the species involved in the reaction reaches the concentration associated with chemical equilibrium. The value that $\boldsymbol{E}$ attains when the cell is at standard conditions is indicated as $\boldsymbol{E}°$.

When the external circuit between the two electrodes is closed, electric current flows through the solution inside the cell. This produces a drop in $\boldsymbol{E}$ due to the internal electric resistance of the cell's solution. Because $\boldsymbol{E}$ refers to open-circuit conditions, its value represents the maximum electric potential difference between the cell's electrodes at the considered state of the cell.

7.1.5 An example: The Daniell cell

The Daniell cell provides a typical example of an electrochemical cell. It is based on the redox reaction (7.1.3), which is rewritten here in the following more general form:

$$\mathrm{Zn(s)} + \mathrm{CuSO_4(aq)} \rightleftarrows \mathrm{Cu(s)} + \mathrm{ZnSO_4(aq)}\,. \qquad (7.1.6)$$

The cell can operate both as a galvanic cell and as an electrolytic cell. In the first case, it drives reaction (7.1.6) in the left-to-right direction. In the second case, it drives the same reaction in the opposite direction.

In order to fully describe the state of the cell, the amounts of the chemical reactants that it contains must be specified at each time of its operation. This is most effectively done by the state variable ξ, called the *extent of reaction* (Appendix C.3). In the present case, all stoichiometric coefficients of reaction (7.1.6) are equal to unity. This means that, at any stage of the reaction, the extent of reaction equals the number of moles of each of the species, whether consumed or produced in the reaction. We denote the number of moles of species A at any stage of the reaction as n_A and the initial value of n_A as n_A^o. We can then write:

$$n^{o}_{Cu^{2+}} = n^{o}_{CuSO4(aq)},$$

$$n^{o}_{Zn^{2+}} = n^{o}_{ZnSO4(aq)},$$

$$\begin{aligned} n_{Cu^{2+}} &= n_{CuSO4(aq)} = n^{o}_{CuSO4(aq)} - \xi\,, \\ n_{Zn^{2+}} &= n_{ZnSO4(aq)} = n^{o}_{ZnSO4(aq)} + \xi\,, \\ n_{Cu(s)} &= n^{o}_{Cu(s)} + \xi\,, \\ n_{Zn(s)} &= n^{o}_{Zn(s)} - \xi\,, \end{aligned} \tag{7.1.7}$$

which makes eq. (C.11) of Appendix C specific to the present case. The above equations show that variable ξ alone suffices to determine the chemical composition of the cell's content at any time of its operation.

Galvanic cell mode

As observed above, under standard conditions of temperature and pressure, reaction (7.1.6) proceeds spontaneously from left to right. The energy liberated per mole of reactant (or, more generally, per mole of extent of reaction, ξ) is $-\Delta_r G°$. Refer to eq. (C.21) and the comments following it for more details. If not otherwise used, the energy $-\Delta_r G°$ is dissipated into heat in the reacting mixture. The Daniell cell transforms this energy into usable electric energy, thus avoiding its immediate dissipation. When operating in this way, the Daniell cell acts as a galvanic cell. Fig. 7.1.1 presents an outline of the cell in one of its classical settings as a galvanic cell.

The details of the chemical process that occurs within the cell are described as follows. Some Zn(s) atoms leave the zinc electrode to enter the solution as Zn^{2+} ions. In doing so, each atom surrenders two electrons to the electrode, thereby making the $ZnSO_4$(aq) solution assume a positive charge. The liberated electrons flow from the Zn electrode and through the external circuit to reach the Cu electrode at the other side of the cell, where the electrons are absorbed by the Cu^{2+} ions in the solution of the Cu-electrode compartment. The Cu^{2+} ions are thus reduced to neutral Cu(s) atoms, which are deposited on the copper electrode. The consequent loss of Cu^{2+} ions from the solution charges the same solution negatively. This draws to that solution some Zn^{2+} ions from the solution at the other side of the semi-permeable barrier, thus restoring the electrical neutrality of both compartments of the cell. Then, a new Zn^{2+} ion can leave the Zn electrode, and the process starts again.

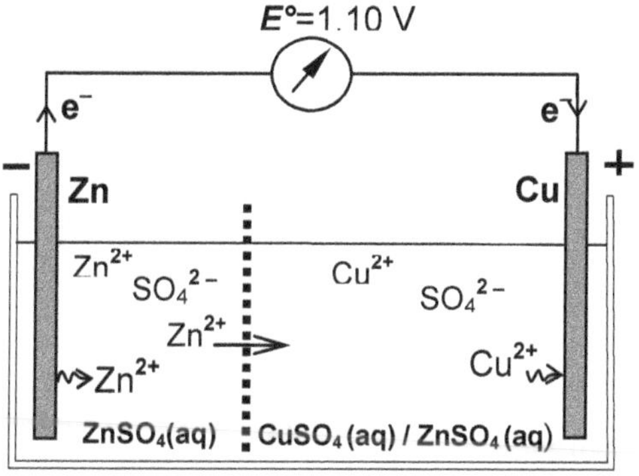

Fig. 7.1.1. Daniell cell in galvanic mode of operation

The process produces a flow of electrons through the external electric circuit, which can be exploited to do work. When the Daniel cell operates in this mode, its Zn electrode is the negative electrode, while the Cu electrode is the positive electrode. Eventually, all of the Zn(s) atoms from the Zn electrode are consumed, or all of the available $CuSO_4$(aq) is transformed into $ZnSO_4$(aq). The last occurrence means that all of the Cu^{2+} ions in the original $CuSO_4$(aq) solution have been depleted as the result of their transformation into Cu(s) atoms that are deposited on the copper electrode.

In the present analysis, we neglect any dissipation that the process may involve, because we are interested in determining the maximum amount of energy that a cell is capable of storing, apart from how that energy is expended. If the cell is rechargeable, as is the case of the version of the Daniell cell considered here, the maximum amount of energy that can be stored depends on how far the recharging process can go. The end of the recharging process need not coincide with the exhaustion of some reactants. The present chapter will show that there is a thermodynamic limit to the energy that a cell can store, which cannot be exceeded, no matter how well the cell is constructed.

Electrolytic cell mode

With the external circuit still closed, the chemical reaction in the cell can be stopped at any time by applying a potential difference $\Delta \boldsymbol{V}$ to its electrodes that is opposite to the value of $\boldsymbol{E}$ at the considered time. That is, $\Delta \boldsymbol{V} = -\boldsymbol{E}$. The applied potential makes the voltage difference between the two electrodes drop to zero, thus stopping the electron current between them. On the other hand, if $\Delta \boldsymbol{V}$ is more negative than $\boldsymbol{E}$, i.e., if $\Delta \boldsymbol{V} < -\boldsymbol{E}$, then an electric current will flow through the electrodes in the opposite direction to that of the electric current in the galvanic mode of operation. Under these conditions, the cell will operate as an electrolytic cell.

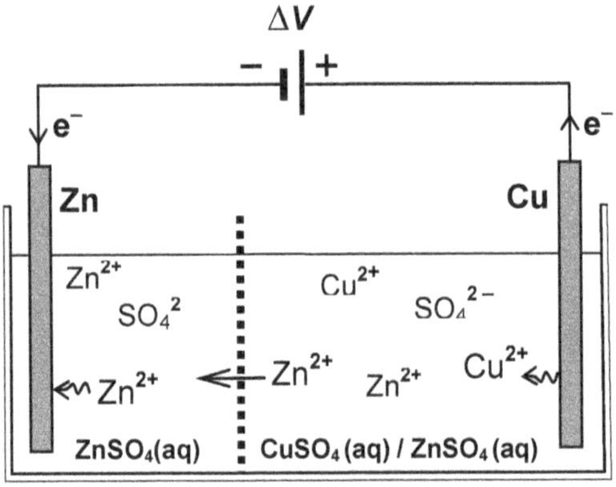

Fig. 7.1.2. Daniell cell in electrolytic mode of operation

Fig. 7.1.2 illustrates the process that takes place in a Daniell cell in the electrolytic mode. The applied potential $\Delta \boldsymbol{V}$ pushes the electrons to the Zn electrode and thence to the adjacent $ZnSO_4$(aq) solution. Once in the solution, these electrons reduce the Zn^{2+} ions into Zn(s) atoms, which are insoluble, so they precipitate. As a result of the loss of positive ions, the $ZnSO_4$(aq) solution becomes negatively charged. This draws Zn^{2+} ions from the other compartment of the cell through the semi-permeable barrier, provided that there are any Zn^{2+} ions available there (possibly from a previous operation in galvanic mode). This, in turn, makes the $CuSO_4$(aq) solution of the Cu electrode compartment acquire a negative charge. As a result, some Cu(s) from the copper electrode will leave the electrode as Cu^{2+} ions. The latter go into the solution of the Cu-electrode compartment to neutralize it. In doing so, each Cu atom surrenders two electrons to the Cu electrode.

These electrons are drawn by the $\Delta \boldsymbol{V}$ generating device and transferred to the zinc electrode. Thus, the process can continue.

From the chemical standpoint, the process corresponds to driving reaction (7.1.6) from right to left, i.e., opposite to its spontaneous direction. To drive the cell this way, the $\Delta \boldsymbol{V}$ generating device must expend energy. Reaction (7.1.6) *absorbs* the energy $-\Delta_r G° > 0$ per mole of Cu(s) or Zn^{2+} as it proceeds from right to left. The absorbed energy is stored in the cell as an increase in the free energy of the solution. It is returned when the cell operates in galvanic mode.

If the difference $(\Delta \boldsymbol{V} - \boldsymbol{E})$ is small enough, the process approximates a sequence of equilibrium states. As observed in Section 1.5, this makes the process thermodynamically reversible. Therefore, an electrochemical cell provides a means of driving a chemical process reversibly in either direction, which has a number of important consequences. In particular, it enables us to establish the relation between $\Delta_r G$ and $\boldsymbol{E}$, which we discuss in the next section.

Under equilibrium conditions, the process is also homogeneous. The reason is that, under these conditions, the variables that describe the state of the system are independent of the points of the system at which they are calculated. (Of course this is quite different from stating that the cell itself is a homogeneous system, which obviously it is not.)

7.2 Relation between Gibbs Energy and Emf

As previously mentioned, the reaction Gibbs energy, $\Delta_r G$, is the energy that a chemically-reacting system absorbs per unit extent of reaction (i.e., per unit ξ) at constant pressure and temperature. Its opposite, $-\Delta_r G$, is the energy that the system librates per unit ξ. As the reaction proceeds, $\Delta_r G$ changes and vanishes at chemical equilibrium. (See Appendix C.4, for more details.)

A relation between $\Delta_r G$ and $\boldsymbol{E}$ can be established by recalling that the work needed to displace a positive electric charge, q^e, across a potential difference, $\Delta \boldsymbol{V}$, is given by $q^e \Delta \boldsymbol{V}$. Therefore, the quantity $\Delta \boldsymbol{V}$ represents the work needed to displace a unit positive charge from potential $\boldsymbol{V}$ to potential $\boldsymbol{V} + \Delta \boldsymbol{V}$. In reversible conditions, the difference between the electric potential of the cell's positive electrode and its negative electrode is $\Delta \boldsymbol{V} = \boldsymbol{E}$. Thus, the work that must be expended to move the infinitesimal charge dq^e from the negative to the positive electrode is given by:

$$\mathrm{d}W_{\mathrm{in}}^{\mathrm{e}} = \mathrm{d}q^{\mathrm{e}}\boldsymbol{E} . \tag{7.2.1}$$

In particular, if $\boldsymbol{E} > 0$ and the charge $\mathrm{d}q^{\mathrm{e}}$ is negative, then the transfer of the charge from the negative to the positive electrode produces a negative value of $\mathrm{d}W_{\mathrm{in}}^{\mathrm{e}}$. This means that the cell does work on the surroundings, since $\mathrm{d}W_{\mathrm{out}}^{\mathrm{e}} > 0$ according to eq. (1.3.2). In particular, if the process takes place reversibly, then eq. (1.11.11) applies. Therefore, from eq. (7.2.1), we have

$$\mathrm{d}G\big|_{p,T} = \mathrm{d}q^{\mathrm{e}}\boldsymbol{E} . \tag{7.2.2}$$

Now, let $\mathrm{d}q^{\mathrm{e}}$ be the negative charge liberated at the cell's anode as the cell's reaction advances by $\mathrm{d}\xi$. Let, moreover, F denote the so-called Faraday's constant, i.e., the charge of an Avogadro number of elementary positive charges (protons). Expressed in coulombs per mole, we have $F = 9.6485\ 10^4\ \mathrm{C\ mol^{-1}}$. Since the charge of the electron is opposite to that of the proton, we can write

$$\mathrm{d}q^{\mathrm{e}} = -zF\,\mathrm{d}\xi . \tag{7.2.3}$$

Here, z indicates the number of moles of electrons that are produced by the oxidation part of the cell's reaction per 1 mol increase in the extent of reaction ξ. The number z is a dimensionless quantity. Its value can be obtained by inspection from the balanced reaction formula, once the oxidized components are singled out. It depends on the stoichiometric coefficients of the reaction and on the valence of the oxidized ions. For instance, in the case of reaction (7.1.3), we have $z = 2$, as can easily be inferred from the same reaction and from eqs (7.1.5) and $(7.1.7)_6$.

By introducing eq. (7.2.3) into eq. (7.2.2), we obtain:

$$\left(\frac{\partial G}{\partial \xi}\right)_{p,T} = -zF\boldsymbol{E} . \tag{7.2.4}$$

As discussed in Appendix C.4, the quantity on the left-hand side of this equation is nothing but the reaction Gibbs energy. It is the same quantity that we previously denoted by the symbol $\Delta_{\mathrm{r}}G$, according to the notation that is more frequently used. With that notation, eq. (7.2.4) rewrites as:

$$\Delta_r G = -zF\boldsymbol{E}. \tag{7.2.5}$$

This is the fundamental equation of electrochemistry. It relates the emf of a cell to a basic thermodynamic property of the chemical reaction that takes place within the cell.

Equation (7.2.5) applies, in particular, to the special case in which both the reactants and the products of reaction are in their respective standard states, i.e., the state that is also assumed as the reference state for their Gibbs free energy (Appendix C.5). In that case, the electrochemical cell is said to be in standard condition, and eq. (7.25) becomes

$$\Delta_r G^o = -zF\boldsymbol{E}^o, \tag{7.2.6}$$

where $\boldsymbol{E}^o$ is the standard emf defined in Section 7.1.4.

Equation (7.2.6) affords a means of calculating $\boldsymbol{E}^o$ from the thermodynamic tables on the standard Gibbs energy of formation. The quantity $\Delta_r G^o$ can be obtained from these tables by subtracting the energy of formation of the products of reaction from that of the reactants, cf. eq. (C.22). For a Daniell cell, $\boldsymbol{E}^o = 1.10$ V.

In order to determine how $\boldsymbol{E}$ varies with the concentrations or, more correctly, with the activities a_i of the species involved in the reaction, we can introduce eqs (7.2.5) and (7.2.6) into eq. (C.55) to obtain

$$\boldsymbol{E} = \boldsymbol{E}^o - \frac{RT}{zF}\ln Q, \tag{7.2.7}$$

where Q is the reaction quotient. By means of eq. (C.56), we can write eq. (7.2.7) in the more explicit form below:

$$\boldsymbol{E} = \boldsymbol{E}^o - \frac{RT}{zF}\ln\frac{\left(\prod_h a_h^{\nu_h}\right)_{\text{products}}}{\left(\prod_k a_k^{|\nu_k|}\right)_{\text{reactants}}}. \tag{7.2.8}$$

Eqs (7.2.7) and (7.2.8) are two equivalent expressions of the *Nernst equation*. Refer to Appendix C.6 for a complete explanation of the symbols used.

If we apply eq. (7.2.8) to reaction (7.1.3), we obtain:

$$\boldsymbol{E} = \boldsymbol{E}^o - \frac{RT}{zF}\ln\frac{a_{Zn^{2+}}}{a_{Cu^{2+}}}. \tag{7.2.9}$$

7.3 The Electrochemical Cell's Free Energy

Let us first recall a few basic formulae from physical chemistry that will be used in this section. More details concerning these formulae are provided in Appendix C.

The Gibbs free energy of a solution of r different components A_i ($i = 1,2, \ldots r$), including solvent, can be expressed as

$$G = G(p,T,n_1,n_2,\ldots,n_r) = \sum_{i=1}^{r} n_i \mu_i + C , \tag{7.3.1}$$

cf. Appendix C, eq. (C.4). In this equation, n_i indicates the number of moles of solute A_i, while μ_i is the *partial molar free energy* of that component. The quantity μ_i is also referred to as *chemical potential*. It represents the contribution to G from 1 mol of A_i in the solution. At variance with eq. (C.4), eq. (7.3.1) contains the arbitrary constant C that originates from the definition of internal energy, since we are considering the Gibbs equation in its general form (B.19). Although C is inessential and in many cases omitted, it must be considered explicitly here because of its role in the arguments that follow.

The chemical potential of each component A_i is usually expressed in the form:

$$\mu_i = \mu_i^o(p,T) + RT \ln a_i , \tag{7.3.2}$$

where R is the universal gas constant, a_i is the *activity* (or *effective concentration*) of component A_i, and $\mu_i^o(p,T)$ is the value assumed by μ_i when A_i is in its standard state of concentration. In many practical applications, constant pressure and constant temperature are considered. In that case, the dependence of $\mu_i^o(p,T)$ on p and T is of little relevance, and the quantity is simply written as μ_i^o.

A widely-used measure of concentration of a solute is its *molality*. This is defined as the ratio of the moles of solute to the mass of solvent. Thus, the molality, b_i, of n_i moles of solute A_i in a mass, m_{solv}, of solvent is given by:

$$b_i = \frac{n_i}{m_{solv}} , \tag{7.3.3}$$

cf. Appendix C, eq. (C.41). In terms of b_i, the activity is usually expressed as

$$a_i = \gamma_i \frac{b_i}{b_i^o}, \tag{7.3.4}$$

where b_i^o denotes the molality of solute A_i in its standard state, and γ_i is the *activity coefficient* of the solute. For ideal solutions, $\gamma_i = 1$. In general, however, the activity coefficient is a function of the state of the solution, i.e.,

$$\gamma_i = \gamma_i(p, T, n_1, n_2, \dots n_r). \tag{7.3.5}$$

In principle, this function should be determined experimentally. However, in some important instances, its expression can also be inferred theoretically, as discussed briefly in the next section. Of course, there are different, equally-acceptable ways of defining the concentration of a solute and thus its activity. Each choice leads to different definitions of γ_i and different functions (7.3.5), but the form of eq. (7.3.2) remains unaltered.

Unless otherwise stated, the standard state of a species in solution is taken as the state of unit molality; that is, $b_i^o = 1$ molal. This enables us to omit the denominator b_i^o from eq. (7.3.4) while leaving the division by 1 molal understood. It also facilitates the use of the modern thermodynamic data tables, since these tables refer to unit molal concentrations. Clearly, a different value of μ_i^o would be obtained if we chose, for example, the molar fraction x_i as a measure of concentration and $x_i = 1$ as the standard state for solute *i*. This would be perfectly admissible and would lead to similar formulae as formula (7.3.6) below, although with different values for μ_i^o and γ_i. For more details on how activities are defined and standard states are chosen, refer to Appendix C.5.

From eqs (7.3.2)-(7.3.4), we can express μ_i as

$$\mu_i = \mu_i(p,T,n_1,n_2,\dots,n_r) = \mu_i^o(p,T) + RT \ln \frac{\gamma_i}{m_{solv}} n_i. \tag{7.3.6}$$

Accordingly, the solution's free energy (7.3.1) can be written as

$$G = G(p,T,n_1,n_2,\dots,n_r) = \\ \sum_{i=1}^{r} n_i \, \mu_i^o(p,T) + RT \sum_{i=1}^{r} n_i \ln \frac{\gamma_i}{m_{solv}} n_i + C. \tag{7.3.7}$$

If the solutes are involved in a chemical reaction, their mole number n_i will vary as the extent of reaction, ξ, increases. More precisely,

$$n_i = \nu_i \xi + n_i^o , \qquad (7.3.8)$$

where n_i^o denotes the value of n_i at the beginning of the reaction (see Appendix C.2.) Note that the superscript ^{o} attached to n_i^o refers to the initial value of n_i. The quantities ν_i that appear in eq. (7.3.8) are the stoichiometric coefficients of the balanced reaction equation that is taken into consideration. They are assumed to be negative for the species on the left-hand side of the reaction equation (the reactant side) and positive for the species on the right-hand side of the same equation (the product side).

In the present chapter, we are concerned with the values of G irrespective of the direction in which the cell reaction proceeds. Pressure and temperature are assumed to be constant. In this case, functions (7.3.7) and (7.3.5) reduce to $G = G(\xi)$ and $\gamma_i = \gamma_i(\xi)$, because all variables n_i are functions of ξ only. However, as shown by eq. (7.3.8), these functions depend on the values of ν_i, and hence on the considered reaction. This makes the functions $G(\xi)$ and $\gamma_i(\xi)$ different for different reactions. By contrast, when expressed in the forms (7.3.7) and (7.3.5), the quantities G and γ_i depend only on the concentration of the solutes (i.e., on the values of n_1, n_2, ... n_r) apart from the reaction that is taking place in the system.

By introducing eq. (7.3.8) into eq. (7.3.3), we obtain

$$b_i = \frac{\nu_i}{m_{solv}} \xi + \frac{n_i^o}{m_{solv}} . \qquad (7.3.9)$$

When introduced into eq. (7.3.4), this yields:

$$a_i = \frac{\gamma_i(\xi)}{m_{solv}} (\nu_i \xi + n_i^o) . \qquad (7.3.10)$$

Remember that division by $b_i^o = 1$ molal is understood in the right-hand side of this equation. By introducing eq. (7.3.8) into eqs (7.3.5) and (7.3.6), we obtain:

$$\mu_i = \mu_i(\xi) = \mu_i^o + RT \ln[(\nu_i \xi + n_i^o)\gamma_i(\xi)] - RT \ln m_{solv} . \qquad (7.3.11)$$

Again, since the argument of a logarithm must be dimensionless, the arguments of the logarithms appearing in this formula are understood to be divided by a unit quantity of the appropriate dimension to make them dimensionless.

A general expression for $G = G(\xi)$ follows from eq. (7.3.1) after eqs (7.3.11) and (7.3.8) are introduced into it. Though not particularly complex, the resulting expression would be rather lengthy. Moreover, it would contain the functions $\gamma_i = \gamma_i(\xi)$, which can only be specified when the reactions that are occurring in the system are given. For this reason, it is far more practical to determine $G(\xi)$ in each particular case, once that the system and the chemical reaction occurring in it have been specified. A further advantage of this way of proceeding is that, for constants T and p, the contribution to G from all species that do not take part in the reaction is a constant. That contribution can therefore be included in the arbitrary constant C that appears in eq. (7.3.1), so that there is no need to calculate it explicitly.

7.4 Activity Coefficients of Ions in Water Solution

The activity coefficient, γ_i, of a solute of a non-ideal solution depends on the concentration of the solute and must be determined experimentally. Various approximate formulae are available to evaluate these coefficients theoretically. For solutions that contain ionic solutes, however, it is not possible to change the concentration of one single ionic species without also changing the concentration of the other ions which result from the dissociation of the neutral compound that produced the ionic species. The reason is that all of the different ionic species that are produced by the dissociation of a given compound enter into solution simultaneously as the compound dissociates. This means that we cannot measure the activity coefficient of just one single species of ions; instead, we measure the so-called *mean ionic activity*. The latter is an appropriate geometric mean of the activities of the individual ionic species of all ions that are formed as a compound dissociates in solution.

Several formulae are available to determine the mean ionic activity coefficients of an ion in aqueous solutions at any concentration. They aim at extending the range of validity of the well-known Debye-Hückel limit law, a theoretically derived formula that is only applicable to very low concentrations. Most of these formulae express γ_i as a function of the total content of all ionic charges of the solution, as represented by the solution's ionic strength, I, which is defined as

$$I = \frac{1}{2} \sum_{\rm i} z_{\rm i}^{2}\, b_{\rm i} \,. \tag{7.4.1}$$

Here the summation is understood to include all ionic species contained in the solution, b_i is the molality of ionic species i, and z_i is the valence number of the same species, i.e., a positive or negative integer equal to the number of elementary charges of the considered ion.

For low values of ionic strength, e.g., for I less than about 0.1 molal, many of these formulae can be written as

$$\log \gamma_{\rm i} = -0.51 \left| z_{\rm i}\, z^{\rm c}{}_{\rm i} \right| \left(\frac{\sqrt{I}}{1 + {\rm A}\sqrt{I}} - {\rm B}\, I \right), \tag{7.4.2}$$

where 'log' indicates the common logarithm (base 10), and $z^{c}{}_{i}$ is the valence of the conjugated ion that results from the dissociation of the parent compound from which ion i originated [66]. Of course, the coefficients that appear in eq. (7.4.2) are not dimensionless, because I, as defined by eq. (7.4.1), is expressed in molality. Eq. (7.4.2) reduces to the extended Debye-Hückel equation when A = 0.33 a_o and B = 0, where a_o is the effective diameter of the ion in Å. For A = 1.0 and B = −0.2 or B = −0.3, equation (7.4.2) becomes the so-called Davies equation, which is often used in geophysics and biology.

Another formula is the Pitzer equation. This is a particularly versatile and accurate semi-empirical formula, but it is by no means the simplest. It can be expressed as

$$\ln \gamma_{\rm i} = \left| z_{\rm i}\, z^{\rm c}{}_{\rm i} \right| f^{\gamma} + b_{\rm i}\, B^{\gamma} + 3\, b_{\rm i}^{2}\, C^{\phi}/\, 2\,, \tag{7.4.3}$$

where

$$f^{\gamma} = -\, A_{\phi} \left[\frac{\sqrt{I}}{1 + 1.2\sqrt{I}} + \frac{2}{1.2} \ln\left(1 + 1.2\sqrt{I}\right) \right]. \tag{7.4.4}$$

Here, A_ϕ is the so-called Debye-Hückel coefficient. For aqueous solutions at 298.15 K, we have A_ϕ = 0.3919 (kg mol^{-1})$^{1/2}$. Coefficient B^γ in eq. (7.4.3) is given as

$$B^{\gamma} = \beta^{(0)} + \left(2\beta^{(1)}/\, \alpha^{2} I\right)\left[\, 1 - \left(1 + \alpha\sqrt{I} - \alpha^{2} I/\, 2\,\right)\exp\left(-\alpha\sqrt{I}\right) \right], \tag{7.4.5}$$

or

$$B^{\gamma}=2\beta^{(0)}+\left(2\beta^{(1)}/\alpha_1^{2}I\right)\left[1-\left(1+\alpha_1\sqrt{I}-\alpha_1^{2}I/2\right)\exp\left(-\alpha_1\sqrt{I}\right)\right]$$
$$+\left(2\beta^{(2)}/\alpha_2^{2}I\right)\left[1-\left(1+\alpha_2\sqrt{I}-\alpha_2^{2}I/2\right)\exp\left(-\alpha_2\sqrt{I}\right)\right], \quad (7.4.4)$$

for univalent-bivalent electrolytes or for bivalent-bivalent electrolytes, respectively [24]. For other kinds of electrolytes, the reader is referred to the vast literature on the subject, e.g., [73]. In the above equations, the parameters C^{ϕ}, $\beta^{(0)}$, $\beta^{(1)}$, and $\beta^{(2)}$ vary according to the electrolyte being used. They must be evaluated experimentally if they are not available from thermodynamic tables. In the above formula, $\alpha = 2$ (kg mol^{-1})$^{1/2}$, $\alpha_1 = 1.4$ (kg mol^{-1})$^{1/2}$, and $\alpha_2 = 12$ (kg mol^{-1})$^{1/2}$.

Incidentally, formula (7.4.2) gives the same value of γ_i irrespective of the number of solutes that contribute to the ionic strength of the solution. In fact, the value of γ_i will generally be affected (though usually to a minor extent) by the number and kind of the species in solution. This is one of the reasons why eq. (7.4.2) only provides an approximation. Formula (7.4.3) is free from that short-coming, in that its parameters are experimentally determined to adapt it to each specific solution rather than assuming the same values for entire classes of electrolytes.

7.5 The Free Energy of the Daniell Cell

Let us now calculate the free energy function, $G = G(\xi)$, of an electrochemical cell at constant temperature and pressure. We refer, in particular, to a Daniell cell in the configuration and modes of operation described in Section 7.1.5. However, as Lewis and Randall noted in their classic treatise [35, p. 400] dating back to almost a century ago, the emf and the free energy of an electrochemical cell are properties of the reaction that occurs within a cell, not properties of the cell itself. The cell is only the means by which the chemical energy produced by the reacting substances is tapped. Thus all cells that use the same chemical reaction will share the same thermodynamic limitations that apply to that reaction.

The reaction that takes place in a Daniell cell is expressed by eq. (7.1.3). We assume that, initially, the solution in the cell is 0.2 molal $ZnSO_4$(aq) and 0.2 molal $CuSO_4$(aq) in the Zn-electrode compartment and in the Cu-electrode compartment, respectively. Each compartment is assumed to contain 1 kg of aqueous solvent (i.e., $m_{solv} = 1$ kg for each compartment). Temperature and pressure are assumed to be constant at 298.15 K and 1 bar, respectively.

It is not difficult to verify that, for the kind of Daniell cell we are considering, the concentration of the solution in the Zn-electrode compartment does not change during the cell's operation. Indeed, when the cell operates in the galvanic mode, any Zn^{2+} ions that leave that compartment are replaced by similar ions from the Zn electrode. Conversely, when the cell operates in the electrolytic mode, for every Zn^{2+} ion that enters the Zn-electrode compartment, a similar ion is precipitated as a solid Zn atom. This effectively removes the Zn^{2+} ion from the solution. As a result, the concentration of the solution in that compartment, and thus the ionic strength of that solution, remains constant in both cases.

Likewise, but for a different reason, the ionic strength of the solution in the Cu-electrode compartment remains constant. In this case, the reaction in the cell affects the relative amount of Cu^{2+} and Zn^{2+} ions in the solution, while leaving their total number unchanged, irrespective of the mode in which the cell operates. Consequently the solution's ionic strength, I, stays constant.

In both compartments, the number of SO_4^{2-} ions also remains constant. Since the total concentration of ionic charges (both positive and negative) is the same in the two compartments, the two solutions share the same value of I. Accordingly, the latter can be calculated by referring to the solution in the Zn-electrode compartment. In that case, we obtain

$$\begin{aligned} I &= \frac{1}{2}\left(z_{Zn^{2+}}^{\;2}\; b_{Zn^{2+}} + z_{SO_4{}^{2-}}^{\;2}\; b_{SO_4{}^{2-}}\right) \\ &= \frac{1}{2}\left[2^2\,0.2 + (-2)^2\,0.2)\right] = 0.8 \text{ molal}. \end{aligned} \tag{7.5.1}$$

This value of I will be constant and the same for both compartments, irrespective of how the cell operates.

The reaction that occurs in the cell is an example of a reaction that takes place in an aqueous solution at constant ionic strength. Chemical reactions of this kind are important because they are simple and because they are comparatively frequent in nature. The chemical processes that occur in biology, for instance, quite often take place in aqueous solutions of constant ionic strength.

As observed, the solution in the Cu-electrode compartment is the only one that changes in composition as the cell operates. Therefore, it alone is responsible for the changes in the cell's free energy. Accordingly, we must refer to that solution when applying eq. (7.1.7) to the present case. We assume that, initially, the solution is a pure-component solution, 0.2 molal $CuSO_4$(aq) in 1 kg of water. This means that $n^o_{Cu^{2+}}=n^o_{CuSO4(aq)}=0.2\,\text{mol}$ and $n^o_{Zn^{2+}}=n^o_{ZnSO4(aq)}=0$. From eqs $(7.1.7)_3$ and $(7.1.7)_4$ we then obtain

$$n_{Cu^{2+}}=n^o_{Cu^{2+}}-\xi=0.2-\xi \qquad (7.5.2)$$

and

$$n_{Zn^{2+}}=n^o_{Zn^{2+}}+\xi=\xi\,. \qquad (7.5.3)$$

The Zn electrode and the Cu electrode are assumed to have sufficient mass to avoid becoming exhausted during the whole reaction, no matter its direction. Under these conditions, the quantities $n^o_{Zn(s)}$ and $n^o_{Cu(s)}$ play no role in what follows and therefore need not be specified.

At constant T and p, the Gibbs free energy of the solvent is a constant, because the amount of solvent does not change during the cell's operation. On the other hand, the cell's electrodes are pure metals; as such, their formation energy is zero, and their contribution to the system's free energy is constant. For the considered cell, therefore, eq. (7.3.1) becomes

$$G=G(\xi)=n_{Cu^{2+}}\,\mu_{Cu^{2+}}+n_{Zn^{2+}}\,\mu_{Zn^{2+}}+C\,, \qquad (7.5.4)$$

where $n_{Cu^{2+}}$ and $n_{Zn^{2+}}$ are given by eq. (7.5.2) and eq. (7.5.3), respectively.

The partial molar energies $\mu_{Cu^{2+}}$ and $\mu_{Zn^{2+}}$ entering eq. (7.5.4) have the following forms:

$$\mu_{Cu^{2+}}=\mu_{Cu^{2+}}(\xi)=\mu^o_{Cu^{2+}}+RT\ln[(n^o_{Cu^{2+}}-\xi)\,\gamma_{Cu^{2+}}(\xi)] \qquad (7.5.5)$$

and

$$\mu_{Zn^{2+}}=\mu_{Zn^{2+}}(\xi)=\mu^o_{Zn^{2+}}+RT\ln[(n^o_{Zn^{2+}}+\xi)\,\gamma_{Zn^{2+}}(\xi)]\,. \qquad (7.5.6)$$

These expressions follow directly from eq. (7.3.11) and from $\nu_{Zn^{2+}}= -\nu_{Cu^{2+}}=1$. The latter is a consequence of eq. (7.1.3) and of the sign convention relative to the stoichiometric coefficients in eq. (7.3.8). The quantities $\mu^o_{Cu^{2+}}$ and $\mu^o_{Zn^{2+}}$ that appear in the above equations are the values of μ^o_i for Cu^{2+} and Zn^{2+}, respectively. When dealing with ions, these values usually refer to the so-called *Henry's law standard state* of the considered ion in solution (see Appendix C.5 for details).

By introducing eqs (7.5.5) and (7.5.6) into eq. (7.4.4) and by using eqs $(7.1.7)_3$ and $(7.1.7)_4$, we can write G in the following form:

$$G = G(\xi) = G_o(\xi) + G_1(\xi)\,, \tag{7.5.7}$$

where

$$G_o(\xi) = (n^o_{Cu^{2+}} - \xi)\,\mu^o_{Cu^{2+}} + (n^o_{Zn^{2+}} + \xi)\,\mu^o_{Zn^{2+}} \tag{7.5.8}$$

and

$$\begin{aligned} G_1(\xi) = (n^o_{Cu^{2+}} - \xi)\,RT\,\ln[(n^o_{Cu^{2+}} - \xi)\,\gamma_{Cu^{2+}}(\xi)] + \\ (n^o_{Zn^{2+}} + \xi)\,RT\,\ln[(n^o_{Zn^{2+}} + \xi)\,\gamma_{Zn^{2+}}(\xi)]\,. \end{aligned} \tag{7.5.9}$$

To proceed further, we must specify the mean activity coefficients $\gamma_{Cu^{2+}}(\xi)$ and $\gamma_{Zn^{2+}}(\xi)$. In the present case, they are constant, because I is constant. Therefore, we can avoid referring to the formulae presented in the previous section and assume the following values for a solution with an ionic strength $I = 0.8$ molal at $T = 298.15$ K:

$$\gamma_{Zn^{2+}} = \gamma_{Cu^{2+}} = 0.12\ . \tag{7.5.10}$$

These values are taken from the available experimental data and agree with those that can be obtained from formula (7.4.3) for bivalent-bivalent electrolytes [24].

The values of $\mu^o_{Zn^{2+}}$ and $\mu^o_{Cu^{2+}}$ at standard pressure (1 bar) and temperature (298.15 K) can be obtained from the thermodynamic tables (cf., e.g., [71] or [2, Table 2.5]):

$$\mu^o_{Zn^{2+}} = -\ 147.06\ [\text{kJ mol}^{-1}]\quad \textit{and}\quad \mu^o_{Cu^{2+}} = 65.49\ [\text{kJ mol}^{-1}]. \tag{7.5.11}$$

This completes all of the information needed to calculate $G(\xi)$.

For instance, in the particular case in which $n^o_{CuSO_4(aq)} = 0.2$ mol and $n^o_{ZnSO_4(aq)} = 0$, by substituting the value 8.3145 J K^{-1} mol^{-1} for R, from eqs (7.5.7) to (7.5.9), we obtain

$$\begin{aligned} G = G(\xi) = -212.55\,\xi + 2.479\,[(0.2 - \xi)\ln(0.024 - 0.12\xi) \\ + \xi\ln(0.12\xi)] + C \quad [\text{kJ}]. \end{aligned} \tag{7.5.12}$$

Again, all constant terms are assumed to be included in C.

7.6 Solution's Admissible Range and Cell's Admissible Range

Let us refer again to electrochemical cells in general. In order to determine the admissible range of a system, we must first remove from U and S, and hence from Ψ and G, the contributions of any non-thermodynamic component and any pure thermo-entropic component that the system may contain (Section 3.11). Electrochemical cells that do not include active gaseous phases do not undergo appreciable changes in volume during their operation. At constant temperature and pressure, eq. (1.11.13) applies to these cells, so their Helmholtz and Gibbs free energies differ by a constant. Since it is the change in energy that matters in thermodynamics, we can assume, without loss in generality, that the following relation

$$\Psi \equiv G \tag{7.6.1}$$

applies to these cells when isothermal, constant pressure processes are considered. Accordingly, the distinction between Ψ and G is ignored in the rest of this chapter and in the following chapter.

Similar to what we did in Section 3.11, we append a prime (′) to the quantities that refer to non-thermodynamic or pure thermo-entropic components of the system. The double prime symbol (″) refers, therefore, to quantities relevant to the remaining components of the system. With this notation, the free energies Ψ and G of the whole system can be decomposed as follows:

$$\Psi = \Psi' + \Psi'' \tag{7.6.2}$$

and

$$G = G' + G'', \tag{7.6.3}$$

respectively. Of course, when relation (7.6.1) applies, we also have

$$\Psi' \equiv G' \quad \text{and} \quad \Psi'' \equiv G''. \tag{7.6.4}$$

From eqs (7.2.4) and (7.6.3), it follows that the emf of a cell can be decomposed as:

$$\boldsymbol{E} = \boldsymbol{E}' + \boldsymbol{E}'', \tag{7.6.5}$$

where

$$\boldsymbol{E}' = -\frac{1}{zF}\left(\frac{\partial G'}{\partial \xi}\right)_{p,T} \tag{7.6.6}$$

and

$$\boldsymbol{E}'' = -\frac{1}{zF}\left(\frac{\partial G''}{\partial \xi}\right)_{p,T} \quad . \tag{7.6.7}$$

From the energy standpoint, eqs (7.6.3) and (7.6.5) show that any electrochemical cell can be regarded as made of two subunits in series with each other (Fig. 7.6.1). One subunit generates the free energy G'. This is the non-thermodynamic/pure thermo-entropic component of the cell. Its contribution to the cell's total emf is $\boldsymbol{E}'$. From the analysis of Section 3.11, we know that this component does not set any restriction on the admissible range of the cell. The other subunit has the free energy G'' and produces the emf $\boldsymbol{E}''$. Because it is subjected to the thermodynamic limits discussed in Chapter 3, this subunit determines the admissible range of the whole cell.

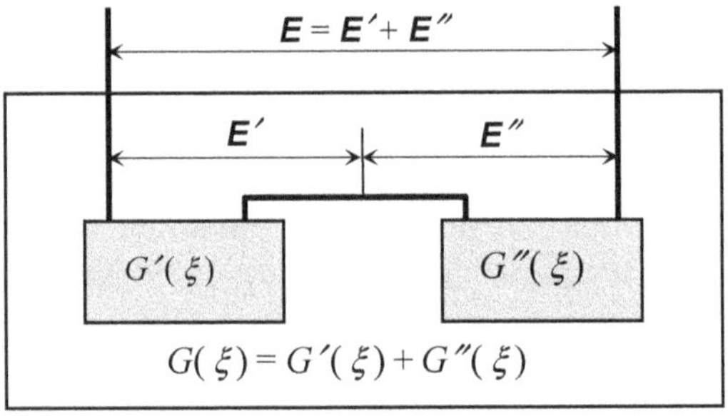

Fig. 7.6.1. Graphic illustration of the meaning of eqs (7.6.3) to (7.6.5)

It is very important to distinguish between the admissible range of the cell and the admissible range of the solution where the cell's reactions occur. The former will be referred to as the *cell's admissible range* or simply the *cell's range*. It is a one-dimensional sub-domain of the *solution's admissible range*, i.e., the admissible range of the cell's solution. The free energy of the solution has the form (7.3.1). The solution's admissible range is a region in a space of r dimensions, where r is the number of the independent variables, n_i, that define the state of the solution.

In ordinary cell operation, the solution's variables n_i cannot vary independently of each other. They are all determined by the extent of reaction ξ, according to eq. (C.11) of Appendix C [cf. eqs $(7.1.7)_{3\text{-}4}$ for the particular case of a Daniell cell]. This makes the cell's admissible range one-dimensional, although the admissible range of the cell's solution is r-dimensional, as observed.

The free energy of any solution, as the free energy of any thermodynamic system, is subjected to the general limitation expressed by relation (3.6.5) or, equivalently, relation (3.7.4). In the r-dimensional plane $(n_1, n_2, \ldots, n_r)$, these relations define an r-dimensional region, which is the admissible range of the solution. The boundary to this region is an $(r-1)$-dimensional curve, which represents the solution's limit surface. If $r = 2$, as occurs in a two-solute solution, the above plane reduces to an ordinary, two-dimensional plane, and the solution's limit surface becomes a one-dimensional curve in that plane. In any case, if the limit surface is continuous, it must be equipotential for the part of the free energy of the system that defines that surface (Section 3.4). This free energy is not the entire free energy of the system if the system contains non-thermodynamic or pure thermo-entropic components.

For instance, in the case we are considering in this section, the free energy part that determines the solution's admissible range is $\Psi'' \equiv G''$. The remaining part of Ψ, i.e., $\Psi' \equiv G'$, does not affect the admissible range of the solution, because it is due to non-thermodynamic or pure thermo-entropic components of the system. The situation is illustrated with reference to a generic cell in Fig. 7.6.2. It is assumed that the chemical reaction that drives the cell involves just two reacting components, the amount of which varies during the reaction. The variables n_1 and n_2 denote the moles of these components at any time of the cell operation. The solution's admissible range is, in this case, a two-dimensional region in the plane of the variables n_1 and n_2. Accordingly, the solution's limit surface coincides with the ordinary (one-dimensional) curve that defines the boundary of the admissible range of the solution.

In contrast, the cell's admissible range is the locus of the values $n_i = n_i(\xi)$ that the variables n_i assume as the cell operates. The actual expression of the functions $n_i = n_i(\xi)$ is obtained by setting $i = 1$ and $i = 2$ into eq. (C.11) of Appendix C and by referring to the stoichiometric coefficients ν_i of the chemical reaction that drives the cell. By eliminating the variable ξ from the equations thus obtained, we have

$$\nu_2 n_1 - \nu_1 n_2 = \nu_2 n_1^{\mathrm{o}} - \nu_1 n_2^{\mathrm{o}}, \tag{7.6.8}$$

where n_1^{o} and n_2^{o} are the values of n_1 and n_2, respectively, at the beginning of the reaction (i.e., for $\xi = 0$). Equation (7.6.8) represents a straight line in plane (n_1, n_2). This is the ξ-axis of Fig. 7.6.2. The cell's admissible range is the segment of that axis that lies within the solution's admissible range.

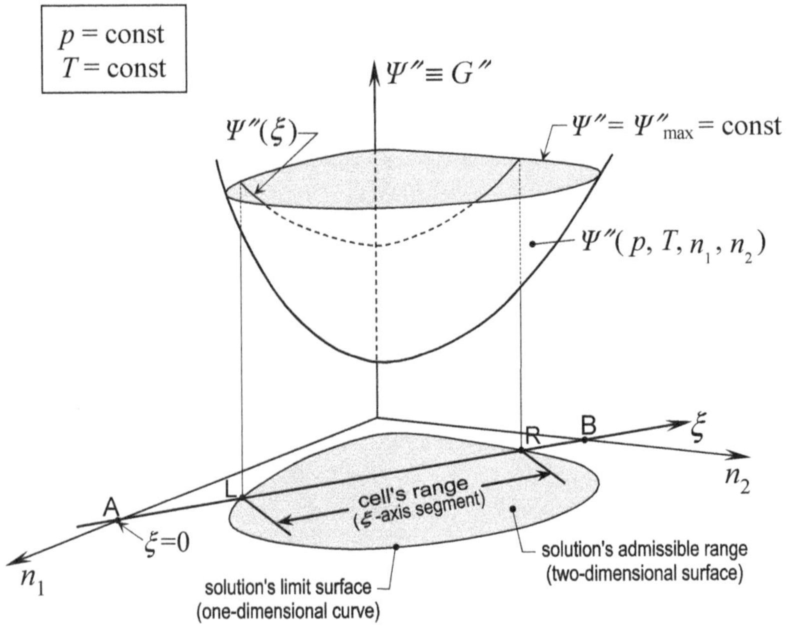

Fig. 7.6.2. Admissible range of an electrochemical cell and that of the solution in which the cell reaction occurs

As is apparent from eq. (7.6.8), the position of the ξ-axis depends on n_1^{o} and n_2^{o}. This suggests that the cell's admissible range can, in principle, be maximized by choosing the values of the initial values of n_1 and n_2 appropriately. This observation should be of practical significance for its implications in the design of the cell.

If the cell's admissible range is exceeded, the cell generally incurs permanent changes, and thus the cell's properties change. In particular, the properties of the cell's solution may change. In the example of the Daniell cell considered in the previous sections, exceeding the admissible range is bound to result in the hydrolysis of the aqueous solvent due to the formation of an excessive electric potential difference between the cell's electrodes (see next section.) The hydrogen and oxygen liberated by hydrolysis reduce the amount of solvent and thus produce a permanent change in the solution's composition. This phenomenon explains, at least partially, the aging of the cell if it is being overcharged.

Curve $\Psi''(\xi)$ in the above figure is the intersection of surface $\Psi''(p, T, n_1, n_2)$ with a vertical plane through the ξ-axis. This curve provides a plot of the part of the cell's free energy that is restricted by thermodynamics. Figure 7.6.3 below shows the same curve in plane (Ψ'', ξ). It also shows the diagram of the contribution $\boldsymbol{E}''=\boldsymbol{E}''(\xi)$ of Ψ'' to the cell's emf, $\boldsymbol{E}$, as obtained from eq. (7.6.7).

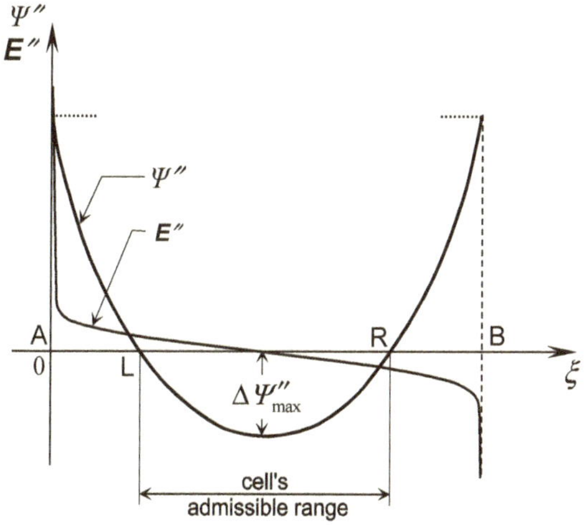

Fig. 7.6.3. Plot of $\Psi''(\xi)$ and of the associated emf, $\boldsymbol{E}''(\xi)$. Although conditioning the cell's admissible range, these functions are often only a minor part of the cell's free energy and emf.

Points L and R of Fig. 7.6.3 denote the left and right limits of the cell's admissible range. At these points, component Ψ'' of the cell's free energy attains its largest value. In the figure, we set that value equal to zero in order to express the cell's admissible range as

$$\Psi''(\xi) \leq 0\,, \tag{7.6.9}$$

cf. eq. (3.7.4). This choice makes Ψ'' negative throughout the admissible range. Its minimum value is then equal to $-\Delta\Psi''_{\max}$.

It occurs frequently in practice that $\Psi''(\xi)$ is only a small fraction of the total free energy Ψ of the cell. One exception are the so-called concentration cells, which are discussed in the next chapter. For the kind of cells considered in the present chapter, Ψ' may be various orders of magnitude greater than Ψ'' and responsible for almost all of the energy of the cell as well as for most of its emf. Yet, Ψ' does not impose any restriction to the cell's admissible range.

The admissible range of the cell is generally only a part of a larger domain of the ξ-axis, where the function $\Psi''(\xi)$ is defined mathematically. At the endpoints of this larger domain, the curve that represents $\Psi''(\xi)$ takes an infinite slope (cf. points A and B in the above figures). This is because functions $\mu_i(\xi)$ defined by eq. (7.3.11) depend on ξ in a logarithmic way. These functions appear in the expression of $G(\xi)$ and thus of $G''(\xi)$ and $\Psi''(\xi)$, cf. eqs (7.3.7), (7.3.8) and (7.6.4). Their derivative tends to infinity as the concentration of the species to which they refer becomes equal to zero. This occurs at the left endpoint of the domain if the species is a product of reaction, and at the right endpoint if it is a reactant, thus making the slope of $\Psi''(\xi)$ vertical at both ends.

A consequence of this is that at the endpoints of the domain of definition of function $\Psi''(\xi)$ the value of the electric potential $\boldsymbol{E}''$ tends to infinity in agreement with eq. (7.6.7). The dielectric capacity of the solvent is therefore exceeded as these endpoints are approached. This produces permanent damage in the solvent or hydrolysis if the solvent is water. It explains why the cell's admissible range is only a part of a larger domain where function $\Psi''(\xi)$ is defined mathematically.

Of course, the cell's solution may contain species other than those involved in the electrochemical reaction of the cell. These additional species contribute to the free energy of the solution, which is bound to affect the maximum amount of energy that the cell can store or supply. The same species may also modify the properties of the solution and hence its admissible range. In what follows, however, the cell's solution will be assumed to be free from any additive.

7.7 The Daniel Cell's Admissible Range

The Daniell cell considered in Section 7.5 does not include any non-thermodynamic components, as defined in Section 3.11. To prove this, let us recall that, as is also apparent from eq. (7.3.2), the standard state free energies μ^{o}_{Cu2+} and μ^{o}_{Zn2+} in eqs (7.5.5) and (7.5.6) are in fact functions of T. That is,

$$\mu^{o}_{Cu2+} = \mu^{o}_{Cu2+}(T) \tag{7.7.1}$$

and

$$\mu^{o}_{Zn2+} = \mu^{o}_{Zn2+}(T), \tag{7.7.2}$$

the dependence on p can be ignored here. The presence of a non-thermodynamic component in the system would show itself in the function U as a temperature-independent, additive term of the form (3.11.9). Of course, the same term would also appear in the system's free energy (whether Helmholtz or Gibbs), because the entropy of a non-thermodynamic component is a constant. However, as is apparent from eqs (7.5.4) through (7.5.6), the expression of G for the Daniell cell lacks any temperature-independent term, apart from the inessential constant C. It must therefore be concluded that the cell does not include any non-thermodynamic component.

A different conclusion holds true for what concerns the presence of pure thermo-entropic components in the cell. The two molar free energies (7.5.5) and (7.5.6) are both of the form (3.12.2). Therefore, the standard state free energies (7.7.1) and (7.7.2) are in fact due to two pure thermo-entropic components of the cell. The molar contribution of these components to S can be calculated from eq. (3.12.5) as given by

$$s'_{Cu2+} = -\frac{\partial \mu^{o}_{Cu2+}(T)}{\partial T} \tag{7.7.3}$$

and

$$s'_{Zn2+} = -\frac{\partial \mu^{o}_{Zn2+}(T)}{\partial T}, \tag{7.7.4}$$

respectively.

The energies (7.7.1) and (7.7.2) of the two thermo-entropic components of the cell contribute to the cell's total free energy with the term

$G_o(\xi)$, defined by eq. (7.5.8). In the notation of eq. (7.6.3) and in view of eq. $(7.6.4)_1$, we can therefore write

$$\Psi'(\xi) \equiv G'(\xi) \equiv G_o(\xi). \tag{7.7.5}$$

As discussed in Section 3.12, this part of the free energy must be ignored when looking for the cell's admissible range. Therefore, according to eq. (7.6.3), the free energy that defines that range is the remaining part of $G(\xi)$, that is $G''(\xi)$. The latter coincides with $G_1(\xi)$, as shown by eqs (7.5.7)-(7.5.9). It is also identical to $\Psi''(\xi)$, according to eq. $(7.6.4)_2$.

For instance, in a Daniell cell in which $n^o_{\mathrm{CuSO4(aq)}} = 0.2$ mol and $n^o_{\mathrm{ZnSO4(aq)}} = 0$, this part of the cell's free energy is given by:

$$\begin{aligned}\Psi''(\xi) \equiv G''(\xi) = 2.479\,[(0.2-\xi)\ln(0.024-0.12\,\xi)+ \\ \xi\ln(0.12\,\xi)]+\mathrm{C} \qquad [\mathrm{kJ}],\end{aligned} \tag{7.7.6}$$

which results from subtracting $G_o(\xi)$ from eq. (7.5.12).

From eqs (7.7.6), (1.11.3) and (C.6), it would not be difficult to derive the expressions of the internal energy U'' and the entropy S'' that correspond to Ψ''. However, explicit knowledge of these functions is not required for the present purpose. The admissible range of the cell can simply be determined by inserting Ψ'' for Ψ into relation (3.7.4):

$$\Psi'' \leq 0. \tag{7.7.7}$$

This is equivalent to applying relation (3.7.1), but it does not require explicit knowledge of U'' and S''.

For relation (7.7.7) to be valid, the reference value of the internal energy of the system must be chosen in such a way that Ψ'' is equal to zero at the boundary of the admissible range. To impose this condition, we should know at least one point of the limit surface. The constant C that appears in eq. (7.7.6) could then be given the value that makes $\Psi'' = 0$ at that point. Alternatively, a point on the limit surface can be determined experimentally by driving the state of the cell towards that surface and crossing it. The procedure entails monitoring the response of the cell to detect any change in its response when the limit surface is crossed and measuring the cell's state variables at the crossing point.

A way to reach the limit surface could be to drive the cell in the electrolytic mode by applying a voltage difference $\Delta\boldsymbol{V}$ to its electrodes. The voltage should then be increased until the limit surface is reached. The process must be slow enough to proceed through equilibrium states, because the admissible range is referred to thermodynamic equilibrium conditions. In these conditions, the applied voltage equals the cell's emf at any time of the process, so that

$$\Delta \boldsymbol{V} = \boldsymbol{E} = \boldsymbol{E}' + \boldsymbol{E}''. \tag{7.7.8}$$

Note that the last equality of this relation is a consequence of eq. (7.6.5).

Of course, the considered process must start from an initial state within the cell's admissible range. To verify if this condition is met, a comparatively small amount of electric energy should be supplied to the cell to change its state. Then, it should be checked whether the cell can recover the original initial state spontaneously, i.e., via a process that does not absorb any energy from the surroundings and gives back the supplied energy (possibly as heat). If this can be done, the considered initial state is not on the boundary of the cell's admissible range. Starting from that state, the quantity $\Delta \boldsymbol{V}$ is then gradually increased until a sudden change in the slope of the voltage/current curve is detected. This indicates the point at which the limit surface of the cell is reached.

In the absence of experimental results, we assume that the cell reaches the boundary of the admissible range when the electric potential difference $\boldsymbol{E}$ between its electrodes reaches the value $\boldsymbol{E}_{\mathrm{hydr}} = \pm 1.23$ V. This is the potential at which the solvent (water) hydrolyzes into its gaseous components, H_2 and O_2. Of course, this assumption should be tested experimentally, as some secondary phenomena may affect the actual limit value of $\boldsymbol{E}$, depending on the cell's architecture.

By assuming that $\boldsymbol{E} = \boldsymbol{E}_{\mathrm{hydr}}$ at a limit point of the admissible range, we can determine the value of the variable ξ at that point by exploiting eqs (7.6.5)-(7.6.7). In the present case, $\boldsymbol{E}'$ is constant, as evident from eqs (7.7.5), (7.5.8), (7.5.11), and (7.6.6). Its value is given by:

$$\begin{aligned} \boldsymbol{E}' &= -\frac{1}{zF}\left(-\mu^{\mathrm{o}}_{\mathrm{Cu}^{2+}} + \mu^{\mathrm{o}}_{\mathrm{Zn}^{2+}}\right) = \\ &-\frac{1000}{2\cdot 96{,}485}(147.06 + 65.49) = -1.09\ [\mathrm{V}]\,, \end{aligned} \tag{7.7.9}$$

since, as already observed, the variable z is equal to 2 in the case of reaction (7.1.3). The endpoints, L and R, of the admissible range (cf. Fig. 7.6.3) are the points where $\boldsymbol{E}$ reaches the value $\boldsymbol{E}_{\mathrm{hydr}} = \pm 1.23$ V. Because $\boldsymbol{E}' = \pm 1.09$ V= const., from eq. $(7.7.8)_2$ it follows that the values of $\boldsymbol{E}''$ at these end points are given by:

$$\boldsymbol{E}''_{\mathrm{L}} = +1.23 - 1.09 = 0.14 \text{ V} \quad \text{and} \quad \boldsymbol{E}''_{\mathrm{R}} = -1.23 - 1.09 = -2.32 \text{ V}, \tag{7.7.10}$$

which make $\boldsymbol{E} = 1.23$ V and $\boldsymbol{E} = -1.23$ V, respectively.

On the other hand, from eqs (7.6.7) and (7.7.6) it follows that:

$$\boldsymbol{E}'' = \boldsymbol{E}''(\xi) = 0.0128 \ln\left(\frac{1-5\xi}{\xi}\right) - 0.0207 \quad [\mathrm{V}] . \qquad (7.7.11)$$

From this and from eq. (7.7.10), we calculate that the values of ξ at the end points of the admissible range are, respectively,

$$\xi_L = 3.7\ 10^{-6} \qquad \text{and} \qquad \xi_R = 0.2 - 3.7\ 10^{-6} . \qquad (7.7.12)$$

These values show that, in the present case, the cell's admissible range does almost coincide with the entire interval [0, 0.2] where the function $\Psi''(\xi)$ is defined mathematically; see eq. (7.7.6) and Fig. 7.7.1 below. A similar coincidence is likely to be the rule rather than the exception for the kind of electrochemical cells considered in this chapter. It is a consequence of the peculiar form of their free energy function [see eqs (7.5.7)-(7.5.9)], and of the u-shaped form of function $\Psi''(\xi)$, the tangent of which diverges abruptly at the ends of the domain of definition (cf. the graph of $\boldsymbol{E}''$ in Fig. 7.7.1).

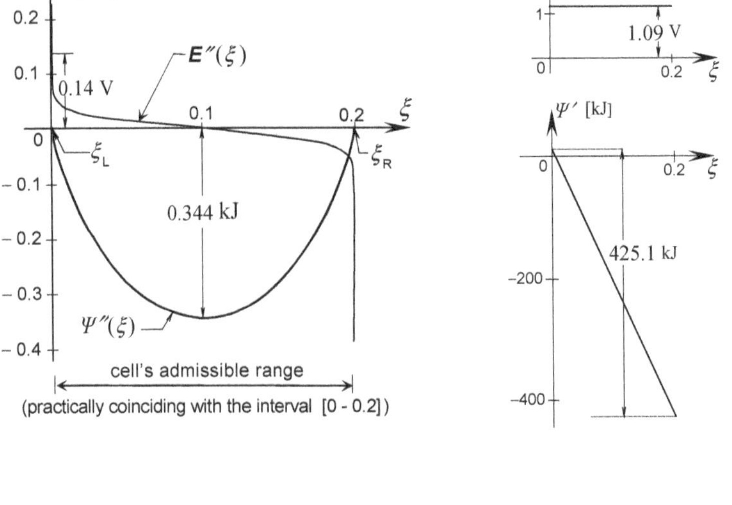

Fig. 7.7.1. The admissible range of the considered Daniell cell as determined by Ψ'' (left diagram). Components Ψ' and $\boldsymbol{E}'$ account for most of the cell's energy and emf, respectively (right diagrams).

By choosing the free energy constant in such a way that Ψ'' vanishes at end of the admissible range, we can determine the maximum contribution of Ψ'' to the cell's free energy. In the present example, this contribution turns out to be a meager Ψ''_{max} =0.344 kJ, i.e., three orders of magnitude less than the maximum contribution coming from $\Psi'(\xi)$. The accuracy of this conclusion can be verified by observing that function $\Psi'(\xi)$ is given by:

$$\Psi' = \Psi'(\xi) = 13.10 - 212.55\,\xi \quad [\text{kJ}], \tag{7.7.13}$$

as follows by inserting eqs (7.5.8) and (7.5.11) into eq. (7.7.5). This function is plotted at the right-hand bottom corner of Fig. 7.7.1, which shows that $\Psi'(\xi)$ contributes up to 425.1 kJ to the cell's free energy.

Because the values of $\Psi''(\xi)$ are so small, the whole cell's free energy $\Psi(\xi)$ practically coincides with $\Psi'(\xi)$. In spite of this, $\Psi''(\xi)$ is responsible for the cell's admissible range. Moreover, $\Psi''(\xi)$ has an important role as far as the chemical equilibrium of reaction (7.1.6) is concerned.

To prove this last point, let us observe that, from eqs (7.6.1) and (7.6.2), we can write the chemical equilibrium condition $\partial G/\partial\xi = 0$ as:

$$\frac{\partial \Psi}{\partial \xi} = \frac{\partial \Psi'}{\partial \xi} + \frac{\partial \Psi''}{\partial \xi} = 0\,. \tag{7.7.14}$$

From this equation and eqs (7.7.6) and (7.7.13), the value of ξ that corresponds to chemical equilibrium can be calculated as

$$\xi_{eq} = 0.2 - 1.16\;10^{-38}. \tag{7.7.15}$$

Although this value is to all practical purposes coinciding with 0.2, the fact that it is not exactly 0.2 is important from the conceptual point of view. It indicates that, within the domain of the definition of $\Psi''(\xi)$, there is actually a point of chemical equilibrium for reaction (7.1.6), in spite of the fact that $\Psi(\xi)$ practically coincides with a straight line in almost all of that domain.

When compared to eq. (7.7.15), equation $(7.7.12)_2$ shows that the thermodynamic limit ξ_R is extremely close to, but still less than, ξ_{eq}. This detail is consistent with the well-known experimental fact that, by electrolytic means, it is virtually impossible to obtain a concentration in the solution as low as the concentration that corresponds to a value of ξ almost equal to 0.2. At these concentrations, the emf $\boldsymbol{E}''$, and hence $\boldsymbol{E}$, would quickly diverge negatively (Fig. 7.7.1). This would result in hydrolysis of the solvent and activate a host of other disturbing phenomena, which would make it impossible for the cell to operate in agreement with eq. (7.1.6), thus preventing ξ from approaching the value of 0.2 any further.

8

Electrochemical Cell's Admissible Range: The Concentration Cells

Concentration-type electrochemical cells, usually referred to as *concentration cells*, take their energy from two solutions that contain the same solute at different concentrations. The process that drives the cells does not produce any chemical change in the solutions involved; rather, it modifies their concentrations.

If two solutions are separated by a semipermeable membrane that allows either the solute or the solvent to pass through it, the concentration difference will drive the solute from the more concentrated solution to the less concentrated solution, or it will drive the solvent in the opposite direction. If the solute is dissociated into ions, the phenomenon can be used to produce an electric current. The process can also be reversed. An electric potential of the appropriate sign can drive the ions or the solvent molecules through the membrane from one solution to the other in the direction that increases the concentration difference between the two solutions.

8.1 Concentration Cells

In a typical concentration cell, the two half-cells that make up the complete cell are identical except for the concentration of their solutions. The solution with the higher concentration has greater molar free energy. Thus, if the two solutions come into contact, there will be a spontaneous migration of the dissolved ions from the higher concentration solution to the lower concentration solution. The process will liberate the free energy ΔG, which, if not otherwise used, will end up as heat.

As is true for transformation-type electrochemical cells, concentration cells operate at constant volume if they do not involve gaseous phases, which we assume to be the case in what follows. Thus, for processes that occur at constant temperature and pressure, relation (7.6.1) also applies to the present type of cells. It follows that there is no need to distinguish between Helmholtz and Gibbs free energy. Accordingly, in the present analysis, as in the previous chapter, the symbols Ψ and G are fully interchangeable.

When operating in the galvanic mode, a concentration cell makes ΔG available as electrical energy. The process ends when the solutions of the two half-cells reach the same concentration. In that state, the cell's free energy reaches a minimum, its emf decreases to zero, and chemical equilibrium is thereby attained. The cell can also operate in the electrolytic mode. In that case, an appropriate electric potential difference must be applied to its electrodes to drive the flow of ions in the opposite direction to the spontaneous direction. The process increases the concentration difference between the two half-cells, thus increasing ΔG and the electric energy that the cell can supply in the reverse mode.

Figure 8.1.1 shows a concentration cell in both modes of operation. Here, A indicates the metal or the hydrogen gas that makes up the cell's electrodes. A salt of substance A, say, salt AB, is dissolved in the aqueous solution of each half-cell. It is partially or completely dissociated into cations $\mathsf{A}^{\alpha+}$ and anions $\mathsf{B}^{\alpha-}$ (the superscript α denoting their valence). The concentrations of these ions are generally different in the two half-cells. In order to indicate that a quantity refers to a specific half-cell, we append the index L or R to it, depending on whether it refers to the left-hand half-cell or to the right-hand half-cell. With this notation, the molality of an ion in the solution of the right-hand part of the cells and in the solution of the left-hand part of the cells is denoted as b_{R} and b_{L} respectively.

The electrochemical reaction of a concentration cell is

$$\mathrm{A}^{\alpha+}(\text{soln}, b_{\mathrm{R}}) \underset{\text{electrolytic}}{\overset{\text{galvanic}}{\rightleftarrows}} \mathrm{A}^{\alpha+}(\text{soln}, b_{\mathrm{L}}), \tag{8.1.1}$$

where the acronym "soln" indicates that $\mathrm{A}^{\alpha+}$ is in solution, not necessarily an aqueous one. We assume that $b_{\mathrm{L}} < b_{\mathrm{R}}$. In this case, reaction (8.1.1) proceeds from left to right when the cell operates in galvanic mode and right to left in electrolytic mode.

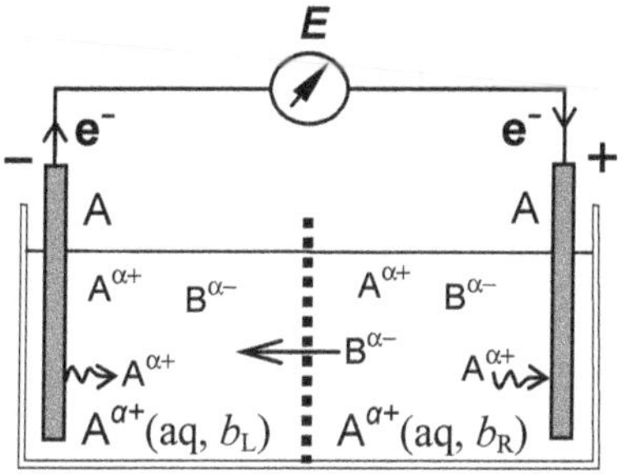

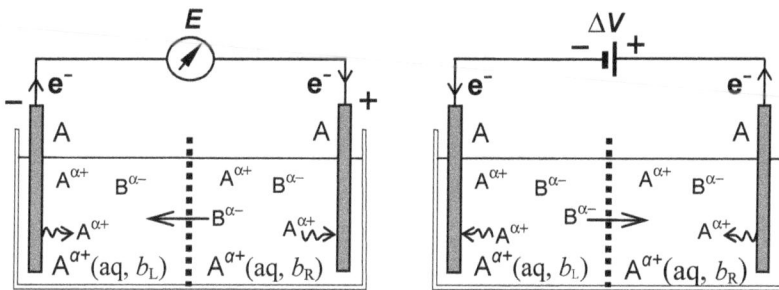

Fig. 8.1.1. Representation of a concentration cell in galvanic mode (*left*) and in electrolytic mode (*right*)

Process (8.1.1) is a redox reaction, as were the processes considered in the previous chapter. When the cell operates in galvanic mode, the A atoms are oxidized into $\mathrm{A}^{\alpha+}$ ions at the cell's negative electrode. The reaction is described by eq. (7.1.1). The electrons collected at that electrode are transferred via the external circuit to the positive electrode, from which they can combine with another $\mathrm{A}^{\alpha+}$ ion in order to transform it into a neutral atom. The last transformation is actually the reduction process (7.1.2). In the electrolytic mode, the above reactions are inverted. The reduction process occurs at the negative electrode, while oxidation occurs at the positive electrode.

As the cell operates, other reactions besides reaction (8.1.1) may take place, including the transfer of $\mathrm{B}^{\alpha-}$ ions from one half-cell to the other. However, the energy that these reactions liberate or absorb does not contribute to the electrical energy produced or absorbed by the cell. For this reason, we ignore any reaction other than (8.1.1) when calculating the Gibbs free energy of the cell, consistently with what we did in the case of an electrochemical cell of the transformation type. This amounts to considering the concentrations of $A^{\alpha+}$ in the two half-cells as the state variables of the system, since temperature and pressure are assumed to be constant. Because G is linear in μ_i [cf. eq. (7.3.1)], and reaction (8.1.1) is solely responsible for the cell's emf, ignoring any reaction other than (8.1.1) will not invalidate relation (7.2.4) or affect the maximum value of the electrical energy that the cell can store or supply.

With the sign convention used when writing a chemical reaction in the form (C.10) (Appendix C.3), the stoichiometric coefficients of eq. (8.1.1) have the following values:

$$\nu_L = 1 \qquad \text{and} \qquad \nu_R = -1 . \tag{8.1.2}$$

Note that the indices L and R refer to the half-cell, not to the respective sides of eq. (8.1.1). By referring to eq. (8.1.1) and by introducing eq. (8.1.2) into eq. (7.3.8), we obtain:

$$\begin{aligned} n_L &= n_L^o + \xi \\ n_R &= n_R^o - \xi , \end{aligned} \tag{8.1.3}$$

where n_L and n_R denote the moles of dissolved $A^{\alpha+}$ ions in the solutions of the left- and right-hand half-cells, respectively, and n_L^o and n_R^o are their respective initial values. For the solution at hand, the moles of $B^{\alpha-}$ and the moles of salt AB are the same as the moles of $A^{\alpha+}$. Therefore, it is unnecessary to identify the particular species in solution to which the above numbers of moles refer. Accordingly, the notation used to designate these numbers of moles does not contain any indication of the particular species in solution to which they refer.

By setting

$$n_{tot}^o = n_L^o + n_R^o , \tag{8.1.4}$$

and by eliminating ξ from equations (8.1.3), we obtain:

$$n_L + n_R = n_{tot}^o . \tag{8.1.5}$$

For the type of concentration cells considered, the quantities n_L^o, n_R^o, and hence n_{tot}^o are fixed at the cell's fabrication time. Therefore, eq. (8.1.5) means that the total amount of $A^{\alpha+}$ ions, i.e., the sum of n_L and n_R, is conserved as the cell operates.

8.2 Cell's Free Energy

As in the case of the transformation-type electrochemical cell, the energy that a concentration cell absorbs or supplies is that of the electrochemical reaction that occurs within it. This energy equals the change in the free energy, G, of the two-solution system contained in the cell. For given values of pressure and temperature, this is a function of the concentrations of $A^{\alpha+}$ ions in the two solutions. That is,

$$G = G(b_L, b_R), \tag{8.2.1}$$

where b_L and b_R are the molalities of $A^{\alpha+}$ in the solutions in the left and right half-cell, respectively.

If $(m_{solv})_L$ and $(m_{solv})_R$ denote the masses of solvent of the above solutions, we obtain from eq. (7.3.3) that

$$b_L = \frac{n_L}{(m_{solv})_L} \quad \text{and} \quad b_R = \frac{n_R}{(m_{solv})_R}. \tag{8.2.2}$$

The amount of solvent in each solution is constant; hence, $(m_{solv})_L$ and $(m_{solv})_R$ also are constant. Therefore, from eq. (8.2.2), we can express eq. (8.2.1) in the alternative form:

$$G = G(n_L, n_R). \tag{8.2.3}$$

Of course, when expressed in this form, function G will be different for different values of $(m_{solv})_L$ and $(m_{solv})_R$.

We now calculate function (8.2.3) explicitly. From eqs (7.3.2) to (7.3.4), we can write the partial molar energy of $A^{\alpha+}$ ions in the left half-cell solution as:

$$\mu_L = \mu^o_{A^{\alpha+}} + RT \ln\left[\frac{b_L\, \gamma(b_L)}{b^o_L}\right], \tag{8.2.4}$$

where $\gamma = \gamma(b_L)$ is the mean activity coefficient of the considered ions (an appropriate function of their concentration, b_L). Likewise, the partial molar energy of $A^{\alpha+}$ in the solution of the right half-cell can be written as:

$$\mu_R = \mu^o_{A^{\alpha+}} + RT \ln\left[\frac{b_R\, \gamma(b_R)}{b^o_R}\right]. \tag{8.2.5}$$

The quantity $\mu^{\text{o}}_{\text{A}^{\alpha+}}$, which appears in eqs (8.2.4) and (8.2.5), is the molar free energy of $\text{A}^{\alpha+}$ in the standard state. This quantity has the same value in both equations, since the standard state is independent of the actual concentration of the solution. Also, for non-gaseous solutes, $\mu^{\text{o}}_{\text{A}^{\alpha+}}$ is almost independent of pressure, as observed previously. Therefore, for all practical purposes, we can write $\mu^{\text{o}}_{\text{A}^{\alpha+}} = \mu^{\text{o}}_{\text{A}^{\alpha+}}(T)$.

To proceed further, we adhere to the usual convention of taking the standard state of a solution as its state at unit molality. With this convention, we have:

$$b^{\text{o}}_{\text{L}} = b^{\text{o}}_{\text{R}} = 1 \;\; [\text{mol} / \text{kg}^{-1}]. \tag{8.2.6}$$

Moreover, to simplify the forthcoming formulae, we refer to the case in which each half-cell contains a unit mass of solvent, i.e.,

$$(m_{\text{solv}})_{\text{L}} = (m_{\text{solv}})_{\text{R}} = 1 \;\; [\text{kg}]. \tag{8.2.7}$$

This makes b_{L} and b_{R} coincide numerically with n_{L} and n_{R}, respectively, as apparent from eq. (7.3.3). Under these conditions, functions (8.2.1) and (8.2.3) become numerically identical, although their independent variables have different meanings and dimensions in the two functions.

With the above assumptions, we can apply eq. (7.3.1) and calculate the free energy of each solution of the cell. By using eqs (8.2.4) to (8.2.6) and by adding together the free energies of the solutions of the two half-cells, we obtain:

$$\begin{aligned} G = G(n_{\text{L}}, n_{\text{R}}) = (n_{\text{L}} + n_{\text{R}})\mu^{\text{o}}_{\text{A}^{\alpha+}}(T) + n_{\text{L}}\, RT \ln[n_{\text{L}}\, \gamma(n_{\text{L}})] \\ + n_{\text{R}}\, RT \ln[n_{\text{R}}\, \gamma(n_{\text{R}})] + \text{C}\,, \end{aligned} \tag{8.2.8}$$

where $\gamma(n_{\text{L}})$ and $\gamma(n_{\text{R}})$ are the functions that are obtained from the analogous functions $\gamma = \gamma(b_{\text{L}})$ and $\gamma = \gamma(b_{\text{R}})$, once we express b_{L} and b_{R} by means of eq. (8.2.2). As usual, the arguments of the logarithms that appear in the above expression are understood to be divided by 1 mol to make them dimensionless. Equation (8.2.8) is the expression of the free energy of the *two-solution system* contained in the cell, or the *cell's solution* for short. Temperature is shown explicitly on the right-hand side of that equation to indicate how that equation varies if a different value of temperature is considered.

8.3 The Admissible Range of Cell's Solution

As repeatedly stated in the previous chapters, to determine the admissible range of a system we must ignore the presence of non-thermodynamic and pure thermo-entropic components in the system (Section 3.12). Accordingly, when it comes to calculating the free energy of the system, we decompose G into two terms, as in eq. (7.6.3), and eliminate the non-thermodynamic/pure thermo-entropic component, G'. The remaining part of G is G'', which is the part of the free energy that controls the system's admissible range (Sections 3.11 and 3.12).

The first term on the right-hand side of eq. (8.2.8) is only a function of temperature, because the sum of $(n_L + n_R)$ is constant, as stated in eq. (8.1.5). This means that the part of G given by

$$G' = (n_L + n_R)\mu^{0}_{A^{\alpha+}}(T) \tag{8.3.1}$$

represents the pure thermo-entropic component of the cell's free energy (Section 3.12). This part should be ignored when seeking the admissible range of the solution (actually, a system of two solutions) that drives the cell. Therefore, from relation (3.7.4), the admissible range of the cell's solution is given by:

$$G''(n_L, n_R) \leq 0 \ , \tag{8.3.2}$$

where

$$\begin{aligned} G'' = G''(n_L, n_R) = n_L\, RT \ln[n_L\, \gamma(n_L)] \\ + n_R\, RT \ln[n_R\, \gamma(n_R)] + \mathrm{C} \ . \end{aligned} \tag{8.3.3}$$

According to relation (8.3.2), the solution's limit surface is the curve $G'' = 0$. Remember that, for relation (3.7.4) and hence relation (8.3.2) to be valid, the constant C that appears in eq. (8.3.3) must be assigned the value that makes $G'' = 0$ at the boundary of the admissible range.

The only solution-specific property that enters the above expression of G'' is the activity coefficient γ of the ions involved in reaction (8.1.1). It follows that the shape of function G'' and the boundary to the solution admissible range are determined by $\gamma(n)$. Figure 8.3.1 shows the equipotential curves of G'' of a silver-electrode concentration cell (henceforth referred to as $Ag|Ag^+$ concentration cell) that operates with silver nitrate ($AgNO_3$) solutions at different concentrations in the two sections of the cell. Thanks to assumption (8.2.7), the concentrations of the solutions in the two half-cells are measured by their Ag^+ molar content, i.e., n_L and n_R, respectively. The inset in Fig. 8.3.1 shows the graph of the mean activity coefficient $\gamma = \gamma(n)$ of the Ag^+ ions at T = 25 °C, as obtained from the available experimental data [25].

At the points of the 45°-sloped dot-dash line of the main diagram of Fig. 8.3.1, the concentration of the two solutions is the same ($n_L = n_R$), since the amount of solvent in each half-cell is assumed to be the same. This is the locus of all possible chemical equilibrium states at zero electric potential between the cell's electrodes. The point at which the same line meets the boundary of the admissible range can be found from the solubility data tables. At that point, $n_L = n_{sat}$ and $n_R = n_{sat}$, where n_{sat} indicates the maximum amount of solute that can be dissolved in the solution at saturation at the considered temperature. For T = 25 °C, the solubility of $AgNO_3$ is 14.16 mol per kg of water. This makes n_{sat} = 14.16 mol, assuming that m_{solv} = 1 kg.

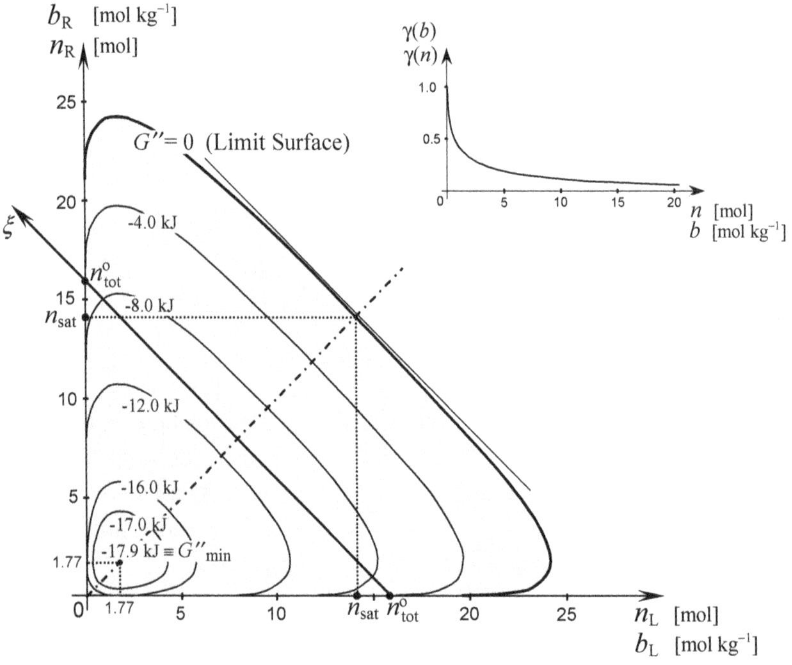

Fig. 8.3.1. Equipotential curves $G''(n_L, n_R)$ = const at 25 °C for the two-solution system of an Ag|Ag^+ concentration cell. Curve G'' = 0 is the limit surface (limit curve) of the cell's solution. The inset shows the plot of $\gamma = \gamma(n)$ for Ag^+ ions at 25 °C. The validity of eq. (8.2.7) is assumed, which also makes the diagrams readable in terms of variables b_L and b_R.

The equipotential curve $G'' =$ const through point ($n_L = n_{sat} = 14.16$, $n_R = n_{sat} = 14.16$) is the boundary to the admissible range, since that boundary is equipotential for G''. We arbitrarily assigned that curve the value of $G'' = 0$ (Fig. 8.3.1). This means that the admissible rage of the cell's solution is represented by relation (8.3.2). The value of C that meets condition $G'' = 0$ is

$$\mathrm{C} = -2 n_{sat} RT \ln[n_{sat}\, \gamma(n_{sat})], \qquad (8.3.4)$$

as follows from eq. (8.3.3). When applied to the present case, eq. (8.3.4) yields the value C = −13.01 kJ. The total free energy of the two-solution system considered in this example can then be calculated from eq. (8.2.8), once we set $\mu^{o}_{A\alpha+} = \mu^{o}_{Ag+} = 77.11$ kJ mol^{-1}, as obtained from the thermodynamic tables [36].

8.4 The Admissible Range of Concentration Cells

A concentration cell must always comply with eq. (8.1.5) during operation. This means that the variables n_L and n_R must move on a straight line in the (n_L, n_R) plane. This line is referred to as the ξ-axis. Figure 8.3.1 shows a possible placement of this axis. It is orthogonal to the 45°-sloped line through the origin, as can easily be verified. Its distance from the origin depends on the value of n^{o}_{tot}, which depends on how the cell is made. The admissible range of the cell is the portion of the ξ-axis that falls within to the admissible range of the cell's solution. The situation is analogous to the one generally described in Fig. 7.6.2.

From eq. (8.1.5), it follows that component (8.3.1) of the free energy of the cell's solution remains constant during the cell's operation. Therefore, all of the free energy that the cell stores or produces comes from component G'', as defined by eq. (8.3.3). This also implies that all of the cell's emf coincides with component $\boldsymbol{E}''$, defined by (7.6.7).

By using eq. (8.1.5) and eliminating n_R from (8.3.3), the free energy of the considered concentration cells can be expressed as

$$\begin{aligned} G_{cell} = G_{cell}(n_L) = RT\Big\{ & n_L \ln\big[n_L\, \gamma(n_L)\big] \\ & + (n^{o}_{tot} - n_L)\ln\big[(n^{o}_{tot} - n_L)\gamma(n^{o}_{T} - n_L)\big]\Big\}, \end{aligned} \qquad (8.4.1)$$

to within an inessential additive constant. From this and from eq. (7.2.4), the cell's emf as a function of n_L and T follows:

$$\mathbf{E} = \mathbf{E}(n_{\mathrm{L}}) = -\frac{1}{zF}\left(\frac{\partial G_{\mathrm{cell}}}{\partial n_{\mathrm{L}}}\right)_T . \qquad (8.4.2)$$

To exemplify, Fig. 8.4.1 shows the admissible ranges of an Ag|Ag^+ concentration cell for different values of $n_{\mathrm{tot}}^{\mathrm{o}}$. In the present case, we have $z = 1$, since one mole of Ag liberates one mole of electrons as it ionizes into Ag^+ ions. Unless the ξ-axis happens to cross the $G'' = 0$ curve of the (n_{L}, n_{R})-plane (see Fig. 8.3.1), which is the case represented in the top diagram of Fig. 8.4.1, the ends of the cell's admissible range coincide with the intersections of the ξ-axis with the coordinate axes, n_{L} and n_{R}. The positive branches of the latter axes, up to the point where they cross curve $G'' = 0$, are part of the boundary of the admissible range of the cell's solution, simply because $n_{\mathrm{L}} \geq 0$ and $n_{\mathrm{R}} \geq 0$, as follows from the very definition of mole number.

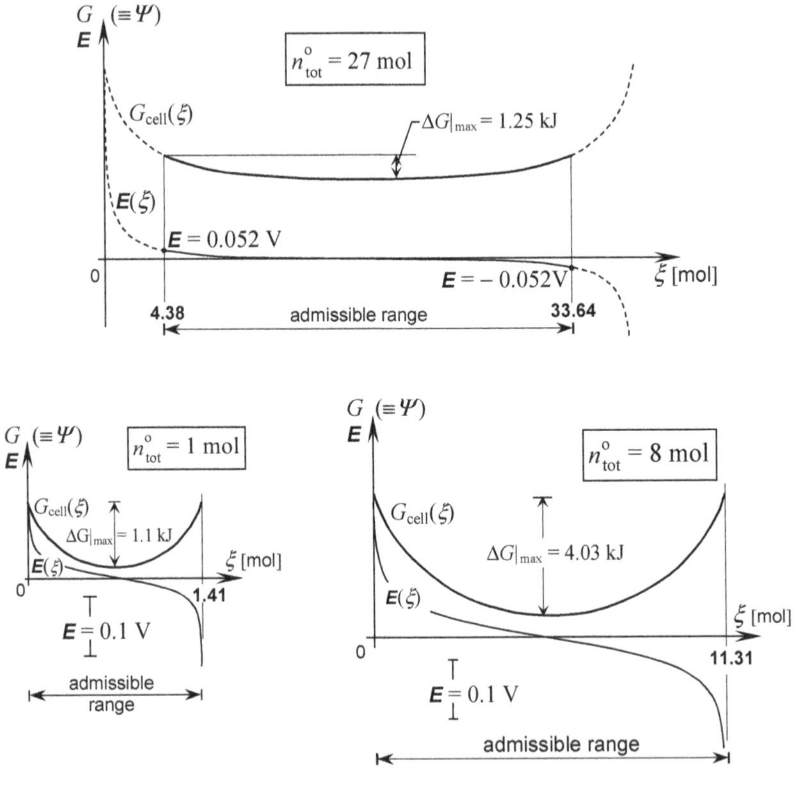

Fig. 8.4.1. Admissible range, free energy, and emf curves of three Ag|Ag^+ concentration cells made with different values of $n_{\mathrm{tot}}^{\mathrm{o}}$. The admissible range of each cell results from the intersection of its ξ-axis with the boundary of the admissible domain of Fig. 8.3.1.

The same diagrams show that ΔG_{max}, i.e., the maximum amount of free energy that the cell can store, depends on n_{tot}^{o}. In every case, ΔG_{max} is much less than the value ΔG = 17.9 kJ, which represents the maximum free energy of the cell's solution (see Fig. 8.4.1.). When configured in this way, a concentration cell is not a very efficient means for tapping the energy available from the cell's solution.

The slope of the $G(\xi)$ curves represented in Fig. 8.4.1 is related to the emf of the cell, cf. eq. (8.4.2), and thus to its electric power. A shallow $G(\xi)$ curve means that the cell is of little use as an energy supplier/absorber, irrespective of the value of ΔG_{max}. Due to the cell's internal electrical resistance, the charge/discharge process of the cell in this case is too slow to be of practical use.

8.5 Concentration Cells with Constant Concentration Half Cell

A more effective way to draw energy from the difference in concentration of two solutions is to keep one of them at a constant concentration. In the main diagram of Fig. 8.3.1, this means that the ξ-axis becomes parallel to the concentration axis relevant to the solution of the half-cell at variable concentration. If the concentration of the constant-concentration half-cell corresponds to the minimum of surface G'', then the energy ΔG_{max} available from the cell will be equal to the maximum amount of free energy that can be obtained from the two-solution system. This is much more than the energy that otherwise can be drawn if the cell is configured as described in the previous sections.

In practice, a constant concentration half-cell can be created by making it much larger than the other half-cell. If the amount of ions that the two half-cells exchange with each other is small in comparison with the total amount of ions contained in the solution of the larger half-cell, then the concentration of the latter will not change significantly as the cell operates.

The constant-concentration condition is rigorously met if the larger half-cell has access to an unlimited amount of solution, a situation that occurs quite frequently in biological systems. In other cases, the volume of biological fluid that surrounds a living cell can be regarded as unlimited, because the various components of that fluid are regenerated continuously by the organism to which the cell belongs or by the complex combination of processes that occur in the biosphere. Under these conditions, the concentration of solutes in the biological fluid surrounding the cell remains constant, even if that fluid has a comparatively small volume.

A concentration cell consisting of a half-cell at variable concentration and a half-cell at constant concentration is shown in Fig. 8.5.1 in both the galvanic and electrolytic modes of operation. The figure refers to the case where the constant concentration half-cell has an unlimited volume. In the same figure, as in the rest of the present section, the index R is appended to the quantities that refer to the constant-concentration half-cell. In the present notation, this corresponds to considering that half-cell as the right-hand half-cell. Accordingly, the variable-concentration cell is considered to be the left-hand half-cell, and the index L is associated with the quantities that pertain to it.

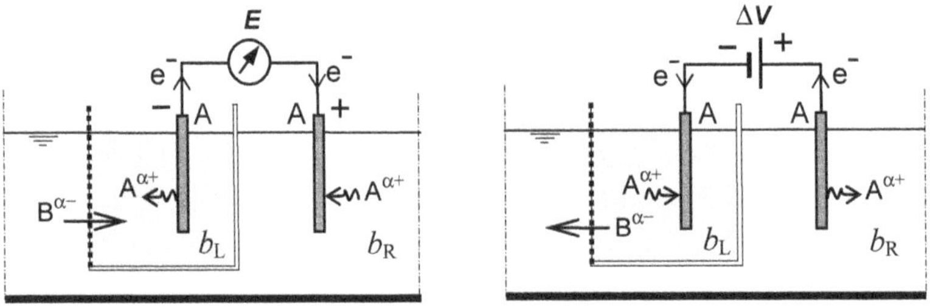

Fig. 8.5.1. Concentration cell consisting of a finite half-cell immersed in an infinite solution at the constant concentration b_R. The complete cell, i.e., the finite half-cell plus the surrounding solution, is represented in the galvanic (*left*) and electrolytic mode (*right*).

As in any concentration cell, the reaction that drives the cell has the form (8.1.1). In the present case, we have

$$b_R = \text{const}, \tag{8.5.1}$$

since the concentration of $A^{\alpha+}$ in the solution of the constant-concentration half-cell does not change. This equation, together with eq. $(8.2.2)_2$, implies that, in any finite portion of volume of the constant-concentration half-cell, we must have

$$n_R = \text{const}\,. \tag{8.5.2}$$

For a half-cell of finite volume, this means that more salt AB must be added to or removed from the solution in order to keep the concentration of that solution constant as the cell operates in the galvanic or electrolytic mode, respectively. On the contrary, the concentration b_L of the dissociated salt in the solution of the variable-concentration half-cell changes as the cell operates. From eq. (8.2.2)$_1$ it follows that n_L must change if, as assumed here, $(m_{solv})_L$ is constant. Therefore, in a concentration cell that includes a constant-concentration half-cell, the following equations hold true:

$$\begin{aligned} n_R &= n_R^o = \text{const} \\ n_L &= \xi , \end{aligned} \qquad (8.5.3)$$

no matter the operation mode of the cell. Incidentally, in the present cell, the variable n_L may substitute ξ as the variable that defines the cell's state during its operation.

Because the constant-concentration part of the cell can be considered to be infinite, there is no limit to the amount of $B^{\alpha-}$ ions available to the finite part of the cell. Thus, the $B^{\alpha-}$ ions are not a limit component of the cell's reaction in the galvanic mode of operation (Fig. 8.5.1, left). In that case, the cell's reaction can proceed until the concentration b_L of the $A^{\alpha+}$ ions in the variable-concentration part of the cell reaches the saturation value, after which point no more $A^{\alpha+}$ ions can leave the electrode of this half-cell, and the process must halt. Here, we assume that the electrode is massive enough not to be exhausted before this state is reached. On the contrary, when the cell operates in the electrolytic mode (Fig. 8.5.1, right), the process ends when the variable concentration half-cell has no more $A^{\alpha+}$ ions to neutralize or, equivalently, no more $B^{\alpha-}$ ions to expel.

The free energy of the cells considered in this section can be obtained from eq. (8.2.8) once the variable n_L is assumed as the cell's state variable. Thus, we obtain:

$$G_{cell} = G_{cell}(n_L) = n_L \mu^o_{A^{\alpha+}}(T) + n_L RT \ln\left[n_L \gamma(n_L)\right] + C . \qquad (8.5.4)$$

This applies to the case in which $(m_{solv})_L = 1$ kg, as does eq. (8.2.8). In this equation, the constant C is assumed to include all terms that contain the variable n_R in the original expression (8.2.8). In the present case, these terms are constants because n_R is constant [see eq. (8.5.3)$_1$].

From eq. (8.2.8) and from eq. (8.2.2)$_1$, the following alternative expression, valid for any value of $(m_{solv})_L$, can also be obtained:

$$G_{cell} = G_{cell}(b_L) = (m_{solv})_L\, b_L \left\{\mu^o_{A^{\alpha+}}(T) + RT \ln\left[b_L \gamma(b_L)\right]\right\}. \qquad (8.5.5)$$

For simplicity, the inessential constant C is omitted from the right-hand side of this equation. The arguments of the logarithms in eqs (8.5.4) and (8.5.5) are understood to be divided by a unit factor with appropriate dimensions to make them dimensionless.

To illustrate some of the basic features of these cells, let us consider the case of an $Ag|Ag^+$ concentration cell, the right-hand half-cell of which is kept at a constant concentration. For simplicity, we refer to the case in which the amount of solvent in each half-cell is given by eq. (8.2.7), so that b_L and b_R coincide numerically with n_L and n_R, respectively. Assume, for instance, that the concentration of $AgNO_3$ in the solution of the constant-concentration half-cell is $b_R = 1.77$ mol kg^{-1}, which corresponds to the value at which surface $G''(b_L, b_R)$ reaches its minimum (see Fig. 8.3.1.). The diagrams of the free energy and the emf of such a cell are shown in Fig. 8.5.2. Here the graph of $G_{cell}(n_L)$ is obtained from eq. (8.5.4) or, equivalently, from eq. (8.5.5) by referring to the values of γ plotted in the inset in Fig. 8.3.1. By taking the derivative of $G_{cell}(n_L)$, as specified by eq. (8.4.2), the graph of the cell's emf, $\boldsymbol{E}(n_L)$, is obtained.

The curves of Fig. 8.5.2 are typical for this kind of cell. Although these curves may correspond to different amplitudes of the admissible range, all of them have similar features irrespective of the electrolytes used and the concentration value of the constant-concentration solution. The only exception is when n_R is so high that the cell's admissible range is reduced to a small segment of the n_L-axis about the point of minimum of the function $G_{cell}(n_L)$. For the example considered here, this would occur for values of n_R that are a little less than 24.2 mol.

Although component G'' of the cell's free energy contributes little to the total free energy of the cell or to its emf, it determines the cell's admissible range, as in the case of a transformation cell (Section 7.6).

An interesting feature of these diagrams is that the cell's free energy is almost linear over a large part of the cell's admissible range. This has the effect of making the cell's emf constant over the same part of the admissible range. Although the value of the emf may be comparatively small ($\boldsymbol{E} = -0.8$ V in the considered example), the fact that it remains constant despite the large concentration changes in the variable-concentration half-cell is a highly desirable feature for an energy supply/storage system. This means that the system can exhibit the same force (e.g., electric, magnetic, or mechanic) irrespective of the state of charge of the cell, provided that it is far enough from the minimum point of the cell's free energy function. A system with this property can supply the same power irrespective of its state of charge.

It can also be charged to the maximum of its capacity without having to accommodate for a high emf, which could otherwise undermine its integrity.

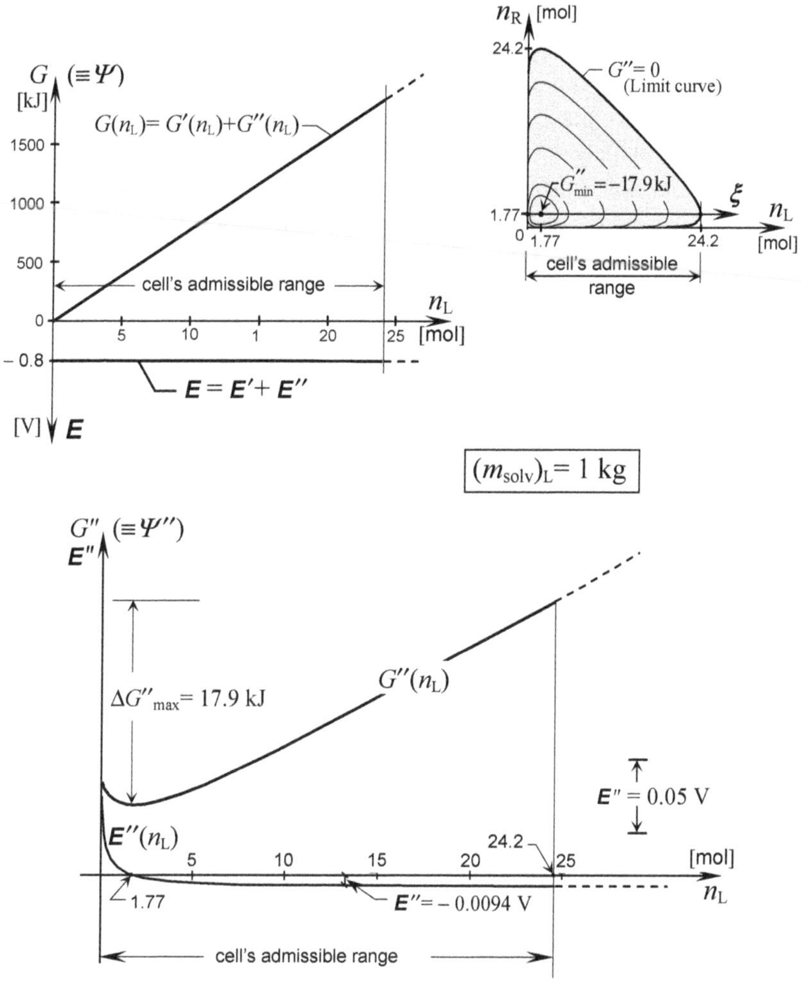

Fig. 8.5.2. Free energy and emf diagrams of an $Ag|Ag^+$ concentration cell with its right-hand half-cell at the constant concentration $n_R = 1.77$ mol/kg. *Top left*: Cell's free energy and emf as functions of n_L. *Top right*: Solution admissible range (shadowed) and cell's range. *Bottom*: Free energy G'' and its contribution, $\boldsymbol{E}''$, to the cell's emf. All diagrams refer to $(m_{solv})_L = 1$ kg.

8.6 Electrochemical Cells and Free Energy of Cell's Solution

In this chapter and in the previous chapter, we were concerned with the free energy that can be extracted from a system of two solutions that differ in composition or concentration and that are separated by a semi-permeable membrane. This system is where the redox reactions that drive the cell take place. Its free energy function can be expressed in the form (8.2.8), which is a particular case of the more general expression (7.3.1). From that energy function, we determined the admissible range of the above-mentioned two-solution system, which we referred to as the *admissible range of the cell's solution* (Sections 7.6 and 8.3).

In the considered cases, the electrochemical cell is the tool by which energy is extracted from the cell's solution. Of course, the energy that the cell can tap from the cell's solution depends on how the cell operates. Different cells working along different paths of the admissible range of the cell's solution have different ranges and thus different capacities for storing energy. The analysis presented should open the way for optimizing the cell's operation based on its intended use.

As the cell operates, its state moves on a segment belonging to an axis on the state variable plane, i.e., the ξ-axis represented in Figs 7.6.2, 8.3.1 and 8.5.2 (inset). The endpoints of that segment determine the cell's admissible range, as distinguished from the cell's solution admissible range. The maximum amount of energy that the cell can store depends on the length of that segment and, above all, on the placement of the ξ-axis to which it belongs.

Although derived for electrochemical cells, the results of the above analysis are general in character. The admissible range and the maximum storable energy of any device that takes its energy from any two-solution system of the kind considered here are determined by the ξ-axis (or, more generally by the ξ-path) over which the device operates. Whether the device produces electrical energy, mechanical energy, heat energy or drives energy-absorbing chemical reactions is of no relevance as far as its admissible range and energy-storage capacity are concerned. Consequently, most of the conclusions obtained here for the electrochemical cells are applicable to any system that is driven by a redox chemical reaction, no matter what kind of energy the system produces.

In particular, the same conclusions apply without essential changes when the energy from the above redox reaction is used by a living cell. This opens a new window into the realm of biology. In a living cell, the two-solution system where the redox reaction takes place may be represented by the fluids inside and outside the cell, or it may simply be the solution (cytosol) inside the cell. In the last case, the semi-permeable membrane that separates the reacting species is replaced by a sophisticated system of macromolecules, proteins, and enzymes that prevent and control the reaction while maintaining the reactants in the same solvent medium. This will be considered in some detail in the next chapter.

9

The living cell's admissible range

When it comes to producing energy, all animals, plants, and—with a few rare exceptions—bacteria use the same strategy. They extract energy chemically from organic nutrients by one or more oxidation-reduction reactions. This energy comes from the potential energy of the valence electrons of the reactants. The reaction displaces the electrons to a lower electric potential than they had before the reaction. The difference in the electric potential is the chemical energy released by the reaction per unit of displaced charge. It can either be harnessed directly as electrical power or used to drive energy-absorbing processes and chemical reactions.

Electrochemical cells do just that. When operating in the galvanic mode, they produce electric power from chemical energy; when operating in the electrolytic mode, they absorb electric energy to reverse the natural course of a chemical reaction (see Chapters 7 and 8).

Living cells use the same principle to move, control their temperature and pressure, communicate, and synthesize the whole array of organic materials they need to grow, reproduce, defend, and repair themselves.

Phototrophic organisms, namely plants and some kinds of bacteria, also use light energy (photosynthesis) to synthesize organic compounds to be used as food. A few bacteria, such as *Halobacterium halobium*, may, under some conditions, transduce light directly into the electric energy they need to maintain the electric gradient between their interior and the surrounding world. However, even light-harvesting organisms resort to redox reactions to access energy from the reserves of nutrients they produce by photosynthesis.

An enormous body of knowledge has accumulated concerning the chemical reactions that living systems actually use to extract energy from nutrients. Most of this knowledge is in the realms of molecular biology and biochemistry. These disciplines are progressing at an astounding pace and have unveiled some of the most intricate details of the chemistry of life, down to the atomic scale.

The scope of the present chapter and the following chapter, however, is more limited. We are interested in the restrictions that the laws of thermodynamics impose on the maximum capacity of a living system to store non-thermal energy under isothermal conditions and hence to expend such energy when no other energy resources are available. This knowledge is macroscopic in character and can be obtained without delving into the details of the biochemistry involved.

9.1 The Living Cell's Free Energy

Living cells are the most complex and cryptic systems in nature. Even the simplest cells comprise a host of different parts, the structures and interactions of which are still not completely understood. Even though the cell machinery is quite complicated, the immediate source of the cell's energy comes from the energy of the aqueous solution that the cell contains. In technical terms, that solution is known as the *cytosol*, and it is enclosed inside the cell, between the external membrane (*plasma membrane*) and the nucleus (if the cell has one). Embedded in the cytosol is a multitude of organelles, complexes, and apparatus devoted to all sorts of specific tasks in order to keep the cell alive and active. Overall, the structure of a cell can be regarded as the engine that draws energy from the cytosol and stores any surplus energy in it or in some other suitable reservoir. The engine in itself, as distinguished from the cytosol, may not possess free energy of its own; in any case, the engine energy is not the energy that the cell uses under normal operating conditions. All the energy available to the cell comes from its cytosol. At constant pressure and temperature, the cytosol's energy is just the Gibbs free energy of a solution, albeit a very particular solution due to the number and kinds of solutes that the cytosol contains.

Of course, the cell may contain reservoirs of nutrients, and hence reservoirs of energy that are not a part of the cytosol. Such reservoirs can be considered as systems in their own right, described by their own state variables. If these variables are thermodynamically orthogonal to the state variables of the cytosol (Section 3.10), they will not affect the thermodynamic limits of the latter. In that case, as far as the Gibbs free energy of the cytosol is concerned, any nutrient reservoir inside the cell should be considered as equivalent to an external source or sink of energy.

On the other hand, if the state of the cytosol is affected by the state variables of the internal reservoirs of nutrients, these variables should enter the expression of the energy of the cytosol. However, there is little evidence that this is actually the case for the living cells we find in nature. Thus, the dependence of the energy of the cytosol on the state variables of any reservoir of nutrients within the cell can accordingly be regarded as having secondary importance. For this reason, we shall henceforth consider the reservoirs of nutrients inside the cell on the same grounds as external sources of nutrients.

Invariably, all living cells are enclosed in a semi-permeable membrane that allows the free passage of water, the cytosol solvent. The membrane is impervious (or almost so) to most of the solutes in the cytosol. Under these conditions, osmosis will produce a pressure differential between the cytosol and the solution surrounding the cell, whenever the *osmolarity* (i.e., the overall concentration of dissolved particles) of the two solutions is different. The expression of the Gibbs free energy of the cytosol must take this pressure differential into account.

For a derivation of the general expression of the Gibbs free energy of a solution at variable pressure, see Appendix D.2. In the case of non-ideal solutions, that energy is a function of the activities of the components of the solution, as specified by eq. (D.25). When referred to the cytosol at a pressure $p = p° + \Delta p$, eq. (D.25) can be written as:

$$G = n_{H_2O}\,\mu^{o}_{H_2O}(p^{o},T) + \sum_{j=1}^{S} n_j\,\mu^{o}_j(p^{o},T) + V\Delta p + RT\left[n_{H_2O}\ln a_{H_2O} + \sum_{j=1}^{S} n_j \ln a_j\right] + C, \tag{9.1.1}$$

where the index "H_2O" has been substituted for the index "solv" that appears in eq. (D.25), given that the cytosol is an aqueous solution. The meaning of the quantities included in the above equation is fully explained in Appendix D.2. The notation here is the same as that used in the previous chapters.

As discussed in Appendix C.5, the actual expression of the activities a_i and a_{H_2O}, in terms of the concentration of the relevant component in the solution, depends on the measure of concentration that we use. If the concentrations of the solutes are measured in molalities and the concentration of the solvent is measured as a mole fraction, then eqs (D.26) and (C.35) are applicable, respectively. In this case, eq. (9.1.1) can be rewritten as:

$$G = n_{H_2O}\,\mu^{o}_{H_2O}(p^{o},T) + \sum_{j=1}^{S} n_j\,\mu^{o}_j(p^{o},T) + V\Delta p + RT\left[n_{H_2O}\ln \frac{\gamma_{H_2O}\, n_{H_2O}}{n_{H_2O} + \sum_{j=1}^{S} n_j} + \sum_{j=1}^{S} n_j \ln \frac{\gamma_j\, n_j}{m_{H_2O}}\right] + C\,. \tag{9.1.2}$$

In both eqs (9.1.1) and (9.1.2), the cytosol is assumed to be a non-ideal solution. There are two good reasons for making this assumption in spite of the fact that the aqueous solvent in a living cell easily makes up 70% or more of the cell's volume and that the concentration of each single solute is fairly low. The first and most obvious reason is that many of the solutes are in ionic form, which imparts non-ideal behaviour even at low concentrations. The other reason depends on a phenomenon known as *macromolecular crowding* ([19], [16], [45]). This phenomenon is typical of living cells, because their cytosol contains thousands of different species of organic macromolecules, mainly proteins. Each macromolecular species may be at low concentrations and inert to most of the other species. However, due to their sheer size, too many macromolecules in a solution hinder the mobility of each. These molecules are thus prevented from coming in contact with a significant portion of the solvent, which increases their effective concentration in the solution. This may strongly affect the behaviour of a macromolecule, say a protein, to the extent that two proteins may react differently in the cytosol and in a test tube, even if their concentrations are the same in both cases. The mere presence of other macromolecules in the solution makes the difference, even if they are inert. Being of geometric origin rather than physical or chemical origin, this phenomenon is classified as a *steric* effect.

Macromolecular crowding is also instrumental in producing interaction between cell volume and cell reactions. This interaction forms the basis for the regulation of the volume of the cell, a phenomenon which is known as *homeostasis* ([44], [65], [33]).

In expressions (9.1.1) and (9.1.2), all of the complication that results from the non-ideal behaviour of the cytosol is embodied in the activities of its components or, with reference to eq. (9.1.2), in the activity coefficients γ_{H_2O} and γ_i. Activities and activity coefficients are functions of the state variables of the cytosol, namely p, T, n_{H_2O}, n_1, n_2, ... and n_s. They can be determined experimentally by standard laboratory procedures. However, the actual expression of these functions is generally of little interest when we look for the general qualitative features of the limit surface of a living cell. Moreover, the range of variability of the state variables of a living cell is in fact rather limited. Thus, in many practical cases, the activity coefficients of the components of the cytosol can be considered constant.

9.2 The Cytosol's Admissible Range

In this section, we consider the cytosol as an ordinary solution separated from the cell that contains it and controls its content. Under these conditions, the composition of the cytosol can be changed freely, simply by adding and removing solutes or allowing them to react with each other.

As for any solution or system, the limit to the maximum free energy that the cytosol can store sets the boundary to the cytosol admissible range. This limit may reveal itself in a number of different ways, one of which is the precipitation of one or more solution components (i.e., the attainment of solubility limit of that component). The maximum free energy limit determines the admissible range of the cytosol and hence the cytosol's limit surface. In Section 9.5, we discuss how the admissible range of a living cell relates to the admissible range of its cytosol.

In order to determine the admissible range of the cytosol, the expression of its Helmholtz free energy must be obtained. In the present case, that energy must be distinguished from the Gibbs free energy, since the cytosol of a cell may undergo both pressure and volume changes. This is due both to osmotic pressure (since the cells are enveloped in a semi-permeable membrane) and the fact that the cell is an open system and, as such, it can exchange material with its surroundings.

The Helmholtz free energy of the cytosol can be calculated from the expression of the Gibbs free energy, G, given in the previous section. The procedure is the same as the one leading to eq. (D.48), presented in Appendix D.4. By applying eq. (D.47) to the case in which G is given by eq. (9.1.2), we obtain:

$$\begin{aligned}\Psi = n_{\mathrm{H_2O}}\,\mu^{\mathrm{o}}_{\mathrm{H_2O}}(p^{\mathrm{o}},T)+\sum_{\mathrm{j=1}}^{\mathrm{s}} n_{\mathrm{j}}\,\mu^{\mathrm{o}}_{\mathrm{j}}(p^{\mathrm{o}},T)-p^{\mathrm{o}}\,V \\ +RT\left[n_{\mathrm{H_2O}}\ln\frac{\gamma_{\mathrm{H_2O}}\,n_{\mathrm{H_2O}}}{n_{\mathrm{H_2O}}+\sum_{\mathrm{j=1}}^{\mathrm{s}} n_{\mathrm{j}}}+\sum_{\mathrm{j=1}}^{\mathrm{s}} n_{\mathrm{j}}\ln\frac{\gamma_{\mathrm{j}}\,n_{\mathrm{j}}}{m_{\mathrm{H_2O}}}\right]+\mathrm{C}.\end{aligned} \tag{9.2.1}$$

It can be observed immediately that this expression does not contain any temperature-independent term, apart from the inessential constant C. Therefore, the cytosol does not include any non-thermodynamic component in the sense specified in Section 3.11. The proof of this statement is the same as the proof referring to the Daniel cell in Section 7.7. It can also be observed that the molar Gibbs free energies of solvent and solutes have the form specified in eq. (3.12.2), that is

$$\mu_{\mathrm{H_2O}}=\mu^{\mathrm{o}}_{\mathrm{H_2O}}(p^{\mathrm{o}},T)+\mathrm{f}_{\mathrm{H_2O}}(p,T,n_{\mathrm{H_2O}},n_1,\ldots,n_{\mathrm{s}}) \tag{9.2.2}$$

and

$$\mu_{\mathrm{i}}=\mu^{\mathrm{o}}_{\mathrm{i}}(p^{\mathrm{o}},T)+\mathrm{f}_{\mathrm{i}}(p,T,n_{\mathrm{H_2O}},n_1,\ldots,n_{\mathrm{s}}), \tag{9.2.3}$$

respectively. These equations follow immediately from eqs (9.1.1) or (9.1.2) once the terms relevant to the solvent and the solutes are extracted from them. Using eqs (D.49) and (D.50) to express V in terms of molar volumes, functions $\mathrm{f}_{\mathrm{H_2O}}$ and $\mathrm{f_i}$, introduced above can be written explicitly as:

$$f_{H_2O}(p,T,n_{H_2O},n_1,\dots,n_s)=$$

$$n_{H_2O}\,RT\ln\frac{\gamma_{H_2O}\,n_{H_2O}}{n_{H_2O}+\sum_{j=1}^{s}n_j}-n_{H_2O}\,\overline{V}_{H_2O}\,\Delta p \qquad (9.2.4)$$

and

$$f_i(p,T,n_{H_2O},n_1,\dots,n_s)=$$

$$n_i\,RT\ln\frac{\gamma_i\,n_i}{m_{H_2O}}-n_i\,\overline{V}_i\,\Delta p\,, \qquad (9.2.5)$$

respectively. Note that the activity coefficients γ_{H_2O} and γ_i that appear on the right-hand side of these equations are functions of the variables p, T, n_{H_2O}, n_1, n_2, …, n_s.

According to what we observed in Section 3.12 it can be inferred from eqs (9.2.2) and (9.2.3) that the contributions of $\mu^o_{H_2O}$ and μ^o_i to G are due to pure thermo-entropic components of the cytosol. Thus, these contributions should be ignored when seeking the cytosol's admissible range (Sections 3.11 to 3.14). Therefore, the part of the Helmholtz free energy that determines that range is:

$$\Psi''=RT\left[n_{H_2O}\ln\frac{\gamma_{H_2O}\,n_{H_2O}}{n_{H_2O}+\sum_{j=1}^{s}n_j}+\sum_{j=1}^{s}n_j\ln\frac{\gamma_j\,n_j}{m_{H_2O}}\right]-p^o\,V+C\ , \qquad (9.2.6)$$

as follows from eq. (9.2.1) after the terms that contain $\mu^o_{H_2O}$ and μ^o_i are dropped. The symbol Ψ'', introduced in eq. (9.2.6), is consistent with the notation we used in a similar context in Section 7.6. The arbitrary constant, C, that appears in the same equation is because the actual value of Ψ'' depends on the reference value that is adopted for internal energy.

Rather than referring to the total free energy, it is more convenient to refer to some specific value of it. To do this, we introduce the new state variables $\overline{n}_i$, which represent the *number of moles of solute per mole of solvent*. Since water is cytosol's solvent, we have that

$$\overline{n}_i=\frac{n_i}{n_{H_2O}} \qquad (i=1,2,\dots,s). \qquad (9.2.7)$$

This obviously means that

$$n_i = \bar{n}_i \, n_{H_2O} \, . \tag{9.2.8}$$

From eq. (C.42) in Appendix C, it follows that

$$m_{H_2O} = \mathsf{M}_{H_2O} \, n_{H_2O} \, , \tag{9.2.9}$$

where M_{H_2O} is the molar mass of water (18.015 g mol^{-1} = 18.015 10^{-3} kg mol^{-1}). Accordingly, using eqs (9.2.7) through (9.2.9), we can put eq. (9.2.1) in the following form:

$$\Psi = n_{H_2O} \, \bar{\Psi}(p, T, \bar{n}_1, \ldots, \bar{n}_s) + C \, , \tag{9.2.10}$$

where $\bar{\Psi}$ is the free energy per mole of solvent. It is given by

$$\bar{\Psi} = \mu^{o}_{H_2O}(p^{o}, T) + \sum_{j=1}^{s} \bar{n}_j \, \mu^{o}_{j}(p^{o}, T) - p^{o}\, \bar{V} + RT \left[\ln \frac{\gamma_{H_2O}}{1 + \sum_{j=1}^{s} \bar{n}_j} + \sum_{j=1}^{s} \bar{n}_j \ln \frac{\gamma_j \bar{n}_j}{\mathsf{M}_{H_2O}} \right] . \tag{9.2.11}$$

The quantity $\bar{V}$ introduced here is the *volume of the cytosol per mole of solvent*:

$$\bar{V} = \frac{V}{n_{H_2O}} \, . \tag{9.2.12}$$

Care should be exercised in distinguishing $\bar{V}$ from the partial molar volumes $\bar{V}_i$ and $\bar{V}_{solv}$. The reader is referred to Appendices D.1 and D.2 for a definition of the latter.

Function $\bar{\Psi}$ can be decomposed into the sum of two functions, i.e., L and $\bar{\Psi}''$, according to the following relationship:

$$\bar{\Psi} = \mathsf{L} + \bar{\Psi}'' \, . \tag{9.2.13}$$

L is linear in the variables $\bar{n}_i$ and is defined by

$$\mathsf{L} = \mathsf{L}(\bar{n}_1, \ldots, \bar{n}_s; p, T) = \mu^{o}_{H_2O}(p^{o}, T) + \sum_{j=1}^{s} \bar{n}_j \, \mu^{o}_{j}(p^{o}, T) \, . \tag{9.2.14}$$

In contrast, $\overline{\Psi}''$ is non-linear in the same variables and is given by

$$\overline{\Psi}'' = \overline{\Psi}''(\overline{n}_1, \ldots, \overline{n}_s; p, T) =$$
$$RT\left[\ln \frac{\gamma_{H_2O}}{1 + \sum_{j=1}^{s} \overline{n}_j} + \sum_{j=1}^{s} \overline{n}_j \ln \frac{\gamma_j \overline{n}_j}{M_{H_2O}}\right] - p^o \overline{V}. \qquad (9.2.15)$$

Generally speaking, $\overline{\Psi}''$ is a negative-valued function, because the arguments of the logarithms appearing in its expression are, usually, less than 1.

In terms of L and $\overline{\Psi}''$, eq. (9.2.10) can be written as

$$\Psi = n_{H_2O}\left[L(\overline{n}_1, \ldots, \overline{n}_s; p, T) + \overline{\Psi}''(\overline{n}_1, \ldots, \overline{n}_s; p, T)\right] + C, \qquad (9.2.16)$$

so that eq. (9.2.6) becomes

$$\Psi'' = n_{H_2O}\, \overline{\Psi}''(\overline{n}_1, \ldots, \overline{n}_s; p, T) + C. \qquad (9.2.17)$$

Incidentally, it may be useful to observe that, if n_{H_2O} and $\overline{n}_1$, …, $\overline{n}_s$ are taken as independent variables, the quantity $p^o\overline{V}$ that appears in eqs (9.2.11) and (9.2.15) does not depend on n_{H_2O} explicitly, in spite of the fact that the variable n_{H_2O} appears explicitly in definition (9.2.12). The reason is that the ratio V / n_{H_2O}, i.e., $\overline{V}$, can be expressed as

$$\overline{V} = \frac{V}{n_{H_2O}} = \overline{V}_{H_2O} + \sum_{j=1}^{s} \overline{V}_j \overline{n}_j, \qquad (9.2.18)$$

as immediately follows by applying eq. (D.5) to the present case. The asserted independence is a consequence of eq. (9.2.18), because $\overline{V}_{H_2O}$ and $\overline{V}_j$, as in any molar volume, depend on the concentrations of the components of the solution, not on the absolute amount of these components. Of course, $\overline{V}$ will depend on n_{H_2O} explicitly if the set of independent variables is assumed to be n_{H_2O}, n_1, …, n_s rather than n_{H_2O}, $\overline{n}_1, \ldots, \overline{n}_s$.

The expression of the cytosol's admissible range follows from relationship (3.6.7), once the expression of Ψ'' given by eq. (9.2.17) is inserted for Ψ in the same relationship. After some simple algebra, we obtain:

$$\overline{\Psi}'' \leq \overline{\Psi}''_{max}, \qquad (9.2.19)$$

where $\bar{\Psi}''_{max}$ is the value of $\bar{\Psi}''$ at the boundary of the admissible range. In general, $\bar{\Psi}''_{max}$ is negative, since $\bar{\Psi}''$ is negative. By using eq. (9.2.15), we can finally put eq. (9.2.19) in the explicit form:

$$RT\left[\ln\frac{\gamma_{H_2O}}{1+\sum_{j=1}^{s}\bar{n}_j}+\sum_{j=1}^{s}\bar{n}_j\ln\frac{\gamma_j\bar{n}_j}{M_{H_2O}}\right]-p^o\,\bar{V}\le\bar{\Psi}''_{max}\,. \qquad (9.2.20)$$

Relationship (9.2.20) is the analytical expression of the admissible range of the cytosol. The cytosol's limit surface is the boundary of that range. Its expression is therefore:

$$RT\left[\ln\frac{\gamma_{H_2O}}{1+\sum_{j=1}^{s}\bar{n}_j}+\sum_{j=1}^{s}\bar{n}_j\ln\frac{\gamma_j\bar{n}_j}{M_{H_2O}}\right]-p^o\,\bar{V}=\bar{\Psi}''_{max}\,. \qquad (9.2.21)$$

Functions Ψ, Ψ'', and $\bar{\Psi}''$, as defined by eqs (9.2.1), (9.2.6), and (9.2.15), do not depend on Δp directly, but only through the activity coefficients γ_{H_2O} and γ_j. For non-gaseous species, as is the case for the cytosol's components, the dependence of the activity coefficients on pressure is notoriously rather weak and can be neglected unless Δp is very large, which is usually not the case. Therefore, since eqs (9.2.20) and (9.2.21) do not involve p explicitly, we can conclude that the cytosol's admissible range and limit surface are not strongly dependent on pressure.

At any given temperature, the states that belong to the admissible range (9.2.20) are the only states that the cytosol can attain while complying with constitutive equation (9.2.1). Accordingly, crossing the limit surface (i.e., the boundary of that range) produces a permanent change of some sort in the constitutive properties of the cytosol. The kind of change depends on the process. One possibility is that one or more solutes precipitate because their saturation limit is attained. This would result in new phases—the solid precipitate(s)—which did not exist in the original system. Another possibility is the activation of new reactions. This would lead to the formation of new species, the reaction products, that were not included in the original expression of Ψ. Events of this sort are likely to have major consequence for the system and produce permanent changes in it.

9.3 Simplified Expression of the Cytosol's Admissible Range and Limit Surface

The solutes account for about 30% of the weight of the cytosol. Since the number of species in the solution is very large, the concentration of each single solute is fairly low. Most of the solutes are organic compounds with molecular weights that easily exceed ten times the molecular weight of water, which is the solvent in the cytosol. All of these facts come together to make the moles of each solute per mole of solvent quite low, perhaps in the range of 10^{-3} or less. In a living cell this is the order of magnitude of the quantities $\bar{n}_i$ that appear in the cytosol's free energy expressions given above.

From the above remarks and from the fact that most of the $\bar{n}_i$ do not undergo significant changes during the cell's life processes, it follows that the quantity $1+\sum \bar{n}_j$ essentially remains constant. The same is true for the activity coefficient γ_{H_2O}, since the total amount of both ionic and non-ionic solutes is not expected to undergo drastic changes in ordinary conditions.

Under these conditions, the first term within the brackets of eqs (9.2.11) and (9.2.15) is almost constant, and, for isothermal processes, will remain such after multiplication by *RT*. At constant *T*, therefore, this term can be dropped from eq. (9.2.11) and included in the constant C that appears in eq. (9.2.10). When we do that, function $\bar{\Psi}$ will be given by:

$$\bar{\Psi} = \mu^{o}_{H_2O}(p^o,T) + \sum_{j=1}^{s} \bar{n}_j \mu^{o}_{j}(p^o,T) - p^o\,\bar{V} + RT\sum_{j=1}^{s} \bar{n}_j \ln\frac{\gamma_j \bar{n}_j}{M_{H_2O}} \qquad (9.3.1)$$

rather than by eq. (9.2.11). From eqs (9.2.13) and (9.2.14), it then follows that under the same conditions function $\bar{\Psi}''$ can be approximated to

$$\bar{\Psi}'' = RT\sum_{j=1}^{s} \bar{n}_j \ln\frac{\gamma_j \bar{n}_j}{M_{H_2O}} - p^o\,\bar{V}\,, \qquad (9.3.2)$$

which can be used instead of eq. (9.2.15) above. With this expression for $\bar{\Psi}''$, the expressions of the cytosol's admissible range and limit surface simplify to

$$RT\sum_{j=1}^{s} \bar{n}_j \ln\frac{\gamma_j \bar{n}_j}{M_{H_2O}} - p^o\,\bar{V} \le \bar{\Psi}''_{max} \qquad (9.3.3)$$

and

$$RT\sum_{j=1}^{s} \bar{n}_j \ln\frac{\gamma_j \bar{n}_j}{\mathrm{M}_{H_2O}} - p^o \bar{V} = \bar{\Psi}''_{max} , \qquad (9.3.4)$$

which replace eqs (9.2.20) and (9.2.21), respectively. The quantity $\bar{\Psi}''_{max}$ in the last two equations is understood to be the value attained by $\bar{\Psi}''$, as defined by eq. (9.3.2), at the boundary of the admissible range. In general, therefore, this value is different from that of the homologous quantity $\bar{\Psi}''_{max}$ in eqs (9.2.20) and (9.2.21), which refers, instead, to the value calculated at the same boundary through the rigorous expression (9.2.15). In all of the above equations, T should be considered as a constant parameter, because we are confining our attention to isothermal processes.

As stated before eq. (9.1.2), the activity coefficients γ_i that appear in the formulae of the present chapter are relevant to solute concentrations expressed in molalities. In terms of these concentrations, coefficients γ_i are related to the activities of the cytosol's solutes by eq. (D.26), which we rewrite as

$$\gamma_i = \frac{a_i}{b_i} , \qquad (9.3.5)$$

where b_i is the molality of solute i. In contrast, the activity coefficient of water (the cytosol's solvent), γ_{H_2O}, refers to solvent concentrations in expressed in mole fractions. Its relationship with the activity of water is given by eq. (D.27). That is

$$\gamma_{H_2O} = \frac{a_{H_2O}}{x_{H_2O}} , \qquad (9.3.6)$$

where x_{H_2O} is the mole fraction of water contained in the cytosol.

9.4 Some Basic Features of the Cytosol's Admissible Range

As apparent from eq. (9.2.15), function $\bar{\Psi}''$ can be interpreted as a surface of $(s+1)$ dimensions embedded in space $\{\bar{\Psi}'', p, \bar{n}_1, \bar{n}_2, \ldots, \bar{n}_s\}$ of $(s+2)$ dimensions. This is a formidably large number of dimensions, because the cytosol contains thousands of different solutes. While most solutes can be ignored because they provide a negligible or a constant contribution to $\bar{\Psi}''$, there are still hundreds of species that provide significant, variable contributions to $\bar{\Psi}''$.

To gain some insight into the structure of surface $\bar{\Psi}''$, let us confine our attention to processes that occur at constant $\bar{V}$. In this case, the contribution to the variable part of $\bar{\Psi}''$ that comes from each solute is given, to a good approximation, by

$$(\bar{\Psi}'')_{\bar{n}_i} = RT\bar{n}_i \ln(k_i \bar{n}_i), \qquad (9.4.1)$$

as immediately follows whether we refer to the approximate expression (9.3.2) or to the more rigorous equation (9.2.15). The variable $\bar{n}_i$ in eq. (9.4.1) stands for any of the variables $\bar{n}_1, \bar{n}_2, \ldots, \bar{n}_s$, while k_i is an appropriate factor that depends on the activity coefficient of the solute to which $\bar{n}_i$ refers. It is given by

$$k_i = \frac{\gamma_i}{M_{H_2O}}, \qquad (9.4.2)$$

as can be inferred immediately from eq. (9.2.15).

In principle, factors k_i (i = 1, 2, ..., s) depend on all of the state variables of the system, just as the activity coefficients do. For the time being, however, we shall assume that all k_i are constant, although each one has a different value. This is not unreasonable because the activity coefficients of the solutes are not expected to undergo large changes within the cytosol's admissible range. Under this assumption, each contribution (9.4.1) represents a convex function, involving just one of the variables $\bar{n}_1, \bar{n}_2, \ldots, \bar{n}_s$, and possessing a minimum at a well-defined, positive value of that variable.

The following comments are in order.

A. The first term in the right hand side of eq. (9.3.2) is the sum of convex functions of the form (9.4.1). It is therefore a convex function in the variables $\bar{n}_1, \bar{n}_2, \ldots, \bar{n}_s$. As a consequence, for a fixed value of $\bar{V}$, function $\bar{\Psi}''$ represents a convex (hyper-)surface in space $\{\bar{\Psi}'', \bar{n}_1, \bar{n}_2, \ldots, \bar{n}_s\}$. Being the sum of independent functions of different variables, function $\bar{\Psi}''$ has a minimum at the point of coordinates $\bar{n}_1, \bar{n}_2, \ldots, \bar{n}_s$, at which each of the component functions has its minimum. The minimum of $\bar{\Psi}''$ is an absolute minimum, and no other maxima or minima exist since $\bar{\Psi}''$ is convex. Fig. 9.4.1 represents surface $\bar{\Psi}''$ symbolically as a set of convex curves. Each curve refers to a different constant value of $\bar{V}$.

The importance of the location of the point of minimum of $\bar{\Psi}''$ is due to the fact that, as is apparent from eq. (9.2.17), Ψ'' also reaches its minimum at that point. Therefore, because Ψ'' rather than Ψ defines the admissible range of the system, the point of minimum of $\overline{\Psi}''$ is also the exhaustion state of the system. Of course, the functions Ψ and Ψ'' coincide if the system does not contain non-thermodynamic or pure thermo-entropic components, cfr. Section 3.14.

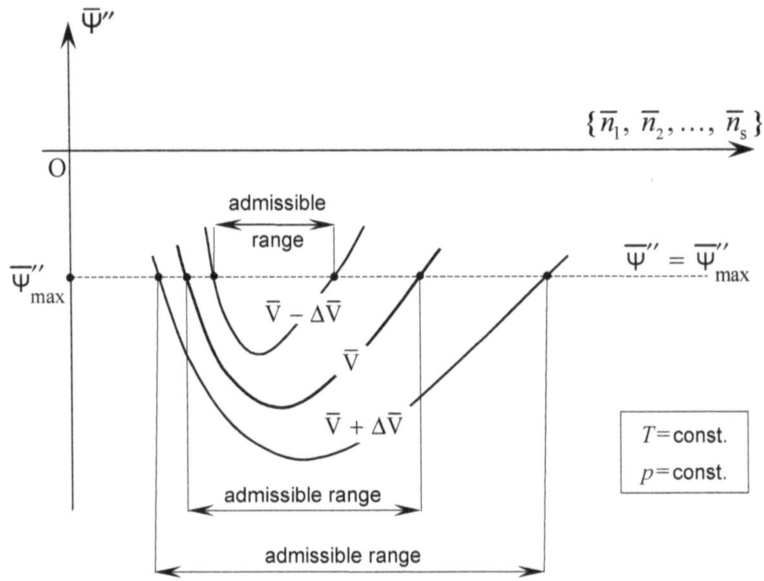

Fig. 9.4.1. Symbolic representation of surface $\overline{\Psi}''$ as a set of convex curves. Each curve refers to a constant value of $\overline{V}$. As $\overline{V}$ increases, the corresponding curve moves down, in agreement with eq. (9.2.15), while the width of the admissible range increases.

B. Condition $\overline{V} = \text{const}$ does not, in any way, mean that the cytosol has a constant volume. It simply means that the volume of cytosol per mole of solvent remains constant. This is not incompatible with a change in V, because the content of both solvent and solutes may change while leaving $\overline{V}$ unaltered. More generally, eqs (9.2.12) and (9.2.18) show that condition $\overline{V} = \text{const}$ can be met at various cytosol compositions, provided that the sum of all terms on the right-hand side of eq. (9.2.18) remains constant. Of course, this is not the case for the processes that normally occur in the cell. However, condition $\overline{V} = \text{const}$ is useful when analyzing the properties of surface $\overline{\Psi}''$ from a qualitative standpoint, as Fig. 9.4.1 shows. The same condition is assumed to hold true in the more detailed representation of the admissible range to be given below. The effect of a change in $\overline{\mathrm{V}}$ is dealt with briefly at point ***H*** later in this section.

C. At constant $\overline{\mathrm{V}}$, the variable term in the expression of the cytosol's admissible range (9.3.3) reduces to the sum of *s* addends of the kind (9.4.1). The same holds true in the case of limit surface (9.3.4). Each addend involves only one of the variables $\overline{n}_1, \overline{n}_2, \ldots, \overline{n}_s$, because the coefficients k_i are constant.

Consider a cross-section of the cytosol's admissible range parallel to coordinate plane $(\overline{n}_A, \overline{n}_B)$. Here, $\overline{n}_A$ and $\overline{n}_B$ are any two of the variables $\overline{n}_1, \overline{n}_2, \ldots, \overline{n}_s$. Analytically, the cross-section plane is defined by

$$\overline{n}_i = \overline{n}_i^{\,o} \quad \textit{for} \quad i \neq A \textit{ or } B \quad (\overline{n}_A \text{ and } \overline{n}_B \text{ unrestricted}), \qquad (9.4.3)$$

where the quantities $\overline{n}_i^{\,o}$ are given constants. From eqs (9.3.3), (9.4.2) and (9.4.3), it follows that the cross-section in question is expressed by

$$RT\overline{n}_A \ln(k_A\overline{n}_A) + RT\overline{n}_B \ln(k_B\overline{n}_B) + \mathsf{R}_{AB} - p^o\,\overline{\mathrm{V}} \leq \overline{\Psi}''_{max}, \qquad (9.4.4)$$

where the quantities k_A and k_B are given by

$$k_A = \frac{\gamma_A}{\mathsf{M}_{H_2O}} \qquad \textit{and} \qquad k_B = \frac{\gamma_B}{\mathsf{M}_{H_2O}}, \qquad (9.4.5)$$

while

$$\mathsf{R}_{AB} = RT\sum\nolimits_{i\neq A,B}\left(\overline{n}_i^{\,o} \ln\frac{\gamma_i\,\overline{n}_i^{\,o}}{\mathsf{M}_{H_2O}}\right). \qquad (9.4.6)$$

If the activity coefficients are variable within the admissible range, then the values of γ_A and γ_B to be introduced in eq. (9.4.5) should be appropriate mean values.

The quantity R_{AB} defined by eq. (9.4.6) is the contribution to $\overline{\Psi}''$ from the variables (9.4.3). The notation used in eq. (9.4.6) emphasizes that the terms relevant to $i = A$ and $i = B$ should be excluded from the summation. Although R_{AB} is constant in each cross-section parallel to plane $(\overline{n}_A, \overline{n}_B)$, the quantity R_{AB} assumes different (negative) values in different parallel planes, i.e., for different values of constants $\overline{n}_i^{\,o}$. Thus, R_{AB} can be taken as a parameter of the family of all of the cross-sections that are parallel to plane $(\overline{n}_A, \overline{n}_B)$. Fig. 9.4.2 shows some them.

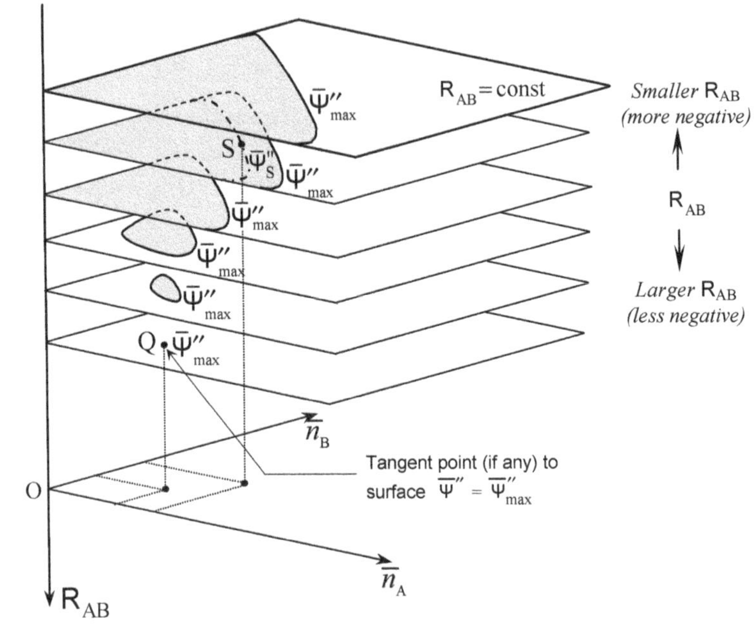

Fig. 9.4.2. Cross-sections of the cytosol's admissible range parallel to coordinate plane ($\bar{n}_A$, $\bar{n}_B$) for different values of R_{AB} , cf. eq. (9.4.4).

D. The boundaries of all of the above cross-sections belong to the limit surface. Therefore, they are all at the same potential $\overline{\Psi}''_{max}$. The value of $\overline{\Psi}''$ at a generic state S within the admissible range is denoted as $\overline{\Psi}''_S$. Clearly, $\overline{\Psi}''_S < \overline{\Psi}''_{max}$. The equipotential curve passing through S is $\overline{\Psi}'' = \overline{\Psi}''_S$ and is represented as a dash-dotted line in Fig. 9.4.2. Point Q of the same figure indicates a state on the limit surface where the tangent plane parallels plane ($\bar{n}_A$, $\bar{n}_B$). At Q, therefore, the considered cross section reduces to a point. Because Q belongs to the limit surface, we have that $\overline{\Psi}'' = \overline{\Psi}''_{max}$ at it. Of course, depending on the shape of the limit surface and the pair of variables $\bar{n}_A$ and $\bar{n}_B$ that are considered, there may not be points like Q.

E. In the space of variables $\bar{n}_1, \bar{n}_2, \ldots, \bar{n}_s$, there are κ different coordinate planes ($\bar{n}_A$, $\bar{n}_B$) . Hence, there are κ different families of parallel cross-sections of the kind (9.4.4), one family of cross-sections for each coordinate plane. The number κ equals the number of different pairs $\{\bar{n}_A, \bar{n}_B\}$ that can be drawn from set $\{\bar{n}_1, \bar{n}_2, \ldots, \bar{n}_s\}$. This is the so-called two-combination number associated with that set (cf. e.g. [32, §24.4]). In the present case, therefore, κ is given by

$$\kappa = \binom{s}{2} = \frac{s!}{2(s-2)!}, \tag{9.4.7}$$

where s is the number of elements of the set.

Fig. 9.4.3 represents four possible cross-sections parallel to four different coordinate planes through the same generic state S. The dashed curves in these cross-sections are the intersections of the considered planes with the equipotential surface $\bar{\Psi}'' = \bar{\Psi}''_S$ through S. Point O denotes the exhaustion state. It is the point of minimum of $\bar{\Psi}''$ and thus of Ψ''. It is shown in grey, because it generally does not belong to the considered planes. Clearly, point O is quite another thing than point Q of Fig. 9.4.2. (Cf. also the example considered in Fig. 9.4.4, which gives a three-dimensional representation of the admissible range in the hypothetical particular case of three independent state variables.)

Of course, the position of points S and O is different in different cross-sections, because each cross-section is relevant to a different pair of variables $\bar{n}_i$. Accordingly, the distance of state S from the limit surface may be different depending on the variable $\bar{n}_i$ along which such distance is measured. For a given state S, therefore, this representation allows us to determine at a glance the most critical variable $\bar{n}_i$ as far as the attainment of the thermodynamic limit is concerned.

F. An increase in a coordinate $\bar{n}_i$ of a given state S does not necessarily displace that state nearer to the limit surface. In the case depicted in Fig. 9.4.3, for example, an increase in the variable $\bar{n}_q$ will increase the distance of point S from the limit surface, while a similar increase in the variables $\bar{n}_7$ or $\bar{n}_k$ will bring the same point nearer to that surface. This is consistent with the fact that adding a solute to a solution will affect the solubility of the other solutes differently, depending on the added solute and which other solute is considered. This phenomenon is well known but rather complicated. It is of significance in many fields, including pharmacology. It depends on the effect that the added solute has on the interactions between solvent and solutes and between the solutes themselves. In other words, it all depends on the effect of the added solute on the free energies of the other components of the solution. It is reassuring that the present theory accounts for this phenomenon.

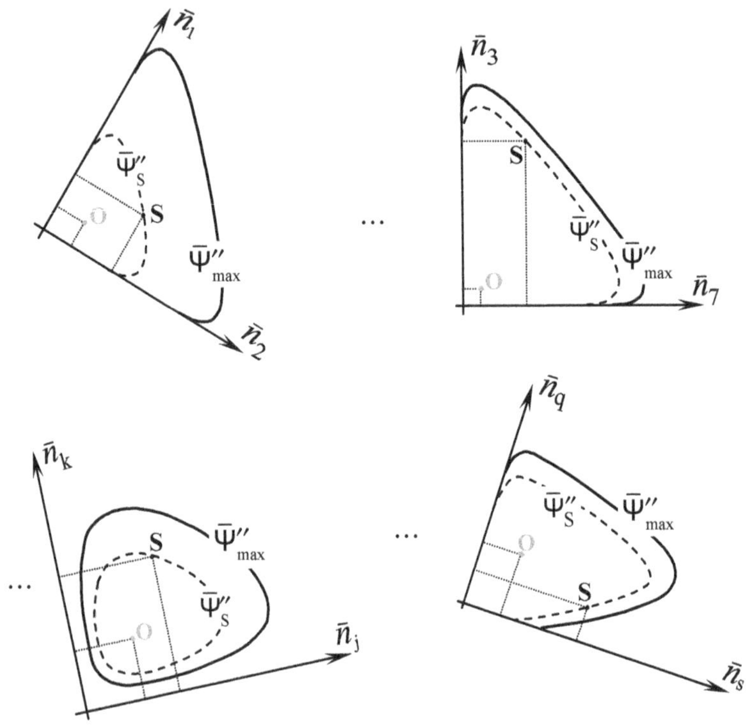

Fig. 9.4.3. Four different cross-sections of the cytosol's admissible range, relevant to four different pairs of variables $\bar{n}_A$ and $\bar{n}_B$. All the cross-sections are assumed to pass through state S.

G. To illustrate what the complete *s*-dimensional admissible range of the cytosol might look like, in Fig. 9.4.4 we represent the case where the state of the cytosol is defined by just three state variables: $\bar{n}_1$, $\bar{n}_2$, and $\bar{n}_3$. Although the state of minimum of $\bar{\Psi}''$ (point O in the above figures) may coincide with the state at which the cytosol's free energy, Ψ , is at its minimum value, this need not be so in general. Thus, depending on the system and on the process, Ψ may decrease as $\bar{\Psi}''$ increases, or vice versa.

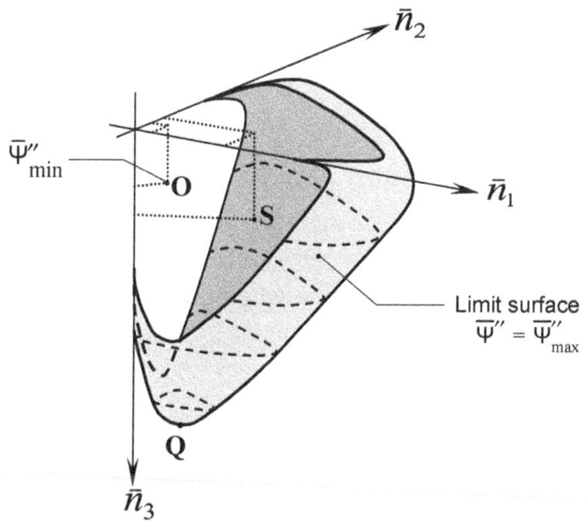

Fig. 9.4.4. Example of an admissible range of a cytosol with just three independent state variables, $\bar{n}_1$, $\bar{n}_2$ and $\bar{n}_3$. At point Q, the tangent plane to the limit surface parallels plane ($\bar{n}_1$, $\bar{n}_2$).

H. The present representations of the cytosol's admissible range apply at constant $\bar{V}$. From relation (9.4.4), we see that, as far as this range is concerned, a decrease in $\bar{V}$ has the same effect on that of making R_{AB} larger (i.e., less negative). In both cases, the admissible range will shrink as a result (cf. Fig. 9.4.1 and 9.4.2). This will reduce the distance of any state S from the limit surface and may even make that state attain the limit surface. In contrast, an increase in $p°$ has the same effect as an increase in $\bar{V}$. Thus, an increase in ambient pressure will enlarge the cytosol's cross-sections, thus increasing the distance of any state from the limit surface.

I. If the dependence of the activity coefficients γ_{H_2O} and γ_j on the variables $\bar{n}_i$ cannot be ignored, the quantities k_A and k_B (and, more generally, k_i) cannot be treated as constants. As a consequence, the shape of admissible range and limit surface will change with respect to the one described above. With due changes and unavoidable complications, however, the main results of the present section still apply to this case, provided that the activity coefficients do not vary too extensively within the cytosol's admissible range, which is a reasonable expectation in practice.

9.5 Admissible Range and Limit Surface of the Living Cell

According to relation (9.3.3), the admissible range of the cytosol is determined by its solutes. It does not matter whether the cytosol is a closed or an open system. Since the admissible range is the set of all compositions or states that the cytosol can attain at the considered temperature, it is of no relevance whether any particular composition is reached by internal transformations or by exchanging material with the surroundings.

The free energy of a living cell is essentially that of its cytosol. Of course, the other parts of the cell also possess free energy. That energy, however, can be ignored, because it is either negligible with respect to that of the cytosol or hardly interacts with the surroundings. This applies, in particular, to the free energy of cell's skeleton and cell's membrane, which constitute the cytosol's container. As a consequence, the primary state variables of the cell are those of its cytosol. Thus, the cell's admissible range must be consistent with that of its cytosol.

However, when confined within the cell, the cytosol is not free to cover all of the admissible states that it can attain as a free solution. The only states that the cell's cytosol can reach are the ones that are compatible with the physical-chemical machinery of the cell and with the permeability properties of the cell's membrane. This reduces the cell's admissible range to a subset of the analogous range of the cytosol. Consequently, the limit surface of a living cell will coincide with that part of the limit surface of the cytosol that is attainable by the cytosol when confined within the cell.

Fig. 9.5.1 illustrates this point by referring to a generic $(\overline{n}_A, \overline{n}_B)$ cross-section of the cytosol's admissible range, relevant to solutes A and B. If the amounts of these solutes change as a result of a chemical reaction, then $\overline{n}_A$ and $\overline{n}_B$ will change in proportion to the extent of reaction, ξ (cf. Appendix C.3). In the plane of the considered cross-section, the point representing the values of $\overline{n}_A$ and $\overline{n}_B$ will move along a straight line. Axes ξ_1 and ξ_2 of Fig. 9.5.1 represent two different instances of these lines. Their slope is determined by the stoichiometric coefficients of the reaction to which they refer. Their actual position depends on the initial values of $\overline{n}_A$ and $\overline{n}_B$. The points at which the each axis intercepts the boundary of the cross-section belong to the cell's limit surface. In the examples considered in the figure, the intersection points are points M and N for the ξ_1-axis and points P and R for the ξ_2-axis.

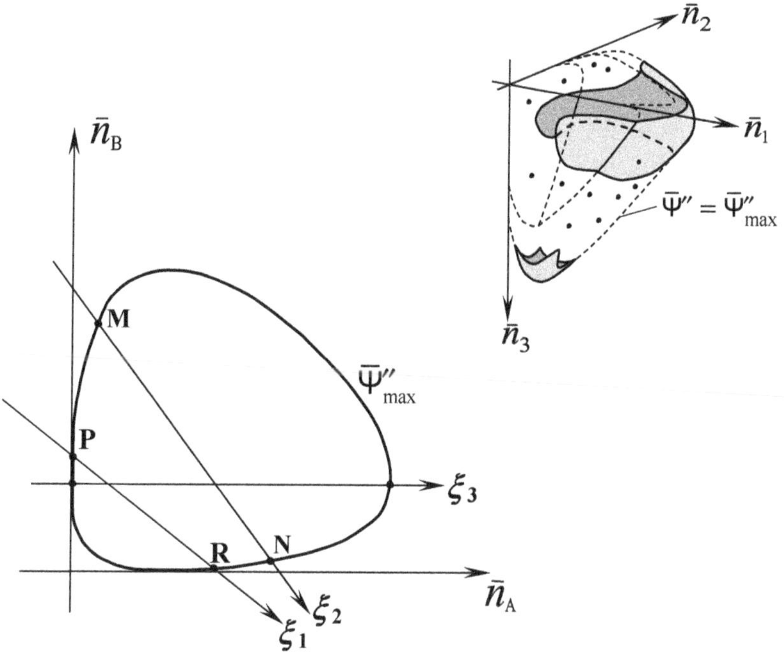

Fig. 9.5.1. ($\bar{n}_A$, $\bar{n}_B$) cross-section of the cytosol's admissible range. Axes ξ_1 and ξ_2 are two examples of lines along which the concentration of solutes A and B varies during a reaction. The intersection points of these axes with the boundary of the cross-section belong to the cell's limit surface. For the three-dimensional case considered in Fig. 9.4.4, the cell's limit surface (which is part of the cytosol's limit surface) may look like the top right diagram.

A different example is represented by axis ξ_3 in the same figure. It refers to the case in which $\bar{n}_A$ varies while $\bar{n}_B$ remains constant. This occurs when component A reacts with some other solute, say solute C (the concentration of which cannot be represented in the considered plane) or when solute A, but not solute B, is exchanged across the cell's membrane. A more general case occurs when the concentrations of A and B change as a result of exchange of material with the surroundings. In that case, the variables $\bar{n}_A$ and $\bar{n}_B$ may vary along any curve in the cross-section plane.

As observed, when the cytosol is confined within the cell, not all of its admissible states can be accessed. Some of these states cannot be reached by the cell and others are either not used or better avoided. For this reason, most of the cytosol's admissible states are out of the reach of the cell. In general, this will make the cell's limit surface a subset of the cytosol's limit surface. The cell's limit surface may consist of different separate regions, some of which are just one single point (the limit points of the reaction lines). The diagram at the upper right-hand corner of Fig. 9.5.1 illustrates this situation with reference to the hypothetical case already considered in Fig. 9.4.4.

In spite of all its discontinuities and irregularities, the cell's limit surface is a part of the cytosol's limit surface. Thus, since the latter is equipotential for $\bar{\Psi}''$, the limit surface of the living cell also must be equipotential for $\bar{\Psi}''$. At the cell's limit surface, function $\bar{\Psi}''$ attains the largest of all the values it attains in the admissible range of the cell, because $\bar{\Psi}'' = \bar{\Psi}''_{max}$ at the cytosol's limit surface. This applies to any point on that surface, even if it is an isolated point. Therefore, the quantity $\bar{\Psi}''_{max}$ can be regarded as a property of the cell. Interesting enough, its value can be calculated as in any chemical solution, by referring to the cytosol's limit surface rather than the living cell itself.

Different regions of the cell's limit surface refer to different modes of the cell's operation. For instance, a muscle cell can exert force quickly, e.g., during a weight-lifting exercise, or maintain the force for a long time, e.g., when sustaining the body's weight. The two alternatives are powered by different chemical processes, and they involve and produce different substances in the cell's cytosol. Fast muscle action, which is powered by *anaerobic glycolysis*, does not require oxygen and produces lactic acid as a by-product. Prolonged exertion of force, which depends on *aerobic glycolysis*, consumes oxygen and produces CO_2. Both processes can start from any state in the admissible domain of the cell if the right reactants are available. The maximum free energy output is obtained if they end at the exhaustion state and start at the point where the reaction line from the exhaustion state intercepts the cell's limit surface. In the case of anaerobic and aerobic glycolysis considered above, the reaction starting points on the limit surface lie on quite different regions of that surface because the relevant reaction lines leading to the exhaustion state are quite different.

The existence of a limit to the free energy that the cell can store makes all the processes and chemical reactions of the cell energetically interrelated. It does not matter if these processes are independent of each other when they occur outside the cell, e.g., when they take place *in vitro*. As the cell approaches its energy-storage limit, the progress of any process that increases the cell's energy requires that another process of the cell releases energy to the surroundings at the same time. If that does not happen, then the free energy of the cell soon exceeds the cell's energy-storage limit. This indirect energetic interaction between otherwise independent processes is likely to play a primary role in the cell's life, due to all the processes that teem in its restricted space.

9.6 Determining the Maximum Storable Free Energy of a Living Cell

According to relation (9.2.19), the admissible range of a living cell is determined by function $\overline{\Psi}''$ once the value $\overline{\Psi}''_{max}$ is given. However, as should be apparent from eq. (9.2.16), if the system includes non-thermodynamic and/or pure thermo-entropic components, the state at which the cell's free energy, Ψ, reaches its minimum is different than the analogous state of function $\overline{\Psi}''$. To calculate the minimum value of Ψ and the corresponding state, we first observe that the variables $\overline{n}_i$ take comparatively small values in the living cell. Accordingly, we can assume that the quantity of solvent in the cell, i.e., n_{H_2O}, is independent of $\overline{n}_i$, which is strictly valid for diluted solutions.

The cell's free energy minimum

With the above assumption, by taking the derivative of eq. (9.2.16) with respect to $\overline{n}_i$ and by using eqs (9.4.2), (9.2.14), and (9.2.15), we can infer that the following relation

$$\frac{\partial L}{\partial \overline{n}_i} + \frac{\partial \overline{\Psi}''}{\partial \overline{n}_i} = \mu_i^o(p^o, T) + RT\left[\ln(k_i\, \overline{n}_i) + 1\right] = 0 \qquad (9.6.1)$$

or, equivalently,

$$\ln(k_i\, \overline{n}_i) = -\frac{\mu_i^o(p^o, T)}{RT} - 1 \qquad (9.6.2)$$

should hold true at the point of minimum of Ψ. Accordingly, the coordinates of that minimum will be given by:

$$\overline{n}_i\Big|_{min} = \frac{1}{k_i \exp\left(\dfrac{\mu_i^o}{RT} + 1\right)} \qquad (i = 1, 2, \ldots, s). \qquad (9.6.3)$$

These coordinates may differ from those of the cell's exhaustion state. In the presence of non-thermodynamic and/or pure thermo-entropic components, the exhaustion state corresponds to the point of minimum of $\bar{\Psi}''$ rather than Ψ.

The value of the minimum of the cell's free energy can be calculated from eqs (9.6.3) and (9.2.10):

$$\Psi_{\min} = n_{H_2O}\,\bar{\Psi}(\bar{n}_1|_{\min},\ldots,\bar{n}_s|_{\min};p,T)+C. \qquad (9.6.4)$$

From eqs (9.2.15), (9.2.14), and (9.2.15), we can write the above equation more explicitly as

$$\Psi_{\min} = n_{H_2O}\left[\mathsf{L}(\bar{n}_1|_{\min},\ldots,\bar{n}_s|_{\min};p,T)+\bar{\Psi}''(\bar{n}_1|_{\min},\ldots,\bar{n}_s|_{\min};p,T)\right]+C$$

$$= n_{H_2O}\left[\mu^{o}_{H_2O}(p^{o},T)+\sum_{j=1}^{s}\bar{n}_j|_{\min}\,\mu^{o}_{j}(p^{o},T)\right.$$

$$\left.+RT\left(\ln\frac{\gamma_{H_2O}}{1+\sum_{j=1}^{s}\bar{n}_j|_{\min}}+\sum_{j=1}^{s}\bar{n}_j|_{\min}\ln\frac{\gamma_j\,\bar{n}_j|_{\min}}{\mathsf{M}_{H_2O}}\right)-p^{o}\,\bar{\mathrm{V}}\right]+C.$$

(9.6.5)

Note that the exponential function that appears in eq. (9.6.3) is expected to assume quite large values, because the standard Gibbs free energies, μ^{o}_{i}, of the cytosol's solutes are in the range of tens to hundreds of kJ/mol; at room temperature, the factor RT is just about 2.48 kJ/mol (cf. e.g. Tables 2.5 and 2.6 of [2]). This locates the point of coordinates ($\bar{n}_1|_{\min},\ldots,\bar{n}_s|_{\min}$), i.e., the point of minimum of Ψ, near the origin of the $\bar{n}_i$ axes. For $\bar{n}_j|_{\min}\cong 0$, we can approximate eq. (9.6.5) as

$$\Psi_{\min} \cong n_{H_2O}\left[\mu^{o}_{H_2O}(p^{o},T)-p^{o}\,\bar{\mathrm{V}}\right]+C, \qquad (9.6.6)$$

because in this case $\gamma_{H_2O}=1$. This also means that in this case $\Psi_{\min}$ coincides with the free energy of the pure solvent, as eq. (9.6.6) shows.

Determining a point of the limit surface

To determine the limit surface, we need the coordinates of one of its points. Any point will do, since that surface is equipotential for $\bar{\Psi}''$. The value of $\bar{\Psi}''_{\max}$ can then be obtained from eq. (9.2.15) as the value of $\overline{\Psi}''$ at that point. The experimental procedure to find a point on the limit surface may start from any state of the cell's admissible range. From that point, we must make the cell follow a process that leads farther and farther from the exhaustion state. The occurrence of some permanent changes in the cell's constitutive properties will indicate that the limit surface has been crossed. The transition may be more or less conspicuous, depending on the process. Death of the cell and its splitting into two daughter cells are among the most dramatic and common outcomes.

Most cells do not tolerate desiccation. They can only withstand moderate dehydration, e.g., water loss of 20 to 30%. For these cells, removing water from the cytosol is a fairly simple method to bring them to the limit surface. In practice, a desiccation process can be implemented by embedding the cells in a hypertonic solution, thus withdrawing water from them through osmosis. Every cell has some osmoregulatory capacity to oppose the loss of water. Dehydration, however, is bound to ensue if the hypertonic environment is strong enough and the cell is in contact with it for a sufficiently long time. As the cell dehydrates, the variables $\overline{n}_{\mathrm{i}}$ increase, in agreement with eq. (9.2.7). Eventually, the state of the cell reaches the limit surface.

From that point on, further dehydration will bring the cell to a state of free energy in excess of the maximum free energy that the cell can store. Therefore, if the limit surface is crossed, a change in the cytosol's constitutive properties must occur that makes the new state of the cell compatible with the energy increase that the process entails. The kind of change depends on the state at which the limit surface is crossed. Generally speaking, it may be a sudden change in the constitutive response due to phase changes or the activation of new state variables, energetically orthogonal to the original ones. This may be due to the activation of new reactions and/or the formation of new aggregations of solutes. In any case, since the constitutive response has changed, the cell will be incapable of recovering its original state on its own (i.e., through an unloading process without absorbing non-thermal energy from the surroundings while keeping its constitutive equations unchanged). In the case of dehydration processes, this means that, once the limit surface is crossed, the cell cannot be brought back to its original state simply by rehydrating it.

The values of the variables $\overline{n}_{\mathrm{i}}$ that correspond to the point at which the cell's limit surface is crossed can be determined by monitoring the composition of its cytosol in a sequence of dehydration/rehydration cycles at increasing levels of dehydration. The limit surface is reached when additional dehydration produces changes in the cytosol's properties that are not reversed by subsequent rehydration. By inserting the coordinates $\bar{n}_i$ corresponding to this limit into eq. (9.2.15), the value $\bar{\Psi}''_{max}$ of the considered cell is obtained. Of course, experiments are generally performed on a whole colony of identical cells, since working with a single cell is impractical.

For desiccation-tolerant cells, the above procedure is inapplicable. These cells use strategies that prevent them from reaching the limit surface when they are desiccated. Thus, they can survive a water loss of 90% or more. The argument will be briefly considered in Section 9.10.

Free energy at the points of the limit surface

Although the limit surface is equipotential for $\bar{\Psi}''$, the cell's free energy Ψ will vary from point to point on that surface. This can be seen from eq. (9.2.16). When applied to a point of the limit surface, let's say point $(\bar{n}_1|_{Lim}, \ldots, \bar{n}_s|_{Lim})$, the equation yields:

$$\begin{aligned}\Psi_{Lim} &= \Psi(\bar{n}_1|_{Lim}, \ldots, \bar{n}_s|_{Lim}) = \\ &n_{H_2O}\left[\mathsf{L}(\bar{n}_1|_{Lim}, \ldots, \bar{n}_s|_{Lim}) + \bar{\Psi}''_{max}\right] + \mathrm{C},\end{aligned} \tag{9.6.7}$$

where Ψ_{Lim} is the value of Ψ at the considered point. In the above equation, the dependence of Ψ_{Lim} on p and T is ignored, because we are considering processes at constant temperature and pressure.

As observed in Section 9.4, the function $\bar{\Psi}''$ is convex. From eq. (9.2.16), it then follows that function Ψ, which is the sum of a convex function ($\bar{\Psi}''$) and a linear function (L), also is convex. The fact that Ψ is convex implies that the largest value of the cell's free energy occurs at the boundary of the admissible range, i.e., at the points of the limit surface. Function $\bar{\Psi}''$ determines this surface but not the free energy that the cell can store. However, once the limit surface is known, the values of Ψ at the points of that surface can be obtained from eq. (9.6.7).

Maximum amount of free energy of a cell

The maximum amount of free energy that a cell can release in a process that starts from the limit surface is given by

$$\Delta\Psi_{max} = \Psi_{Lim} - \Psi_{min}\,. \tag{9.6.8}$$

If Ψ does not coincide with $\bar{\Psi}''$, then $\Delta\Psi_{max}$ may differ for different points of the limit surface. More explicitly, by subtracting eq. (9.6.5) from eq. (9.6.7), we obtain

$$
\begin{aligned}
\Delta\Psi_{max} &= \Delta\Psi_{max}(\bar{n}_1|_{Lim},\dots,\bar{n}_s|_{Lim}) \\
&= n_{H_2O}\,[\mathsf{L}(\bar{n}_1|_{Lim},\dots,\bar{n}_s|_{Lim}) - \mathsf{L}(\bar{n}_1|_{min},\dots,\bar{n}_s|_{min}) \\
&\qquad\qquad + \bar{\Psi}''_{max} - \bar{\Psi}''(\bar{n}_1|_{min},\dots,\bar{n}_s|_{min})] \\
&= n_{H_2O}\Bigg\{\sum_r (\bar{n}_r|_{Lim} - \bar{n}_r|_{min})\,\mu_r^o(p^o,T) \\
&\qquad + RT\Bigg[\ln\frac{\gamma_{H_2O}}{1+\sum_{j=1}^{s}\bar{n}_j|_{Lim}} - \ln\frac{\gamma_{H_2O}}{1+\sum_{j=1}^{s}\bar{n}_j|_{min}} \\
&\qquad + \sum_r\left(\bar{n}_r|_{Lim}\ln\frac{\gamma_r\,\bar{n}_r|_{Lim}}{\mathsf{M}_{H_2O}} - \bar{n}_r|_{min}\ln\frac{\gamma_r\,\bar{n}_r|_{min}}{\mathsf{M}_{H_2O}}\right)\Bigg]\Bigg\}.
\end{aligned}
\tag{9.6.9}
$$

Here, eq. (9.2.21) has been used to express $\bar{\Psi}''_{max}$ in terms of the variables $\bar{n}_i|_{Lim}$. Equation (9.6.9) gives the free energy, $\Delta\Psi_{max}$, that the cell can release in an unloading process (actually a chemical reaction) that starts from any given point of the limit surface and ends at the point of the cell's minimum free energy. The quantity $\Delta\Psi_{max}$ is also the maximum amount of free energy that the cell can store in the reverse process.

The first and the last summations within the curled brackets in eq. (9.6.9) are confined to the cytosol components that participate in the considered process. Generally, the chemical reaction that produces the process will involve only a few of the cytosol's solutes. For this reason, the index, r, of the above summations will run over just a few terms, thus making the practical use of eq. (9.6.9) much simpler than it first appears.

Moreover, if the point of minimum of Ψ happens to be near the origin of the $\bar{n}_i$ axes, then approximation (9.6.6) applies. In this case, from eqs (9.6.8), (9.6.7), and (9.6.6), we can express $\Delta\Psi_{max}$ in the following simpler form:

$$\Delta \Psi_{max} = n_{H_2O} \left[\sum_r \bar{n}_r|_{Lim} \, \mu_r^o(p^o, T) + RT \left(\ln \frac{\gamma_{H_2O}}{1 + \sum_{j=1}^{s} \bar{n}_j|_{Lim}} + \sum_r \bar{n}_r|_{Lim} \ln \frac{\gamma_r \, \bar{n}_r|_{Lim}}{M_{H_2O}} \right) \right]. \tag{9.6.10}$$

The role of $\Delta \Psi_{max}$ *in the cell's life*

Frequently, living systems must expend energy on very short notice. They must do so to catch food, avoid predators, and process whatever information happens to hit their senses. Any delay in these activities could mean death. Reacting to external stimuli requires the expenditure of energy. To react quickly enough, the cells cannot take energy from their internal energy reserves (mainly glycerol deposits inside the cell). The energy needed for the cell's instantaneous response is taken directly from the free energy of the cytosol itself. In this case, the cytosol does not behave much differently from the electrolytic solution of a rechargeable battery, which is ready to supply electric energy as soon as the circuit to which it is connected is closed.

To recharge the free energy of the cytosol, the energy reserves inside the cell are first summoned. The energy transfer process from these reserves to the cytosol is a fast process, but not as fast as the above mentioned process that takes the energy directly from the cytosol. Anyway, the recharging of the cytosol consumes the cell's internal reserves. Thus, the cell starts to take energy from the surroundings to replace the energy that it loses. This part of the recharging process involves a complex of chemical transformations from the external sources of nutrients, and is referred to as cellular respiration. The energy transfer by cellular respiration takes more time than the transfer of energy from the cell's internal reserves to the cytosol. However, the combination of all the above processes enables the cytosol not only to supply the almost instantaneous amount of energy that the cell needs for its immediate response, but also to keep supplying energy to the cell for a comparatively long time, or even indefinitely. In this way the cell manages to obtain from the cytosol all the energy that it needs to operate, in spite of the fact that the maximum amount of free energy that the cytosol can store is rather limited and cannot exceed $\Delta \Psi_{max}$.

The energy $\Delta\Psi_{max}$ is what gives the cell the capacity to act instantaneously. It also gives the cell the time to activate the other energy sources that allow the cell to obtain all the energy that it needs. The value of $\Delta\Psi_{max}$ may be different for different processes and for cells of different kind and size. For example, nerve cells and muscle cells may well possess different values of $\Delta\Psi_{max}$.

Spending all of $\Delta\Psi_{max}$ may make the cytosol's state reach the limit surface. This may mean death for the cell, if further energy is drown from it. It is expected, therefore, that a cytosol will normally use only a fraction of the energy $\Delta\Psi_{max}$. The greater the value of $\Delta\Psi_{max}$, the larger its usable fraction will be. The energy $\Delta\Psi_{max}$ can thus be viewed as an indicator of the maximum instantaneous power with which the cell can respond. It is also an indicator of the cell's ability to deal with quick, repetitive stresses. Formulae (9.6.9) and (9.6.10) show that $\Delta\Psi_{max}$ can be determined from the cytosol's composition. This provides a way to evaluate the state of charge (or state of health) of a living cell from the composition of its cytosol.

9.7 Approximate Formulae for $\Delta\Psi_{max}$

In order to assess the relative contribution of the various terms involved in eq. (9.6.9), let us first observe that at ambient temperature the quantities μ_r^o relevant to most of the solutes involved in the cell's reactions are one or two orders of magnitude greater than the factor RT.

Let us focus our attention on the first two terms of the part in square brackets of eq. $(9.6.9)_3$, i.e., the part that is multiplied by RT. These two terms are logarithms of a quantity that is not too far from unity and have opposite signs. Therefore, they do not provide much contribution to $\Delta\Psi_{max}$ and tend to cancel each other.

Now consider the last summation in eq. $(9.6.9)_3$, within the same square brackets. It consists of the sum of pairs of terms of the general form

$$x \ln(a\,x) - y \ln(b\,y), \qquad (9.7.1)$$

where a and b are appropriate constants. The two terms of each pair tend to cancel each other. Moreover, each term approaches zero as the relevant variable, x or y, does. In the case in question, x and y stand for the variables $\bar{n}_r|_{Lim}$ and $\bar{n}_r|_{min}$, respectively. In a living cell, their values are usually comparatively small.

From all of these observations, we conclude that the terms that are multiplied by RT in eq. (9.6.9)$_3$ generally contribute much less to $\Delta\Psi_{max}$ than the other terms in the same equation. Therefore, in many practical cases, eq. (9.6.9) can be approximated as

$$\Delta\Psi_{max} = n_{H_2O} \sum_{r} (\bar{n}_r|_{Lim} - \bar{n}_r|_{min})\,\mu_r^o(p^o,T), \qquad (9.7.2)$$

where the coordinates $\bar{n}_r|_{min}$ are given by eq. (9.6.3). When the latter coordinates are near enough to the origin of the $\bar{n}_i$ axes, then formula (9.7.2) can be further simplified as

$$\Delta\Psi_{max} = n_{H_2O} \sum_{r} \bar{n}_r|_{Lim}\,\mu_r^o(p^o,T). \qquad (9.7.3)$$

The above approximations are consistent with what we found in Section 7.7 concerning the Daniell cell. In that case, the terms of the cell's free energy expression that were multiplied by RT [i.e., function Ψ'', eq. (7.7.6)] afforded a negligible contribution to the total free energy of the cell. This is also apparent from the values of Ψ'' and Ψ' in Fig. 7.7.1. Though concerning a non-biological system, the result is essentially the same.

Comparison of the right-hand side of eqs (9.7.2) and (9.7.3) with definition (9.2.15) shows that, in a living cell, the contribution that $\bar{\Psi}''$ makes to Ψ may be so small that it can be ignored altogether when evaluating its free energy. However, no matter how small its contribution, $\bar{\Psi}''$ determines the limit surface of the cell and thus the coordinates $(\bar{n}_1|_{Lim}, \bar{n}_2|_{Lim}, \ldots, \bar{n}_s|_{Lim})$ of the points of this surface. The important point to note here is that by limiting the cell's admissible range, function $\bar{\Psi}''$ also limits the whole free energy, Ψ, of the cell.

9.8 Power Regulation: How Cells Make Choices

As previously discussed, the maximum amount of energy that any system can store under isothermal conditions is determined by the laws of thermodynamics. It depends on system's internal energy and entropy functions or, equivalently, on Helmholtz free energy function. In the last two sections, the maximum amount of storable energy was calculated, in particular, for a living cell.

Although the rate at which a system absorbs thermal energy cannot exceed the rate of its entropy change multiplied by the system's absolute temperature (second law, Section 1.8), there is no limit to the time rate at which a system can give off its internal energy. Living systems exploit this fact to regulate to their advantage the rate at which they spend energy. Inanimate systems lack this regulatory capacity; which puts their energy exchanges at the mercy of their surroundings.

To control the time rate of their energy exchanges, living cells regulate the time rates of the reactions that they use to produce energy. To do so, they synthesize special organic catalysts (*enzymes*) that accelerate specific reactions. Or they use other substances (*inhibitors*) that inhibit the enzymes, thereby damping the reactions that the enzymes catalyze. Without enzymes and inhibitors, a cell could not wait for favourable changes in the environment or oppose hostile situations. Before long, the cell would be driven to its minimum free energy state and die.

Under isothermal conditions, the non-thermal energy, W_{out}, that a system supplies to the surroundings must meet limitation (1.11.6). By dividing it by dt, the latter becomes

$$\dot{W}_{\text{out}} \le -\frac{\mathrm{d}\Psi}{\mathrm{d}t}, \tag{9.8.1}$$

where $\dot{W}_{\text{out}}$ is the (non-thermal) power released by the system. For simplicity, the indication $T = \text{const}$ is omitted from the above relation and from the formulae that follow.

For isothermal processes at constant pressure and volume, the change in Helmholtz free energy coincides with the change in Gibbs free energy (Section 1.11). For the same processes, therefore, relation (9.8.1) can also be written as

$$\dot{W}_{\text{out}} \le -\frac{\mathrm{d}G}{\mathrm{d}t}. \tag{9.8.2}$$

In the case of cytosol, the Gibbs free energy depends on variables p, T, n_1, n_2, ... and n_s, cf. eqs (9.1.1) or (9.1.2). Therefore, under isothermal conditions at constant p and V, we have that:

$$\frac{\mathrm{d}G}{\mathrm{d}t} = \sum_{\mathrm{i}=1}^{s} \frac{\partial G}{\partial n_{\mathrm{i}}} \frac{\mathrm{d}n_{\mathrm{i}}}{\mathrm{d}t} = \sum_{\mathrm{i}=1}^{s} \mu_{\mathrm{i}} \frac{\mathrm{d}n_{\mathrm{i}}}{\mathrm{d}t}, \tag{9.8.3}$$

the last equality in this sequence of equations following from eq. (C.8).

Because the cytosol contains hundreds of different components, the number (s) of terms entering eq. (9.8.3) is quite large. Only a few of these components, however, are involved in each single chemical reaction that takes place in the cytosol. In the notation of Appendix C.1, each of these reactions can be written as

$$\sum_{\alpha} \nu_{\alpha} \mathrm{A}_{\alpha} = 0 , \tag{9.8.4}$$

where the quantities A_{α} indicate the cytosol components that are involved in the reaction. The stoichiometric coefficients, ν_{α}, entering the above equation are understood to be negative or positive depending on whether component A_{α} is a reactant or a product. In the same equation, index α is assumed to run over the particular subset of values of i which refers to the cytosol components that take part in the reaction.

Each variable n_{α} is related to the extent of reaction, ξ, as specified by eq. (C.11). This means that

$$\frac{\mathrm{d}n_{\alpha}}{\mathrm{d}t} = \nu_{\alpha} \frac{\mathrm{d}\xi}{\mathrm{d}t} . \tag{9.8.5}$$

In a homogeneous system, we have that $n_{\alpha} = [\mathrm{A}_{\alpha}]\ V$, where $[\mathrm{A}_{\alpha}]$ is the molar concentration (mol/dm^3 or mol/L) of component A_{α} and V is the system's volume. For homogeneous systems, therefore, eq. (9.8.5) can be written as:

$$\frac{1}{\nu_{\alpha}} \frac{\mathrm{d}[\mathrm{A}_{\alpha}]}{\mathrm{d}t} = \frac{1}{V} \frac{\mathrm{d}\xi}{\mathrm{d}t} . \tag{9.8.6}$$

The left-hand side of eq. (9.8.6) represents the *reaction rate* and will be denoted by r. That is:

$$r = \frac{1}{\nu_{\alpha}} \frac{\mathrm{d}[\mathrm{A}_{\alpha}]}{\mathrm{d}t} . \tag{9.8.7}$$

For a given reaction at a given pressure, temperature, and cytosol composition, the quantity r is independent of the particular component A_{α} to which we refer when applying eq. (9.8.7). The value of r, however, depends on pressure, temperature, and cytosol composition, non-reacting solutes included. Observe that, due to the sign convention we adopted for the stoichiometric coefficients, r is a non-negative quantity, because the time derivatives $\mathrm{d}[\mathrm{A}_{\alpha}]/\mathrm{d}t$ are positive for products and negative for reactants.

If we consider a single chemical reaction, all of the cytosol components that are not involved in the reaction are to be considered as constant. Thus, their time rates, $\mathrm{d}n_i/\mathrm{d}t$, is to be put equal to zero. In this case, from eq. (9.8.3) we obtain

$$\frac{\mathrm{d}G}{\mathrm{d}t} = r\,V \sum_{\alpha} \nu_{\alpha}\,\mu_{\alpha}\,, \tag{9.8.8}$$

as immediately follows from eqs (9.8.5)-(9.8.7). The summation on the right-hand side of eq. (9.8.8) represents the reaction's Gibbs energy, as defined in Appendix (C.18). Thus, eq. (9.8.8) can be written as

$$\frac{\mathrm{d}G}{\mathrm{d}t} = r\,V\,\Delta_{\mathrm{r}}G\,. \tag{9.8.9}$$

When introduced into relation (9.8.2), this equation yields

$$\dot{W}_{\mathrm{out}} \leq -r\,V\,\Delta_{\mathrm{r}}G\,, \tag{9.8.10}$$

which sets a limit to the maximum power that the cell can supply from the considered reaction.

In terms of *power per unit volume*, $\dot{w}_{\mathrm{out}}$, this limitation can be written as

$$\dot{w}_{\mathrm{out}} \leq -r\,\Delta_{\mathrm{r}}G\,. \tag{9.8.11}$$

Reaction rates in vivo

The rates of most of the chemical reactions taking place in a living cell are controlled by the cell itself or by the organism to which the cell belongs. In some kinds of cells, e.g., skeletal muscle cells, certain reactions are controlled at a distance via the organism's nervous system. Depending on this control, the cell synthesizes, absorbs, liberates, or expels appropriate material to accelerate or inhibit the appropriate reaction(s) as needed to carry out its function in its ever-changing environment.

Of course, the laws of physics apply to living cells just as they do to any other system in nature. These laws, however, do not suffice to predict how the cell will react to external actions, unless there is complete microscopic knowledge of cell's state and properties, which is clearly not possible. Therefore, on the macroscopic scale that we are using in the present thermodynamic approach, the cell's response to external inputs is not always predictable.

As discussed in Section 2.3, both physical and subjective variables should be known in order to define the state of a living system in a more complete way. However, the subjective variables are not addressed in the present purely physical approach, because they are not relevant to it. The values of the subjective variables are controlled by special molecules, proteins, and organs inside the cell and by the organism's nervous system if the cell belongs to an organism. These variables are responsible for what will be referred to as the 'cell's choices', i.e., the particular way that a cell responds to a given stimulus among the many alternative ways that are compatible with the values of the physical state variables used to describe the cell itself. To condition cell's choice, the subjective variables selectively activate, increase, or reduce the reaction rates of some of the biochemical reactions that take place within the cell. For this reason, the value of the *in vivo* time rate of these reactions cannot generally be predicted by the present physico-chemical model, as it can when the same reactions take place in inanimate systems, e.g., *in vitro*.

A cell cannot boost or reduce the rate of a reaction unconditionally, though. It takes energy for the cell to produce a catalyst or to acquire/expel selected materials. Moreover, the amount of additional substances that can be accumulated within the cell is limited. This restricts the extent to which a cell can accelerate or inhibit a reaction. Within these limits, the time rates of the reactions rest upon the cell's subjective variables. For this reason, the value of the reaction rate of most of the biochemical reactions that occur within the cell cannot be predicted from theory, and must be determined experimentally.

The value of $\Delta_r G$ for reactions taking place in the cytosol

The situation concerning the value of $\Delta_r G$ is different from that concerning the reaction rate. The quantity $\Delta_r G$ can be determined quite accurately from eq. (C.18) once reaction and concentration of reacting components are specified. The environment in which the reaction takes place may affect the reaction rate, but it is of no relevance at all as far as the value of $\Delta_r G$ is concerned.

The determination of $\Delta_r G$ for a given chemical reaction is a classical exercise of chemistry. It can be calculated from eq. (C.18) once the chemical potentials of the reacting components are expressed in the form (C.23). Alternatively, we can apply eq. (C.52) or eq. (C.53), once the standard molar reaction Gibbs energy, $\Delta_r G°$, is calculated from thermodynamical data tables, cf. eqs (C.21) or (C.22). The most delicate part in this kind of calculations is assigning the correct values to the effective concentrations (*activities*) of the reacting components in such a complex environment as the cytosol. In practice, however, $\Delta_r G$ may not be strongly dependent on them, so that even a rough estimate of their values is often acceptable.

Of course, the concentration of the reactants of any reaction that occurs in the cell must be small enough to keep the cytosol state within the cell's admissible range. The greater the amount and number of solutes, the closer the value of $\bar{\Psi}''$ to the limit value $\bar{\Psi}''_{max}$ [cf. eqs (9.2.15) and (9.2.19)], and the smaller the concentration increase that is allowed to a solute during a reaction. Accordingly, the value of $\Delta_r G$ of any reaction that takes place in the cytosol's admissible range is limited by the presence of other solutes that are not involved in the reaction. As relation (9.8.10) makes apparent, this limit to $\Delta_r G$ entails a limit to the maximum power that the cell can supply, because r is finite.

Example. The ATP cost of muscle power

Almost all mechanical energy produced by animal muscles comes from the hydration of adenosine triphosphate (ATP) into adenosine diphosphate (ADT) according to reaction:

$$\text{ATP (aq)} + \text{H}_2\text{O (l)} = \text{ADP (aq)} + \text{Pi (aq)} + \text{H}_3\text{O}^+\text{(aq)}, \qquad (9.8.12)$$

where Pi denotes an inorganic phosphate group, such as $H_2PO_4^-$. In the standard state [i.e., at 25 °C (298.15 K) and unit concentration of all reaction components], the Gibbs free energy of this reaction is $\Delta_r G^\circ$ = +10 kJ/mol. This means that the reaction absorbs energy, rather than releasing it as it does in physiological conditions.

Let us refer to the following molal concentration [moles per kg of solvent (cytosolic water)]: [ATP] = 5 10^{-3} mol/kg; [H_2O] = 55 mol/kg; [ADP] = 0.6 10^{-3} mol/kg; [Pi] = 10 10^{-3} mol/kg; [H_3O^+] = 10^{-7} mol/kg (corresponding to pH = 7). These are indicative values extrapolated from [5], and they refer to the cytosol in human skeletal muscles (see [11, p. 502] for other kinds of cells). The relevant reaction quotient, as obtained from eq. (C.56), is:

$$\mathrm{Q} = \frac{[\mathrm{ADP}][\mathrm{Pi}][\mathrm{H_3O^+}]}{[\mathrm{ATP}][\mathrm{H_2O}]} = 2.18\ \ 10^{-12}.$$

The corresponding value of $\Delta_r G$ can be calculated from eq. (C.55). At T = 310 K (i.e., blood temperature, 37 °C), we obtain:

$$\Delta_r G = \Delta_r G^0 + RT \ln \mathrm{Q} =$$
$$10 + 8.314 \cdot 10^{-3} \cdot 310 \cdot \ln(2.18\ \ 10^{-12}) = -59.2 \quad \mathrm{kJ\ mol^{-1}},$$

since $R = 8.314\ 10^{-3}$ kJ K^{-1} mol^{-1}. In writing this formula, we neglect a small change in $\Delta_r G°$ due to the increase in temperature from 25 °C to 37 °C.

According to the data reported in [1], the mechanical power per body weight that an athlete expends in level running varies from 20 W/kg to 70 W/kg, corresponding to velocities from 2.5 m/s to 6.5 m/s, respectively. For example, let us refer to an athlete who weighs 75 kg. The total mechanical power that he or she must expend in order to run varies from 1.50 kW to 5.25 kW depending on velocity. Most of this power is produced by the extensional muscles of the athlete's knees, mainly the *quadriceps femoris* of both legs. This amounts to the involvement of a muscle mass of about 3.2 kg per leg, if we accept the anthropometric data reported in [5]. Since the density of skeletal muscle averages about 1.1 g/cm^3 (cf. [43]), it follows that the volume of the muscle cells producing that power is about $V = 2.91$ dm^3.

From these data, we can calculate the power output per unit volume of muscle cells as

$$\dot{w}_{\text{out}} = 0.52 \sim 1.80 \quad \text{kW dm}^{-3}.$$

From the value of $\Delta_r G$ given above and from relation (9.8.11), we infer that, in the velocity range 2.5 m/s ~ 6.5 m/s, the time rate of ATP hydrolysis in the athlete's muscles must meet the following relation:

$$r \geq \frac{0.52}{59.2} \sim \frac{1.80}{59.2} = 8.78 \cdot 10^{-3} \sim 30.40 \cdot 10^{-3} \quad \text{mol s}^{-1}\ \text{dm}^{-3}.$$

By multiplying these values by muscle volume and by recalling that the ATP molar mass is 507.18 g mol^{-1}, from eq. (9.8.10) we can conclude that the minimum amount of ATP that the athlete must consume in his running varies from 12.95 to 44.87 grams per second depending on athlete's velocity. This is the ATP cost of the mechanical power that the muscles must supply. It does not include the ATP that the muscle cells consume to perform the ordinary biological tasks that keep them alive. These tasks are estimated to require an additional expense of 0.7 grams of ATP per second [9].

The value $r = 30.49 \cdot 10^{-3}$ mol s^{-1} dm^{-3} calculated above provides an indication of the maximum value that r may attain in a muscle cell. From this value, the maximum power that a muscle can release can be determined from eqs (9.8.10) or (9.8.11). In these equations we must insert the largest value that $\Delta_r G$ can reach in the considered cells. Similarly to what we did in the example, this value of $\Delta_r G$ can be obtained by inserting into eq. (C.55) the value of Q that corresponds to the state at which the concentration of ATP in the cell reaches its largest value. This ATP concentration can be measured experimentally by taking samples from a well trained muscle. Also, by setting a limit to the overall content of solutes, eq. (9.3.4) can help to find out the best cytosol composition that allows a muscle cell to have the largest possible concentration of ATP.

9.9 Dissolved Gases and Pressure Changes

The amount of gas dissolved in the cytosol is quite low. This applies to both polar and non-polar gases. Polar gases, such as ammonia (NH_3) and hydrogen sulfide (H_2S), are highly soluble in aqueous solvents and thus in the cytosol. The cell membrane, however, is impervious to these gases, which hinders their access to the cytosol. In contrast, non-polar gases, such as O_2, CO_2, and N_2 are allowed free passage through the cell's membrane. However, they are poorly soluble in water. Therefore, the amount of gas dissolved in the cytosol is small in both cases, but for different reasons, and can usually be neglected when determining the free energy of the cytosol.

Many vital reactions that occur within a living cell consume or produce substantial amounts of gases, which the cell takes from or supplies to the medium that surrounds it. To bring gases to/from the cytosol, many organisms use special proteins that bind to the gas, pass through the cell's membranes, and are soluble in the cytosol [11, pp. 47-49]. Examples of this kind of protein in mammals are hemoglobin and myoglobin. They transport O_2 and CO_2 to/from the cytoplasm of red blood cells and muscle cells, respectively [11, Ch.5]. Thus, the molecules of these low-solubility gases become available in the cytosol in the form of protein-gas complexes. This is quite different from having the same gas dissolved in the cytosol, and it is the reason why the contribution of these gases to the cytosol's free energy can be ignored despite that the cytosol absorbs/produces consistent amounts of them.

Another question concerns the effect of external pressure on the cytosol's free energy and thus the cell's limit surface. The effect of pressure on the free energy of a liquid solution is treated at length in Appendix D. Equation (D.60) shows that that energy is a function of the pressure change, Δp. If dissolved gases are ignored, which in eq. (D.60) means setting both n_{gas} and index g equal to zero, the same equation reduces to eq. (D.48). When applied to the cytosol, eq. (D.48) coincides with eq. (9.2.1), which we considered in the previous sections. Thus, neglecting the dissolved gases makes the explicit dependence on Δp disappear from the cytosol's free energy.

The cytosol's free energy also depends on Δp in an implicit way, because the compressibility of the solution makes the cytosol's volume, V, depend on pressure. The same applies to $\bar{\mathrm{V}}$ due to eq. (9.2.12). Therefore, as far as the cytosol's limit surface is concerned, a change in pressure is tantamount to a change in $\bar{\mathrm{V}}$. This will affect the cytosol's limit surface in a way similar to that depicted qualitatively in Fig. 9.4.1.

However, due to the small compressibility of the aqueous solution in question, pressure changes of several orders of magnitude are required to produce any notable volume effect. The wealth of results available from the literature concerning the capacity of single cells or simple organisms to thrive under elevated changes in pressure appears to support this conclusion (see [6] for a review). Of course, the same conclusion does not necessarily apply to a living organism as a whole. In that case, the amount of gases dissolved in the fluid that surrounds the cells varies with pressure. This may require that the pressure should not change too much or too fast for the organism to function properly.

9.10 Desiccation and Anhydrobiosis

Water is the solvent for almost all biological reactions that occur in a living cell. It also is essential for the maintenance of the lipid bilayer that makes up the membranes of cells and organelles. Dehydration below a few ten percent of normal water content is lethal for most living organisms, so it should be no surprise that living cells counteract desiccation actively. Doing so, however, usually consumes the cell's energy reserves. Thus, if dehydration lasts long enough, most cells are incapable of surviving. In terms of the present theory, this means that dehydration finally succeeds in bringing the cytosol's state beyond the limit surface.

There are remarkable exceptions. Most seeds and pollens, a few plants, some small aquatic animals, and many bacteria, yeasts, and fungi are desiccation tolerant. They can survive almost complete desiccation while remaining in an inert state for years, decades, and, at least in principle, centuries. *Anhydrobiosis* is the state they acquire as a result of partial or total desiccation. In this state, they stop their vital functions almost completely and stay dormant to recover normal life when rehydrated.

While in the dehydrated state, the organism is much like inanimate solid matter. Thus, it shares with solid matter the capacity to resist very harsh environmental conditions as far as pressure, temperature, radiation, and physicochemical attacks are concerned. This confers anhydrobiotic organisms an obvious survival advantage (for a review, refer to [22], [27] and [72]).

In order to interpret desiccation and anhydrobiosis within the framework of the present theory, we first multiply both sides of relation (9.2.20) by n_{H_2O}. By using eqs (9.2.7) and (9.2.12), we obtain:

$$RT\left[n_{H_2O} \ln \frac{\gamma_{H_2O}\, n_{H_2O}}{n_{H_2O} + \sum_{j=1}^{s} n_j} + \sum_{j=1}^{s} n_j \ln \frac{\gamma_j\, n_j}{n_{H_2O}\, M_{H_2O}} \right] - p^{o} V \leq n_{H_2O}\, \overline{\Psi}''_{max} \,. \tag{9.10.1}$$

The above relation expresses the admissible range of the cell in terms of n_j and V rather than $\overline{n}_j$ and $\overline{V}$. The latter are meaningless in the dehydrated state ($n_{H_2O} = 0$), as apparent from (9.2.7) and (9.2.12).

A reduction in n_{H_2O} makes increase the terms within the brackets and the term $-p^{o}V$ in relation (9.10.1); i.e., these terms become less negative. The increase in $-p^{o}V$ is caused by the decrease in V due to the loss of water. As a result the left-hand side of the relation increases as n_{H_2O} decreases, i.e., as the cytosol desiccates. At the same time, the reduction in n_{H_2O} makes the right-hand side of relation (9.10.1) decrease, since, at constant T, the quantity $\overline{\Psi}''_{max}$ is constant. It can be concluded that dehydration, if not stopped in time, will quickly drive the cytosol to a state that is incompatible with relation (9.10.1), i.e., beyond the limit surface. This may permanently damage the cell and even cause it to die.

Moreover, the loss of water produces an increase in the concentration of the cytosol's solutes if the amount of solutes is kept constant. This may in turn activate or accelerate unwanted chemical reactions, the rates of which are negligible or acceptably low at normal concentrations. Furthermore, a reduction in the cell's water content also brings about changes in the cytosol's pH and reduces the polarizing actions that the water molecules exert on the cellular material. This may lead to protein and enzyme denaturation and may damage the cell membranes.

Normal cells use various strategies to counteract desiccation. They expel solutes from the cytosol, thus reducing the contribution of the second addend within the brackets in relation (9.10.1). This keeps the cytosol's state away from the limit surface despite the decreasing values of n_{H_2O} and V. It also helps to maintain the solute concentration within acceptable values. Clearly, there is a limit to this strategy since the cytosol's solutes cannot be eliminated at will, because they are vital to the cell.

The cells also oppose the loss of water by closing the water channels of their membrane or by producing osmoprotectants in the cytosol to reduce the osmotic water transfer to the surrounding medium. All of these measures are at the expense of the energy reserves of the cell (as distinguished from the free energy of the cytosol) and/or reduce the cytosol's functionality. When the energy reserves are exhausted, the cytosol's water content no longer can be preserved. Then, dehydration will drive the cytosol beyond the limit surface and damage the cell.

Knowledge of the cytosol's admissible range helps to single out the solutes that are more critical to the cell's survival during dehydration. As apparent from eq. (9.2.7), desiccation makes the variables $\overline{n}_j$ increase in proportion to the amount of the water loss. The variables $\overline{n}_j$ that are closest to the limit surface will be the first to cross it. By referring to the case depicted in Fig. 9.4.3, from eq. (9.2.7) we can see that variables $\overline{n}_3$ and $\overline{n}_7$ are the closest to the limit surface. Therefore, as n_{H_2O} is reduced, the increase in these variables will drive the state of the cell across that surface, thus making the same variables the first variables that may produce damage in the dehydration process.

Apart from any damage it may produce, desiccation leads the cell to a state that is not dissimilar from that of a mixture of solid components. In such a state, the material that comprises the cell is almost completely inert due to the lack of a medium in which its various components can interact. This means that, in the desiccated state, the activities of the cell's components tend to be rather small, which makes the free energy of the desiccated cell approach the sum of the standard free energy of its pure desiccated solutes, which is similar to what happens in a mechanical mixture. The small reactivity of desiccated biological matter has been known for ages and exploited to preserve food. It is the basis of one of the most successful technologies of the modern food industry.

In general, the free energy of the hydrated cell is less than the free energy of the desiccated cell plus the free energy of the pure water of hydration. Therefore, the cell's rehydration process tends to occur spontaneously. Whether it will bring the cell back to its original state depends on whether the dehydrating process and subsequent rehydration occur entirely within the cell's admissible range, because constitutive changes (i.e. damages) are unavoidable if the limit surface is crossed during any phase of the process. In that case, it will be impossible for the cell to resume its original state through rehydration. The rehydration process would then lead the cell into a new state, usually a lifeless one.

If the damages are comparatively mild, the rehydrated cell may regain some of its original capacity to draw energy from its surroundings and repair itself to its pristine condition. Any repair, however, modifies the constitutive equations of the system. This means that the repair process cannot be an unloading process, because the unloading processes occur entirely within the system's admissible range and thus cannot produce any constitutive change. Therefore, without receiving energy from the outside, the cell cannot recover its original state once the limit surface has been crossed (cf., Sect. 3.8).

To survive dehydration, anhydrobiotic organisms use a special strategy. They replace water with another appropriate substance, usually a non-reducing disaccharide such as trehalose or sucrose [12]. The substance is synthesized by the organism in large amounts, because it has to replace a lot of water. It must be at least as inert as water and possess a polar molecule capable of replacing water in the forces that it exerts on lipids and cell membranes and in the bonds that it makes with proteins and other bio-molecules. Such a water substitute will keep the cellular material stable even when the stabilizing action of water is lost due to dehydration [29].

The substances that anhydrobiontes produce in order to survive desiccation are highly soluble in water, but they are solid in the pure state. As the water content of the cytosol decreases, they solidify to a prevalently amorphous, vitreous state. This avoids damage from crystallization forces, freezes the cellular molecular setup into a glassy matrix, and favours a chemically-inert condition by reducing molecular mobility. When water is available again, the matrix dissolves; water resumes its place in the cell structure; and life resumes its normal pace.

Admissible range of anhydrobiotic cells

The best way to describe cell dehydration and rehydration within the framework of the present theory is to consider the cytosol as a multi-solvent solution. In this case, the solvent is made of two co-solvents:

water and the non-reducing disaccharide that replaces water in the desiccated state. Here, we assume that trehalose is the second co-solvent, but the following analysis applies to any other appropriate co-solvent.

As the cell dehydrates and rehydrates, the composition of the water-trehalose solvent mixture can vary continuously from almost 100% water (the cell's normal state) to about 100% trehalose (the dehydrated state). The general expression of the free energy of a two-solvent solution is discussed in Appendix D, Section D.6. When applied to the present case, the overall mole number of the solvent, as defined by eq. (D.61), is expressed as:

$$n_{\text{solv}}\Big|_{\text{mix}} = n_{\text{H}_2\text{O}} + n_{\text{trh}}\,, \tag{9.10.2}$$

where n_{trh} is the mole number of trehalose in the solvent mixture. In the present case, eqs (D.64) - (D.67) can be rewritten as:

$$\overline{n}_{\text{H}_2\text{O}} = \frac{n_{\text{H}_2\text{O}}}{n_{\text{H}_2\text{O}} + n_{\text{trh}}}\,, \tag{9.10.3}$$

$$\overline{n}_{\text{trh}} = \frac{n_{\text{trh}}}{n_{\text{H}_2\text{O}} + n_{\text{trh}}} = \frac{1}{1 - \overline{n}_{\text{H}_2\text{O}}}\,, \tag{9.10.4}$$

$$\overline{n}_{\text{j}} = \frac{n_{\text{j}}}{n_{\text{H}_2\text{O}} + n_{\text{trh}}}\,, \tag{9.10.5}$$

and

$$\overline{\text{V}} = \frac{V}{n_{\text{H}_2\text{O}} + n_{\text{trh}}}\,. \tag{9.10.6}$$

The variables $\overline{n}_{\text{j}}$ and $\overline{\text{V}}$ defined by eqs (9.10.5) and (9.10.6) replace the analogous variables defined by eqs (9.2.7) and (9.2.12), which apply to the one-solvent solution case. All of the new variables in (9.10.5) and (9.10.6) are meaningful for both $n_{\text{H}_2\text{O}} = 0$ and $n_{\text{trh}} = 0$. This makes them well suited to represent all states that an anhydrobiotic system can attain.

In terms of the new variables, the cell's Helmholtz free energy per overall mole number of solvent can be obtained from eq. (D.69):

$$\bar{\Psi} = \bar{n}_{H_2O}\,\mu^o_{H_2O}(p^o,T) + \bar{n}_{trh}\,\mu^o_{trh}(p^o,T) + \sum_{j=1}^{s} \bar{n}_j\,\mu^o_j(p^o,T) - p^o\bar{V}$$
$$+RT\left[\bar{n}_{H_2O}\ln\frac{\gamma_{H_2O}\,\bar{n}_{H_2O}}{1+\sum_{j=1}^{s}\bar{n}_j} + \bar{n}_{trh}\ln\frac{\gamma_{trh}\,\bar{n}_{trh}}{1+\sum_{j=1}^{s}\bar{n}_j}\right. \tag{9.10.7}$$
$$\left. + \sum_{j=1}^{s}\bar{n}_j\ln\frac{\gamma_j\,\bar{n}_j}{\bar{n}_{H_2O}\,\mathsf{M}_{H_2O} + \bar{n}_{trh}\,\mathsf{M}_{trh}}\right] + C,$$

where M_{H_2O} and M_{trh} denote the molar masses of water and trehalose, respectively. From eq. (9.10.7), by repeating the same analysis as the one leading from eq. (9.2.11) to eq. (9.2.15), we have

$$\bar{\Psi}'' = \bar{\Psi}''(\bar{n}_1,\ldots,\bar{n}_s;p,T) =$$
$$RT\left[\bar{n}_{H_2O}\ln\frac{\gamma_{H_2O}\,\bar{n}_{H_2O}}{1+\sum_{j=1}^{s}\bar{n}_j} + \bar{n}_{trh}\ln\frac{\gamma_{trh}\,\bar{n}_{trh}}{1+\sum_{j=1}^{s}\bar{n}_j}\right. \tag{9.10.8}$$
$$\left. + \sum_{j=1}^{s}\bar{n}_j\ln\frac{\gamma_j\,\bar{n}_j}{\bar{n}_{H_2O}\mathsf{M}_{H_2O} + \bar{n}_{trh}\mathsf{M}_{trh}}\right] - p^o\,\bar{V}.$$

From this and from eq. (9.2.19), the following expression of the cytosol's admissible range is finally obtained:

$$RT\left[\bar{n}_{H_2O}\ln\frac{\gamma_{H_2O}\,\bar{n}_{H_2O}}{1+\sum_{j=1}^{s}\bar{n}_j} + \bar{n}_{trh}\ln\frac{\gamma_{trh}\,\bar{n}_{trh}}{1+\sum_{j=1}^{s}\bar{n}_j}\right. \tag{9.10.9}$$
$$\left. + \sum_{j=1}^{s}\bar{n}_j\ln\frac{\gamma_j\,\bar{n}_j}{\bar{n}_{H_2O}\,\mathsf{M}_{H_2O} + \bar{n}_{trh}\,\mathsf{M}_{trh}}\right] - p^o\,\bar{V} \le \bar{\Psi}''_{max}\,.$$

This expression is equivalent to previous expressions (9.2.20) and (9.10.1), but it is now in terms of the state variables (9.10.3)-(9.10.6). This enables us to follow the state of the cytosol up to complete desiccation, because all state variables remain finite and meaningful until complete desiccation. Note that the condition $n_{H_2O} = 0$ is a rather theoretical one, some trace of water, albeit with reduced mobility, is always present in the cell as a remnant of the hydration layer of some protein.

The activity coefficients γ_j, γ_{H_2O}, and γ_{trh} in expression (9.10.9) must be understood as functions of the state variables (9.10.3)-(9.10.6). Therefore, these functions are different from the analogous functions that enter the expressions of the cytosol's admissible range in terms of other independent variables, namely relations (9.2.20) and (9.10.1).

Relation (9.10.9) helps to clarify the role of trehalose in anhydrobiontes. As n_{H_2O} approaches zero, the first term within brackets in that relation tend to vanish, while the term $-p^0\,\overline{V}$ outside brackets increases due to the volume reduction due to the loss of water. Trehalose contrasts the overall increase of the left-hand side of relation (9.10.9), thus keeping the cytosol state within its admissible range. It does so by introducing a negative term in the left-hand side of the above expression (the second term within the brackets), by limiting the contributions to $\overline{\Psi}''$ coming from the cytosol's solutes (the third term in the brackets) and, finally, by replacing the lost water, thus limiting the reduction of $\overline{V}$ and slowing the increase of $-p^0\,\overline{V}$.

In addition, by vitrifying the cytosol, trehalose hampers the molecular mobility of its solutes and hence their reactivity. In the dehydrated state, the solute activity coefficients also are expected to decrease, because trehalose is not as reactive as water. This helps to reduce the left-hand side of relation (9.10.9) in spite of the increased concentration of the solutes and to preserve the cell's status quo thanks to the low chemical reactivity of trehalose.

10

The Formation of Living Systems

The gap between the functions U and TS is generated when the system is formed. In a homogeneous system, this gap determines the system's admissible range and the maximum amount of free energy that the system can store and supply. Similar results apply to each homogeneous part of non-homogeneous systems. In the present chapter, we examine how that gap varies during the process that transforms inanimate matter into a living organism.

The formation of living systems, as it occurs in nature, is invariably an autonomous process. This means that no significant external work—whether generalized or not—is done on the system as it evolves into a living organism. Instead, the process uses energy resources that are freely available to the system. They may either be part and parcel of the system itself, as is usually the case with seeds and eggs, or be at the system's disposal from its maternal or natural environment. This makes the occurrence of life in nature a spontaneous, unavoidable, if quite special, complex process.

In spite of the daunting intricacy of the structure of even the simplest living system, the thermodynamics of the formation of its admissible range is comparatively simple. It affords important clues about the system's capacity to survive and reproduce.

10.1 Formation and Subsistence Processes in Living Systems

Living systems are in constant activity. First, they continuously produce new cells for maintenance and growth. It has been calculated that about 11 million cells grow and replicate each second in a human's body. This is a staggering rate even if it is considered that the human body contains some 10^{14} cells (one hundred trillion cells). Second, living systems constantly exchange energy with their surroundings to move, eat, breathe, cool or heat themselves, etc. Every activity of a living system can be classified either as a formation process or as a subsistence process.

Formation processes. These processes were discussed at length in Chapter 5. They are the processes that modify the constitutive equations of the internal energy and entropy of the system. These two quantities are so fundamental to the system's life that one should expect that any non-trivial change in the properties of the system should also affect them. Any constitutive change that does not affect the constitutive equations for internal energy and/or entropy is of minor relevance to the system—whether living or inanimate—in that it leaves its admissible range unchanged.

Subsistence processes. These are all the processes that do not fall into the previous class. Accordingly, they leave the constitutive equations for U and S unaltered. The class of subsistence processes includes, in particular, all processes in which the system interacts with the surroundings while keeping its constitutive properties unaltered.

By modifying the constitutive equations for U and S, a formation process modifies the admissible range of the system and/or the maximum amount of non-thermal energy that it can store or supply (Chapter 3). In contrast, a subsistence process must take place within the admissible range of the system, because that is the range of all states that the system can attain while complying with its actual constitutive equation.

A living system usually performs both kinds of processes at once; e.g., a lamb keeps growing (formation process) while, at the same time, bleating or suckling at its mother's udders (subsistence processes). However, there may be periods in the life of a system in which it prevalently performs just one of the above kinds of processes. Compare the pupal and the adult stages of a honeybee worker. In the pupal stage, the insect hardly moves. It stays confined to a sealed brood cell while the formation of its organs and tissues is being completed. This is entirely a formation process. The adult worker bee that results from that stage is an accomplished biological engine. It takes energy from the surroundings in the form of food and oxygen and uses it to fly, transport material, and attend to a few other related chores in the hive. At that stage, practically all processes of the bee are subsistence processes.

Formation and subsistence processes can usually be studied separately, even when they take place simultaneously. The reason is that formation processes are generally much slower than subsistence processes. Therefore, the latter can, in practice, be analyzed while considering the formation of the system as frozen. It should be noted that the dependence of a formation process on the concomitant subsistence process is often rather weak. Thus, for many practical purposes, a formation process can be studied while ignoring any concomitant subsistence process. Accordingly, we simplify our approach by considering formation and subsistence processes as independent of each other.

The distinction between formation and subsistence helps dissipate much of the intricacies of the thermodynamics of living systems. Formation processes are not subjected to the limitations related to the admissible range, because they modify that range. Conversely, subsistence processes must take place within the admissible range of the system, because that is the only place where a process compatible with the system's constitutive properties can occur. Although both kinds of processes are fundamental to a living system, the thermodynamic restrictions discussed in Chapter 3 apply only to subsistence processes.

10.2 Life in an Isothermal Environment

Life, as we know it on Earth, thrives under isothermal conditions. Although temperature changes normally take place in the habitat of a living system, they are confined to a few tens of centigrade degrees.

From a thermodynamical standpoint, the temperature difference between a living system and its surroundings is even less significant, and, with rare exceptions, it essentially vanishes in fishes and plants. Moreover, the inside temperature of a living system is almost uniform, although it may undergo some minor changes in time.

This all stems from the fact that the biochemistry of living systems is highly sensitive to temperature. Temperature changes of just a few centigrade degrees make some biologically important proteins fold or unfold, thus triggering different vital processes. For instance, avian eggs must be brought to their mother's temperature to start developing, and then they have limited resistance to being kept below or above that temperature. To give another example, a cell culture in the lab must be at a constant temperature to grow—typically around 37 °C for animal cells and 25 °C for plant cells. When brought to a temperature that is too far above or below its natural temperature, every living system stops activity or dies. No wonder that life thrives at constant temperature or in temperate climates but is almost absent in environments where strong temperature excursions occur. To better adapt to the temperature changes of their environment, many living systems, from some monocellular organisms to the most evolved animals, have developed special regulatory capacities to keep their temperatures at the right values.

The thermodynamic consequences of this state of affairs are profound. In particular, they exclude that a living system could thrive by extracting non-thermal energy from the heat of the surrounding world. The small temperature difference between the system and its surroundings excludes that any significant amount of heat could be transformed into non-thermal energy by the system. Such a transformation requires absorbing heat from a source at a higher temperature and transferring part of it to a sink at a lower temperature. This is next to impossible in the almost uniform temperature environment that life requires.

Granted that it could be done, the efficiency of such a process would be ridiculously low. From Carnot's theorem, we know that the maximum theoretical efficiency of the transformation of heat into work is given by the temperature difference between the heat source and the heat sink divided by the absolute temperature of the source. In the case of terrestrial life, this would mean dividing the already meagre temperature difference available to a living system by a quantity as large as about 300 K, the temperature of the living system's environment. Although low, this efficiency can not even be achieved in practice due to the heat losses and irreversibilities involved in any real transformation process. Therefore, although the mere possibility of an organism thriving on temperature gradients is a physical possibility, this would undoubtedly require a biological setup that is foreign to this world.

It can be concluded that the bias of life toward isothermal conditions makes temperature play a minor role in the thermodynamics of the processes of living systems. Biological processes are isothermal or almost so. Although they can take place at different temperatures, the temperature changes have little effect on the response of living systems, provided that the changes are within the comparatively narrow range within which the systems can operate. This makes temperature an unnecessary state variable in most of the applications concerning living systems.

On the other hand, the isothermal character of biological processes makes the notion of admissible range (Chapter 3) particularly well suited to a living system or part thereof. It refers to systems under isothermal conditions, which is the natural condition for a living system.

10.3 Overall Thermodynamics of the Formation of a Living System

We generally distinguish between two main parts of a living system in formation: the living part and the inanimate part. The former may be a single cell, a multicellular embryo, or a complete organism in transformation. The inanimate part consists of the nutrients that the living part needs for its transformation and growth. In real systems, the nutrients are available in the most disparate forms. They can either occupy a definite portion of space around the embryo, as is the case for eggs and seeds, or be freely available from the natural substrate that surrounds the developing organism, as is usually the case for bacteria, fungi, and yeasts. The pupae of many insects feed on their own cells, while mammalian embryos take the nutrients from maternal blood. In most cases, gaseous phases, such as oxygen, carbon dioxide and water vapour, also should be included among the nutrients.

In defining the materials that make up a living system in formation, one should include at least as much nutrient as is strictly needed for the formation of the organism. This is the stoichiometric minimum. However, the following arguments are still valid if more than that minimum is included in the system. The reason is that any surplus nutrient will remain unaltered and thus will not affect the changes in the internal energy and entropy of the system during the formation process.

The formation process also will produce waste material. Although the waste material will not be part of the final organism, it will contribute to the total internal energy and entropy of the system. Thus, the waste material should be considered as included in that system when considering how its internal energy and entropy change over time.

Living organisms are invariably highly non-homogeneous systems. This means, in particular, that their internal energy and entropy are different for different parts of the system. Internal energy and entropy, however, are extensive quantities. Therefore, it makes sense to speak of internal energy, U, and entropy, S, of the whole system as the sum of the internal energies and entropies of its parts. The following general considerations apply to these overall values of U and S.

In addition to being isothermal, the natural process of formation of a living system invariably is autonomous, i.e., no work is done on the forming material to bring it to life ($\dot{W}_{\text{in}} \leq 0$). However, in the process, the system may supply work to the surroundings. This is essentially volume work and, at atmospheric pressure, it usually can be neglected. Thus, no harm is done if we assume that:

$$\dot{W}_{\text{in}} = 0\,. \tag{10.3.1}$$

From the first law (1.4.8), it then follows that

$$\dot{U} = \dot{Q}\,. \tag{10.3.2}$$

In view of the second law (1.6.6), this implies that the following relation holds true in the formation process of a living system:

$$\dot{U} \leq T\dot{S}\,. \tag{10.3.3}$$

On the other hand, all available evidence indicates that the formation of a living system is an irreversible process. Therefore, the equality sign should be dropped from relation (10.3.3). Accordingly, the latter should be written as:

$$\dot{\Psi} = \dot{U} - T\dot{S} < 0\,, \tag{10.3.4}$$

where Ψ is defined by eq. (1.11.3). Thus, in addition to being autonomous and isothermal, the natural processes of the formation of living systems produce a decrease in the free energy of the system that undergoes the formation process. This means that they are processes that occur spontaneously.

The sign of the entropy changes

It is often maintained that the formation of a living system from inanimate matter should involve a decrease in the entropy of the system that undergoes the formation process. This expectation is related to the idea that entropy can be interpreted as a measure of the microscopic disorder of the system and that a living system, being so well organized, should possess less entropy than the inert material from which it came. Some researchers have questioned the general validity of this standpoint. They have concluded that the formation of a living system may also result in an increase in entropy or no change in entropy at all ([8], [26]). The present analysis, however, is valid irrespective of how this issue may be settled, which requires further experimental work.

If the formation of a living system makes the total entropy of the system decrease, the process must be exothermic ($\dot{Q} < 0$), as follows from relation (1.8.1). Then, from relations (10.3.2) and (10.3.3), it can be inferred that, during the process, the internal energy of the forming material should decrease ($\dot{U} < 0$). However, due to relation (10.3.3), the decrease in entropy cannot exceed the decrease in the ratio of U over T. On the other hand, if the formation process makes S increase, then $\dot{Q}$, and thus $\dot{U}$, can be either negative or positive. However, when these quantities are positive, they cannot exceed the value of $T\,\dot{S}$, as immediately follows from relations (10.3.3) and (10.3.2).

Because of condition (10.3.4), the formation of a living system is always a spontaneous process, irrespective of whether it makes the entropy of the forming material decrease or increase. However, the resulting entropy change is bound to be small, as will be shown in Section 10.7.

10.4 The Routine of Formation: A One-Variable Process

A very important characteristic of all natural processes of formation of living systems is that they take the forming system through a fixed sequence of states, ending up with the complete organism. The sequence is strictly the same for every organism of the same species, but it is different for different species. So much so, that one can tell which parts of the organism have been formed—and to what extent—simply by knowing how much time has elapsed since the formation process began.

It follows that the state of a living organism during its formation can be identified by just a few parameters. One of them defines the progress of the formation process from the time it started. This can be taken as the main variable of the process. The other parameters only have to be considered when the conditions under which the formation occurs diverge from those that are typical for the living species to which the forming system belongs. The conditions may be temperature, pressure, and other physical quantities related to the environmental or pathological conditions of the developing system.

When formation occurs under typical conditions, time can be taken as the variable that measures the progress of the formation. This means that during the process, the values of internal energy and entropy of the forming system are functions of t only. Fig. 10.4.1 shows some possible scenarios for the functions $U(t)$ and $T\,S(t)$ during the formation process of a living system. In the figure, the formation process is assumed to start at time $t = 0$ and end at $t = t_{\mathrm{f}}$.

As shown by these diagrams, the free energy, $\Psi(t)$, of the forming system, i.e., the distance between curves $U(t)$ and $T\,S(t)$, can either be positive or negative. The sign of that distance is of no relevance, because it depends on the reference value we use for U. What matters is the sign of the free energy rate, $\dot{\Psi}$, which must always be negative, in agreement with relation (10.3.4). It follows that Ψ must decrease over time. Therefore, if $\Psi < 0$ at a certain time of the process, then the width of the gap between $U(t)$ and $T\,S(t)$ will increase as the formation proceeds.

The free energy cost, $\Delta\Psi_{\mathrm{out}}$, of the formation process is given by:

$$\Delta\Psi_{\mathrm{out}} = -\Delta\Psi = \Psi(0) - \Psi(t_{\mathrm{f}})\,. \qquad (10.4.1)$$

In general, this quantity is greater than the free energy of formation of the living organism, because some free energy is spent in the metabolism of the living material that is formed before the process is complete. For example, in the process of forming a chick from an egg, $\Delta\Psi_{\mathrm{out}}$ is the overall free energy lost by the egg in the incubation process. That energy is greater than the energy that is needed to produce the material making up the chick, since it includes the energy that the formed cells spend to stay alive before hatching occurs.

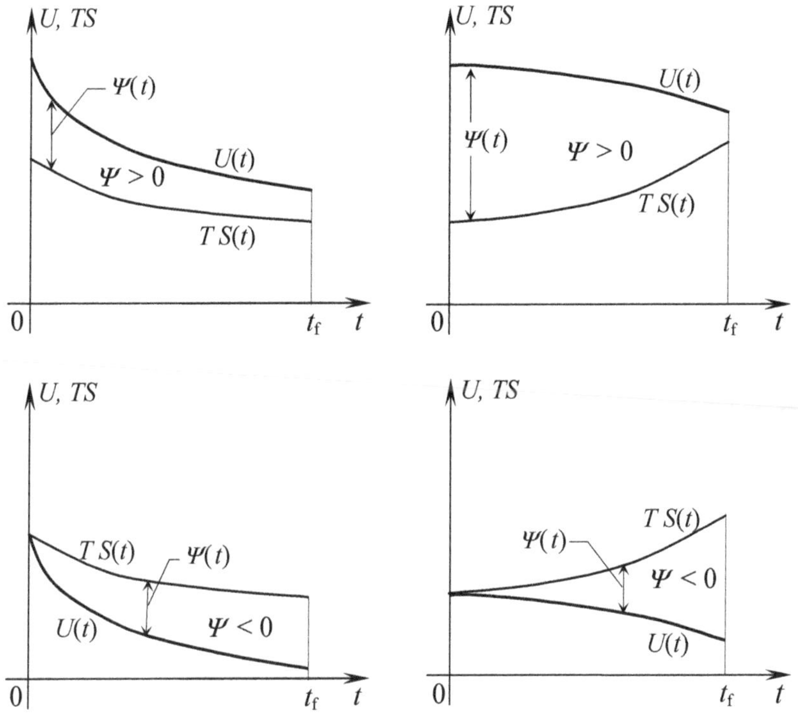

Fig. 10.4.1. Examples of time diagrams for U and $T\,S$ during a spontaneous formation process ($\dot{\Psi} < 0$ in all diagrams). *Top left*: entropy decreasing formation process. *Top right*: entropy increasing formation process. *Bottom diagrams*: same as the top diagrams, but with a reference value of U such that $\Psi = 0$ for $t = 0$ (this makes Ψ negative but leaves $\dot{\Psi}$ unaltered). All diagrams refer to the total values of U, S, and Ψ.

10.5 Heat of Formation

Much of the experimental information concerning the values of U during the formation of a living system comes from calorimetric experiments. A typical experiment is to place the forming system into a calorimetric chamber and measure the amount of heat that is produced by the formation process. Due to the small amounts of heat involved, a differential-type calorimeter usually is used. Thus, what is actually measured is the time rate of heat produced, which also is referred to as thermal power. We denote it as $\dot{Q}_{\text{out}}$, which has a positive value when representing an amount of heat that, per unit time, leaves the system. Since Q denotes an amount of heat absorbed by the system, we have that:

$$\dot{Q} = -\dot{Q}_{\text{out}}\,. \qquad (10.5.1)$$

The curves $\dot{Q}_{\text{out}}(t)$ that are obtained from these experiments are similar to the bottom diagram of Fig. 10.5.1. The formation process starts at time t_i and ends at time t_f. For $t < t_i$, the heat production rate, i.e., $-\dot{Q}_i$, is constant. This represents the metabolic heat, if any, that is needed to keep the living part of the forming system alive during the process. In some cases, the system remains dormant up to time t_i. That is the case, for instance, for seeds before germination or for eggs before incubation. In that case, $\dot{Q}_i = 0$.

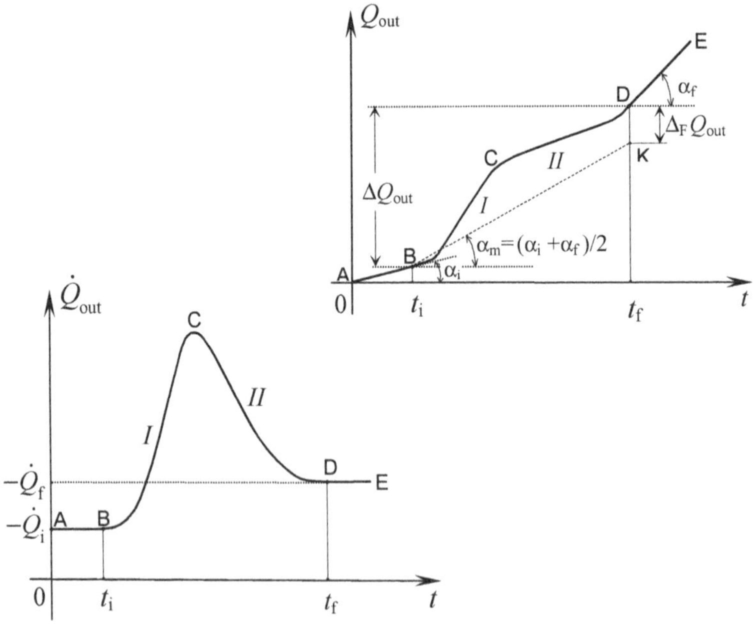

Fig. 10.5.1. *Lower diagram*: Outline of a typical heat power curve, $\dot{Q}_{\text{out}}(t)$, during the formation process of a living system. *Top diagram*: Amount of heat released, $Q_{\text{out}}(t)$, by the forming system obtained by integrating curve $\dot{Q}_{\text{out}}(t)$ in time. The quantity ΔQ_{out} is the total heat released in the interval $[t_i, t_f]$. The quantity $\Delta_F Q_{\text{out}}$ is the part of ΔQ_{out} that is due to the formation of new material.

The process between t_i and t_f usually is composed of two parts, which are indicated as *I* and *II*, respectively, in Fig. 10.5.1. In part *I*, the quantity $\dot{Q}_{out}$ increases and reaches its maximum value at point C. In this part, the increase in $\dot{Q}_{out}$ is due to the heat released by the formation process and to the increase in the heat of metabolism of the living part of the system as new cells are formed.

Part *II* of the formation process takes place from C to D. The heat production rate, $\dot{Q}_{out}$, decreases in this part of the process, because the formation of the system is nearing completion. From point D on, the heat production rate will remain constant at the value of $-\dot{Q}_f$ that is relevant to the metabolism of the formed system at rest.

The top diagram of Fig. 10.5.1 is a plot of the heat, Q_{out}, released by the process. It is obtained from a time-integration of the plot $\dot{Q}_{out}(t)$ represented in the lower diagram of the same figure. For $t < t_i$, and $t > t_f$, the plot of Q_{out} vs. t is a straight line, since $\dot{Q}_{out}$ is constant in these intervals. Of course, the greater the value of $\dot{Q}_{out}$, the steeper the corresponding line of the Q_{out} diagram.

The total amount of heat, $\Delta_F Q_{out}$, due to the formation process is the difference between the total amount of the heat, ΔQ_{out}, that the system releases in the interval $[t_i, t_f]$ and the amount of heat due to metabolism relevant to other concomitant activities that may be taking place in the system. The latter amount of heat varies in time as new cells are formed. It can be evaluated approximately as the product of $(t_f - t_i)$ times the mean value between $-\dot{Q}_i$ and $-\dot{Q}_f$. A similar approximation can be used to evaluate the heat $\Delta_F Q_{out}$. This amount of heat is represented by segment DK in the top diagram of Fig. 10.5.1. Point K of that diagram is obtained as the intersection between the vertical line through t_f and the line starting from B, sloped at the angle $\alpha_m = (\alpha_i + \alpha_f)/2$, which represents the average slope between line AB and line DE.

Depending on the system and on the formation process, the heat power diagram, $\dot{Q}_{out}(t)$, may reduce to part *I* or *II* of the diagram of Fig. 10.5.1. It may also be a sequence of branches in which the above two parts repeat themselves subsequently, one or more times. The amplitude, width, and slope of each part generally will be different for different branches.

A good example of a multi-branched heat power diagram is the $\dot{Q}_{out}(t)$ curve relevant to the formation of the honeybee, *Apis mellifera*. This diagram is shown in Fig. 10.5.2, which was freely adapted from [60]-[62]. For precise quantitative data, the reader is referred to the original papers.

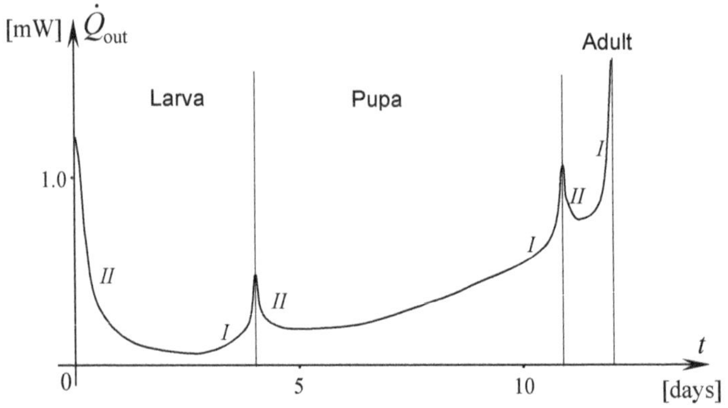

Fig. 10.5.2. Heat production rate during the formation process of one honeybee from the larval stage (immediately after the capping of the brooding cell) to the hatching into an adult worker (adapted from [60]-[62]).

The formation of the chick from an egg affords an instance of a heat power diagram made by just one single branch of the *I* type. Fig. 10.5.3 reports a typical $\dot{Q}_{out}(t)$ diagram relevant to a hen's egg (see, e.g., references [59], [37], and [38] for more details).

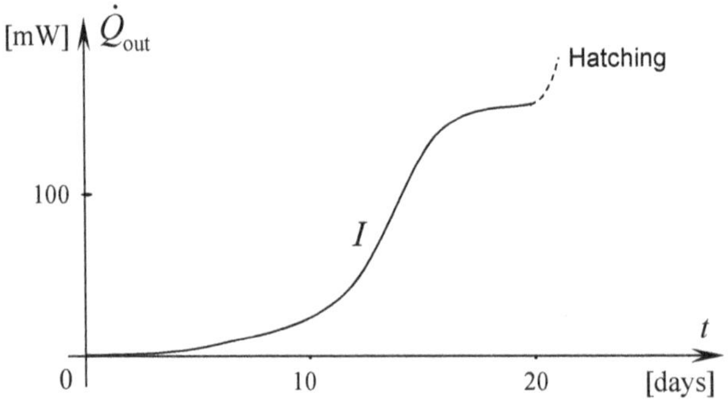

Fig. 10.5.2. Heat production rate during the formation process of a chick from a hen's egg (adapted from [38]).

The germination of seeds affords another instance of an *I*-type $\dot{Q}_{out}(t)$, though with some minor variants. Fig. 10.5.3, taken from the experimental results reported in [58], represents the heat power curves of the germination of malting barley (*Hordeum vulgare L.*) and winterfat (*Krascheninnikovia lanata*) seeds. In the present case, the steepest portion of the diagram is preceded and followed by two low rising parts where $\dot{Q}_{out}$ grows at a rather reduced rate. The low rising part at the end of the diagram indicates that the formation process continues beyond the steepest part of the diagram. This is due to the formation of more cells to complete the seedling.

On the other hand, the increase in $\dot{Q}_{out}$ in the low rising portion at the beginning of the diagram is due to the initial, water-imbibition phase of the germination process. In that phase, the seed embryo is still dormant, and the process concerns the non-living part (nutrients) of the seed. In this part of the process, $\dot{Q}_{out}$ is due to the physical and chemical processes that control the imbibition process rather than the growth of the seed embryo.

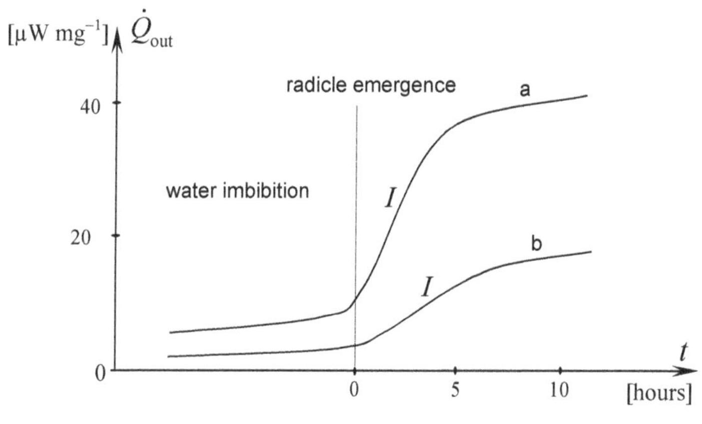

Fig. 10.5.3. Heat production rate (per embryo original weight) during germination of malting barley seeds, curve a, and winterfat seeds, curve b (adapted from [58]).

10.6 The Determination of *U*(*t*)

Equation (1.10.6) applies to all the natural processes of formation of living systems, because these processes occur at constant pressure ($dp = 0$) and do not exchange work with the surroundings ($\dot{W}_{in} = 0$). When the same processes involve gaseous phases, the moles of absorbed gases and the moles of produced gases are consistently the same. Therefore, since isothermal chemical reactions that involve solids and liquids produce quite small changes in volume, we can conclude that the formation of a living system is, to a good approximation, isochoric. As a consequence, eq. (1.10.11) is applicable to them. From that equation and from eq. (1.10.6), it is not difficult to conclude that the equation

$$\dot{U} = -\dot{Q}_{\mathrm{out}} \tag{10.6.1}$$

holds true to a good approximation in any process in which a living system is being formed or, for that matter, in any isothermal biological process that occurs at constant volume and pressure and does not exchange work with the surroundings.

Of course, the constant volume hypothesis in no way means that the volume of the forming system cannot change during the formation process. It simply means that the change in the volume of the system is equal to the volume of the material that the system takes from the surroundings for its formation.

Equation (10.6.1) enables us to interpret the heat power curves described in the previous section as the time-rate of the internal energy of the material involved in the formation process. By time-integrating that equation in the interval $[t_{\mathrm{i}}, t]$, we obtain:

$$U(t) = U(t_{\mathrm{i}}) - [Q_{\mathrm{out}}(t) - Q_{\mathrm{out}}(t_{\mathrm{i}})], \tag{10.6.2}$$

or

$$\Delta U(t) = U(t) - U(t_{\mathrm{i}}) = Q_{\mathrm{out}}(t_{\mathrm{i}}) - Q_{\mathrm{out}}(t). \tag{10.6.3}$$

Expressed in words, these equations state that the heat produced by the system during any time interval of the formation process is equal to the loss of internal energy by the system in the same time interval. This result reduces the experimental determination of $U(t)$ and $\Delta U(t)$ to the experimental determination of Q_{out}.

Both $U(t)$ and $\Delta U(t)$ can also be determined theoretically from the enthalpy, H, of the forming system. To do this, we need to observe that

$$\Delta U(t) = \Delta H(t), \tag{10.6.4}$$

as immediately follows from a time integration of eq. (1.10.11), once it is recalled that, in the present case, $\mathrm{d}p \equiv 0$. The enthalpy change, ΔH, that appears in eq. (10.6.4) can be calculated by well-established procedures of thermochemistry. One of these procedures refers to the equation:

$$\Delta H(t) = \Delta_{\mathrm{f}} H(t) - \Delta_{\mathrm{f}} H(t_{\mathrm{i}}), \tag{10.6.5}$$

where $\Delta_f H(t)$ and $\Delta_f H(t_i)$ are the enthalpies of formation of the considered system at time t and time t_i, respectively.

However, the application of eq. (10.6.5) to the present case tends to be laborious, in that it requires the knowledge of the composition of the forming system at each time, t, at which the above equation is applied. Generally speaking, the procedure requires that the values of the enthalpies of formation of each component must be extrapolated from the standard values taken from thermodynamic tables to account for the actual condition of the components at the time at which their enthalpies of formation are calculated. (Not to mention that the standard values of the enthalpies of many organic compounds of biological interest are not always readily available. In some cases, they cannot even be determined by direct experimentation, because their formation process is too intricate or not completely understood.)

Another way to determine the quantity ΔH appearing in eq. (10.6.5) would be to measure the *heat of combustion* (also called *calorific value*) of the forming system. This is a well-established procedure, which is routinely followed in the laboratory by means of a bomb calorimeter. The difference between the heat of combustion of the forming system measured at time t and the same quantity measured at time t_i is equal to $-\Delta H(t)$. This procedure is especially suitable for systems, such as eggs, that possess a well-defined boundary. The drawback is that, in order to obtain a meaningful graph of $\Delta H(t)$ over the whole formation process, the experiment must be repeated at comparatively short time intervals during the formation process.

10.7 Almost Isentropic Character of Formation Processes

The entropy changes brought about by the formation process of a living system are bound to be rather small. This is a consequence of the fact that the formation of these systems can take place only within a very narrow range of temperatures. (In the case of hens' eggs, for instance, that range is from 37 to 40 °C.) This claim can be proved as follows. First, it must be observed that, in any isothermal, constant pressure, constant volume process, we have:

$$\Delta G|_T = \Delta U|_T - T\Delta S|_T, \tag{10.7.1}$$

as follows from eq. (1.11.14) and eq. (1.11.3). (The notation used here is to emphasize that the increments ΔG, ΔU, and ΔS actually depend on the temperature, T, at which the process occurs.) For a given living system, let (T_{min}, T_{max}) be the temperature interval at which its isothermal formation is possible. If the temperature is outside that interval, the living system will not form. Since the formation of a living system is a spontaneous process, it must lead to a decrease in the free energy of the forming system. Therefore, for values of T inside the above interval, the formation process results in $\Delta G|_T < 0$.

Because no spontaneous formation process can take place outside the temperature interval (T_{min}, T_{max}), it must be argued that for $T > T_{max}$ or for $T < T_{min}$ the same process would make $\Delta G|_T > 0$. Assume then that $\Delta G|_T$ is a continuous function of temperature. Since $\Delta G|_T < 0$ inside the temperature interval (T_{min}, T_{max}), it must vanish for $T = T_{min}$ and for $T = T_{max}$. That is

$$\Delta G|_{T\,min} = 0 \qquad (10.7.2)$$

and

$$\Delta G|_{T\,max} = 0\,. \qquad (10.7.3)$$

From eqs (10.7.3) and (10.7.1), it then follows that

$$\Delta U|_{T\,max} - T_{max}\,\Delta S|_{T\,max} = 0\,. \qquad (10.7.4)$$

Since U and S are state functions, in a simple heating process from T_{min} to T_{max} we have

$$\Delta U|_{T\,max} = \Delta U|_{T\,min} + c\,\Delta T \qquad (10.7.5)$$

and

$$\Delta S|_{T\,max} = \Delta S|_{T\,min} + \frac{1}{\overline{T}}c\,\Delta T\,. \qquad (10.7.6)$$

In these expressions, c is the thermal capacity of the system, $\overline{T}$ is an appropriate value of T in the interval (T_{min}, T_{max}), and ΔT is the amplitude of that interval:

$$\Delta T = T_{max} - T_{min}\,. \qquad (10.7.7)$$

Obviously, all increments in U and S that appear in eqs (10.7.5) and (10.7.6) should be relevant to the temperature change, ΔT, with all the other state variables of the system kept constant. By introducing eqs (10.7.5) and (10.7.6) into eq. (10.7.4), using eq. (10.7.7), applying eq. (10.7.1) to $T = T_{min}$, and recalling eq. (10.7.2), we obtain:

$$c\,\Delta T - \Delta T\,\Delta S\big|_{T\,min} - \frac{T_{max}}{\overline{T}} c\,\Delta T = 0\,. \qquad (10.7.8)$$

That is,

$$\left(1 - \frac{T_{max}}{\overline{T}}\right) c - \Delta S\big|_{T\,min} = 0\,. \qquad (10.7.9)$$

At this point, to show that the entropy change brought about by the formation process of a living system is comparatively small, we may observe that the temperature of formation of most living systems is around 300 K, while the amplitude ΔT of the temperature changes that are tolerated for the formation to take place is a few tens of centigrade degrees at best. Under these conditions, the ratio $T_{max}/\overline{T}$ is approximately equal to unity, given that $T_{max} \geq \overline{T} \geq T_{max} - \Delta T$ and that ΔT is small when compared to T_{max}. Thus, the first term on the left side of eq. (10.7.9) tends to be small. The same equation can therefore be written, to a good approximation, as

$$\Delta S\big|_{T\,min} = 0\,. \qquad (10.7.10)$$

Also, since the value of the ratio $\Delta T/\overline{T}$ is small, we have that $\Delta S\big|_{T\,max} \cong \Delta S\big|_{T\,min}$, as immediately follows from eq. (10.7.6). Thus, for an appropriately small value of ΔT, simple continuity arguments lead us to conclude that

$$\Delta S\big|_{T} \cong 0 \qquad (10.7.11)$$

for every value of T in the temperature interval (T_{min}, T_{max}) in which the formation process can take place. This completes the proof.

In other words, the fact that the natural process of formation of a living system can occur only within a comparatively narrow range of temperatures implies that the entropy changes that the process produces are comparatively small. The formation of living systems that occurs in nature is therefore an almost isentropic process. Accordingly, its driving force is the decrease in the internal energy of the forming system rather than the increase in its entropy.

While the formation process is almost isentropic, it is not adiabatic, as heat is produced in the process (Section 10.5). This is consistent with the fact that the process must be irreversible, in order to produce a non-vanishing admissible range in the organism that is being formed (Section 5.1).

Remarks

A. Because the formation process is approximately isentropic, the following identities hold true to a good approximation:

$$\Delta\Psi \equiv \Delta U \equiv \Delta H \equiv \Delta G \,. \tag{10.7.12}$$

The first of these identities follows immediately from definition (1.13.3). The remaining identities are nothing but a consequence of eqs (1.11.13) through (1.11.16).

B. Though small, the actual values of the entropy changes produced by the formation process can be determined by the usual methods of thermochemistry and, in particular, by the calorimetric method discussed in Section 1.7. That method has the advantage of not requiring that the composition of the system be known. However, it requires cooling the system to temperatures of only a few degrees Kelvin and extrapolating the results to 0 K.

As an alternative, the entropy change could be determined from the relation:

$$\Delta S = \frac{\Delta G - \Delta H}{T}\,, \tag{10.7.13}$$

which follows from definition $(1.11.10)_1$ and from the fact that the process is isothermal. The quantities ΔS, ΔG, and ΔH that appear in eq. (10.7.13) are the differences between the values of S, G, and H at time t and their respective values at the beginning of the process. To apply this method, however, one must know the initial composition of the system and its composition at time t. The values of G and H of each component can be obtained by extrapolating its standard values reported in the thermodynamic tables to the values that correspond to the actual state of the same component at each stage of the formation process. However, as observed in the previous section, the standard values of some organic compounds of biological interest may not be available. Moreover, the need to know the composition of the system at each time of the process makes this procedure quite laborious.

Irrespective of the procedure that is used, the determination of $S(t)$ or $\Delta S(t)$ in the formation process of a living system is a particularly demanding task, because the process is almost isentropic. This requires high precision in measurements and calculations to acquire sufficiently accurate values for the small changes in entropy.

10.8 Unconditional Reserve of Energy: The Quiddity of Life

All living systems, from bacteria to humans, must interact with the surrounding world to survive and reproduce. To do so, they collect information about their environment by means of appropriate receptors. These are particular organs, or even single molecules, that are sensitive to touch, light, special chemicals, and the like. The energy that is required to activate a receptor is extremely small and therefore cannot fuel the complete reaction of the organism in response to the stimulus that it receives. Generally, much more energy is required simply to send the information from the receptors to the appropriate organs within the cell to which the receptor is attached, let alone to produce the complete reaction of the living organism.

Once a signal is elicited from a receptor, a fast reaction is essential to the living system's survival and reproductive success. Fast communication between cells also is required to coordinate even the simplest tasks of a multicellular organism. All of these activities absorb energy. They usually are carried out by chemical or electrochemical means, and, to occur quickly, they demand immediate access to a ready-to-use source of energy.

Every living system carries various reserves of energy. Most of them are in the form of nutrients, such as glycerol and lipids, within the system itself or attached to it in separate reservoirs. However, the chemical reactions required to liberate energy from these energy reserves take more time than the fraction-of-a-second that is needed to a typical instantaneous response of a living system. The nutrient reserves cannot provide that immediate supply of energy. Think about an animal as it reacts to escape a sudden attack by a predator or a predator pouncing upon a moving prey as soon as it comes into reach. In similar situations, the reserves of nutrients are of little use, as there is simply no time to draw energy from them.

To respond promptly, the cells within the system must have access to a source of energy that is capable of being used immediately, with no notice and for whatever task it may be needed. In a living cell, this ready-to-use, unconditional energy is the free energy of its cytosol. In

Chapter 9, we studied this energy at length and determined its main properties and limits. For some important aspects, the cytosol's role in the living cell is analogous to that of an electric battery in an electrical appliance: it stands waiting, ready to power any connected circuitry as soon as the latter is switched on.

To emphasize the importance of the cytosol as a source of energy, it should be added that the process of energy production from nutrients and, in particular, from the reserves of nutrients within the cell, cannot start without some initial expenditure of energy. (For more details on this point, refer to any textbook on biochemistry, e.g. [11, Ch. 14] or [42, Ch. 13]). The energy that the cell uses to start the energy production process is taken from the cytosol's free energy. In the end, the process returns more energy than the energy that is needed to initiate it. However, without the initial investment of energy from the cytosol, the cell could not exploit the reserves of nutrients available to it, no matter how large they are. In other words, it could not be alive.

However, the cytosol does not store unlimited amounts of energy. So, if the external demand of energy from the cell ceases, then the energy produced from the nutrients remains in the cytosol. When the free energy of the cytosol reaches a limit value, which may not coincide with the cytosol's maximum storable energy, the process of producing energy from nutrients must be stopped. Thus, no matter how much energy was supplied by the cell following its instantaneous response, the process ends up by leaving the cytosol recharged with energy, ready for further demands for instantaneous supplies.

It should be clear by now that it is the unconditional, instantaneous source of energy from the cytosol that makes life possible. That energy should, accordingly, be considered as the essence of life. The cell machinery, or the whole organism, for that matter, would simply be inert matter if the cytosol's energy were not available.

Much of the energy cost, $\Delta\Psi_{out}$, to build up a living system is due to the inert machinery that composes it. The unconditional energy that gives life to the system is the free energy of the cytosol of its cells. This energy is only a small fraction of $\Delta\Psi_{out}$. It can be determined from eqs (9.2.10) and (9.2.11) after the cytosol has been isolated from the organism and its composition has been determined.

How much unconditional energy does an animal need? It depends on the quickest event with which it must deal. Being herbivorous and protected by heavy armature, a tortoise can afford to ignore most of the fastest events that threaten other animals. Then, its unconditional energy may be limited to that needed to control metabolism of food and nutrients and to power its brain and nervous system. The amount of unconditional energy contained per unit weight of its muscular tissues would clearly be unfit for the lifestyle of a killer whale or, for that matter, to that of a mosquito.

The analysis in Chapter 9 enabled us to determine the maximum amount of free energy that the cytosol can store and to assess the effect that a change in the cytosol's composition has on its free energy. This should provide an interesting aid to interpret the rich body of knowledge on the molecular and atomic machinery of the living cell that biochemistry is providing at a tremendous pace. It also should be of significance in many fields of life sciences. The mass and composition of the cytosol of the biceps of a weightlifter, for instance, may suffice to predict his maximum lifting power, because his exercise is so fast that it must be done primarily at the expense of the unconditional free energy in the cytosol of the athlete's muscles. Or, the changes in the cell's cytosol composition with age may suggest how to better understand and control the cell's aging and, ultimately, the cell's lifespan.

11

The Onset of Turbulence

An important feature of viscous fluid flow concerns the kind of motion of the fluid particles. If the velocity gradient at the points of the fluid is small enough, no eddies or swirls are formed and the fluid particles move along parallel pathlines. All layers of fluid parallel to the pathlines glide over one another, such that the fluid does not mix. This kind of flow is called laminar.

This situation is bound to change, often abruptly, as the flow velocity and hence the velocity gradient at the points of the fluid increases. Eventually, the flow reaches a threshold velocity, above which the motion of the fluid becomes turbulent. Turbulent flow is characterised by irregular fluctuations, both in direction and intensity, of the velocities of fluid particles. The pathlines become irregular and entangled and the total angular momentum of the fluid increases markedly because of the formation of a host of small vortices. The fluid particles mix and energy dissipation grows due to the increase in the fluid's kinetic energy being turned into heat. In these conditions, to maintain a given flow rate, more energy is expended than in laminar flow conditions. The practical implications of this are evident; hence, the importance of understanding the origin of turbulence in order to achieve a better control of the transition from laminar to turbulent flow.

This chapter shows that the onset of turbulence is a consequence of the limit to the maximum amount of internal energy that a fluid—gas or liquid—can store isothermally. This result is yet another consequence of general relation (3.6.5).

To highlight the essence of the present arguments, the analysis is confined to linearly viscous fluids—also called ideal Newtonian fluids. Moreover, the fluid is assumed incompressible, consistently with what is usually done in classical fluid dynamics. Linearly viscous fluids are important because they represent quite well, under a wide range of conditions, the behaviour of many fluids of practical interest, notably water and air. Although, strictly speaking, all liquids and even more so, all gases are compressible, compressibility proves negligible when the velocity of the fluid does not exceed about 1/3 of the velocity of sound within the same fluid. This makes the assumption of incompressibility a valid approximation in most ordinary applications.

Of course, there are situations where compressibility cannot be neglected. There are also many fluids, such as paints and some kinds of polymer solutions, that are non-Newtonian and thus require a more general analysis than the one presented here. Such an analysis, however, is beyond the scope of this book.

In what follows, the reader is assumed to be familiar with the basic concepts of classical fluid dynamics, as can be learnt, for instance, from Chapters 40 and 41 of [18].

11.1 Viscosity and Viscous Forces

Consider an incompressible, isotropic, homogeneous, nonpolar fluid in steady motion through a certain region of space. Similar arguments can be pursued for more general classes of fluids. Isothermal conditions are assumed. The space in which the fluid flows is referred to as a rectangular Cartesian system of coordinates (x_1, x_2, x_3). These are the spatial coordinates that we adopt to describe the fluid motion. The fluid is acted upon by body forces per unit volume and distributed forces per unit area on the fluid's surface. Body couples and distributed couples per unit surface are absent. Because the flow is steady, the fluid's velocity, $\mathbf{v} \equiv [\mathrm{v}_1, \mathrm{v}_2, \mathrm{v}_3]$, at any point, $\boldsymbol{x} \equiv [x_1, x_2, x_3]$, of space does not depend on time. Under these conditions, the velocity field, $\mathbf{v} = \mathbf{v}(\boldsymbol{x}) = \mathbf{v}(x_1, x_2, x_3)$, of the fluid particles is determined by the following well-known equations of classical fluid dynamics. These equations are the consequence of the general laws of mass conservation, momentum conservation and angular momentum conservation, respectively. They apply to any point of the fluid in steady motion and, in the present form, they are valid in the considered reference system:

$$\operatorname{div}\mathbf{v} = \frac{\partial v_1}{\partial x_1} + \frac{\partial v_2}{\partial x_2} + \frac{\partial v_3}{\partial x_3} = 0\,, \tag{11.1.1}$$

$$\frac{\partial \sigma_{ji}}{\partial x_j} + f_i = \rho\, v_k\, \frac{\partial v_i}{\partial x_k}\,, \tag{11.1.2}$$

$$\sigma_{ji} = \sigma_{ij}\,. \tag{11.1.3}$$

Here, σ_{ij} are the components of the stress tensor $\boldsymbol{\sigma}$, while ρ is mass density and f_i are the components of the body force $\mathbf{f}$ per unit volume of fluid. For a derivation of these equations, the reader is referred to [18] or any other standard textbook on continuum mechanics or fluid dynamics. Equations (11.1.1) - (11.1.3) suffice to determine the velocity field $\mathbf{v}(\boldsymbol{x})$, once appropriate boundary conditions are specified.

To solve the above equations, the relation between $\boldsymbol{\sigma}$ and $\mathbf{v}$ must be specified. This relation describes a constitutive property of the fluid and, therefore, it must be determined from experiment. We may observe that, generally speaking, fluids are characterised by their inability to sustain shear stresses when in equilibrium conditions. Thus, if velocity vanishes at every point of the fluid, the stress tensor must be spherical. That is, $\boldsymbol{\sigma} = -p\,\boldsymbol{1}$ or, in components, $\sigma_{ij} = -p\,\delta_{ij}$. Here, p denotes pressure (assumed positive when representing a compression), $\boldsymbol{1}$ is the unit tensor, and δ_{ij} is the *Kronecker delta.* Outside equilibrium, however, the fluid motion may generate shearing forces on the surface of a fluid element and the stress at that element must balance their action. We want to determine the shearing forces that are created by motion and relate them to the stress at each point of the fluid.

Shearing forces are produced by friction when two adjacent parts of the fluid are in relative motion over their contacting surface. In fluid dynamics, friction forces are known as viscous forces. As with any force resulting from friction, viscous forces are tangential to the surface upon which sliding occurs and they are dependent on the sliding velocity. Viscous forces are measured experimentally by means of appropriate instruments (*viscometers*), all of which produce a shearing flow of some sort in the fluid. The simplest experiment from the conceptual standpoint—although not necessarily the easiest experiment to perform—consists of setting an indefinite layer of the fluid to be tested into a steady homogeneous simple shearing motion. This can be achieved by confining the fluid between two parallel plates maintained in relative motion with respect to each other. The resulting flow is referred to as a *plane Couette flow*. A modern instrument to produce this motion is presented in [67].

Figure 11.1.1(a) provides a sketch of this experiment. The lower plate is stationary, while the upper plate is driven at a constant velocity $\bar{\mathbf{v}}$ in the direction of the x_1-axis by force $\boldsymbol{F}$ in the same direction. Body forces can be neglected because the distance d between the plates is comparatively small. Pressure can have any constant value throughout the fluid, because constant pressure does not affect the motion of an incompressible fluid. The x_3-axis is normal to the plane of the figure.

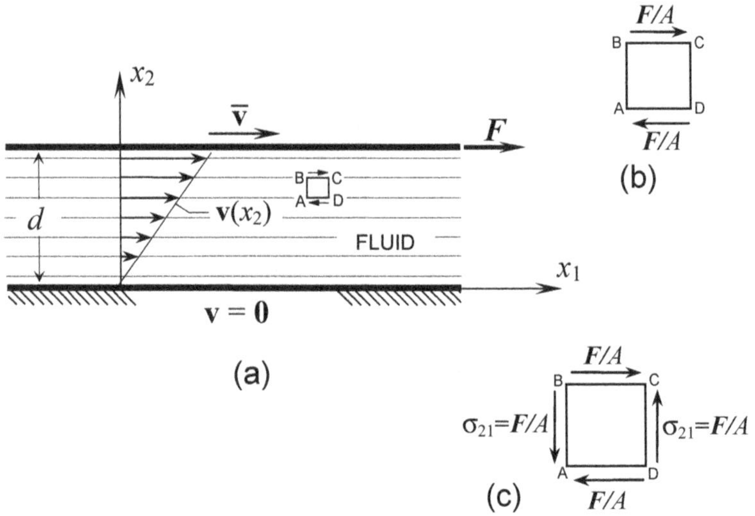

Fig. 11.1.1. (a) Homogeneous simple shearing flow between two parallel plates (*plane Couette flow*). (b) Viscous forces at the surface of a cubic element of fluid. (c) Shearing stress σ_{21} needed to balance the element's angular momentum.

For a large class of fluids, which are referred to as *simple fluids* and include linearly viscous fluids as a particular case, the steady flow that is produced by this experiment is a simple rectilinear shearing. In this flow, the fluid particles move along parallel straight pathlines, directed along the x_1-axis. The speed in each pathline has a constant value that depends on the distance from the lower plate. In the present case, if perfect adherence of the fluid to the plates is assumed, the velocity of the fluid particles varies linearly, from zero at the lower plate to $\overline{\mathbf{v}}$ at the upper plate [see Fig. 11.1.1(a)]. The velocity field of the fluid is, therefore,

$$\mathrm{v}_1 = \frac{\overline{\mathrm{v}}}{d} x_2, \quad \mathrm{v}_2 = \mathrm{v}_3 = 0, \tag{11.1.4}$$

where $\overline{\mathrm{v}}$ denotes the modulus of $\overline{\mathbf{v}}$. This result applies to simple fluids, no matter the particular fluid under test [70]. In this test, the quantity $\overline{\mathrm{v}}$ suffices to determine the velocity at every point of the fluid. The attainment of such velocity field has been verified experimentally (see, e.g., [67]).

The tangential force $\boldsymbol{F}$ that is needed to keep the velocity of the upper plate constant depends on the viscosity of the fluid. Its determination is a major task of the experiment. In linearly viscous fluids, $\boldsymbol{F}$ is found proportional to ratio $\overline{\mathrm{v}}/d$ and, of course, to the area, A, of the sliding plate. The ratio $\overline{\mathrm{v}}/d$ is equal to $\partial \mathrm{v}_1/\partial x_2$, as immediately follows from eq. $(11.1.4)_1$. In steady flow conditions, therefore, the force per unit area that acts in the x_1-direction upon fluid plane $x_2 = d$ can be expressed as:

$$\frac{F}{A} = \eta \frac{\overline{\mathrm{v}}}{d} = \eta \frac{\partial \mathrm{v}_1}{\partial x_2}. \tag{11.1.5}$$

Constant η, introduced in this equation, is the *viscosity coefficient* of the material.

In steady flow conditions, the fluid above any plane normal to the x_2-axis applies to the underlying fluid the same force per unit area as that given by eq. (11.1.5). This can easily be proved by considering a cubic element of material, in motion with the fluid, with sides parallel to the coordinate planes (Fig. 11.1.1). The momentum of such element is constant during the motion, while body forces are zero and pressure is constant or vanishes. From the momentum balance equation it follows that the resultant force acting on the element must be zero. In particular, the x_1-component of the forces acting on the sides normal to the x_2-axis must be equal and opposite.

Let us refer to the above volume element, represented isolated in Fig. 11.1.1(b). The surface forces per unit area acting on the faces of the element are the stress vectors relevant to these faces. Let $\boldsymbol{t}^{(\mathbf{n})}$ be the stress vector relevant to the face of unit normal $\mathbf{n}$. The stress tensor at the point of the fluid where the element is located is related to $\boldsymbol{t}^{(\mathbf{n})}$ by the well-known Cauchy's theorem on stress:

$$\boldsymbol{t}^{(\mathbf{n})} = \boldsymbol{\sigma}\mathbf{n} \qquad \text{or} \qquad t_{\mathrm{r}}^{(\mathbf{n})} = \sigma_{\mathrm{rs}}\, \mathrm{n}_{\mathrm{s}}\,. \tag{11.1.6}$$

In the present case, the stress vectors acting on the faces of the element are known; therefore, we can use eq. (11.1.6) to determine the components of $\boldsymbol{\sigma}$. In particular, if $\hat{\mathbf{j}}$=[0,1,0] denotes the unit vector of the x_2-axis, the component in the x_1 direction of the stress vector on the face of the element normal to the x_2-axis is given by $t_1^{(\hat{\mathbf{j}})} = \eta(\partial \mathrm{v}_1/\partial x_2)$, according to eq. (11.1.5). Therefore, by setting $\mathbf{n} \equiv \hat{\mathbf{j}}$ and $\mathrm{r} = 1$ in eq. $(11.1.6)_2$ we obtain

$$\sigma_{12} = \eta \frac{\partial \mathrm{v}_1}{\partial x_2}\,. \tag{11.1.7}$$

However, as is apparent from eq. (11.1.4), the velocity of the fluid has no tangential components along the faces of the element that are normal to the x_1-axis. This means that no shearing motion occurs along those faces and, therefore, no viscous force acts on them. Thus, if $\hat{\mathbf{i}}$=[1,0,0] is the unit vector of the x_1-axis, the component $t_2^{(\hat{\mathbf{i}})}$ of the stress vector on that face is zero. Therefore, by setting $\mathbf{n} \equiv \hat{\mathbf{i}}$ and $\mathrm{r} = 2$ in eq. $(11.1.6)_2$, we obtain:

$$\sigma_{21} = 0\,. \tag{11.1.8}$$

Results (11.1.7) and (11.1.8) apply irrespective of the value of the pressure that may act on the fluid. They show that viscous forces alone are insufficient to meet eq. (11.1.3) and, thus, the angular momentum balance law in the simple shearing flow considered in the experiment. Other surface forces must come into play to satisfy that law.

11.2 Viscous Stress

To find the missing forces, we begin by observing that, as with any material that allows the propagation of elastic waves, all fluids can exhibit elastic response to stress. In fluids, as in solids, the external actions modify the relative positions of the atoms of the material. The interatomic force that acts between two adjacent atoms changes as a consequence. If the change in the interatomic distance is sufficiently small, the process is elastic. In this case the normal distances are recovered as the external forces are removed. However, if the interatomic distance exceeds a certain limit, the interaction between two adjacent atoms is lost. In solids, this produces rupture or plastic yielding. In fluids, however, the outcome is different. In this case, the interatomic forces are much weaker than they are in solids, permitting greater mobility of atoms and molecules. Thus, any interatomic interaction that breaks is almost immediately replaced by a similar new interaction with nearby atoms. In other words, a fluid repairs itself as it breaks. For this reason, no macroscopic rupture is exhibited by a fluid, unless a portion of it is actually removed from the rest of the fluid.

For sufficiently small stress, the elastic strain of any isotropic material—and of an isotropic fluid, in particular—is related to stress by the same relations (6.1.3) that apply to isotropic linear elastic solids. A way to determine the elastic constants of a fluid is to measure the speed of the different kinds of elastic waves that travel through the fluid. The following well-known equations relate wave speed to the elastic constants of the material. They apply to any medium through which elastic waves can propagate. For every elastic medium, if c_{L} denotes the speed of propagation of pressure waves, we have that

$$c_{\mathrm{L}} = \sqrt{\frac{1}{\rho}(K + \frac{3}{4}G)} = \sqrt{\frac{E(1-\nu)}{\rho(1+\nu)(1-2\nu)}} = \sqrt{\frac{2G(1-\nu)}{\rho(1-2\nu)}}, \qquad (11.2.1)$$

where K is the bulk modulus, E is the Young's modulus, G is the shear modulus and ν is the Poisson's ratio of the material.

Similarly, the speed c_{T} of transverse waves is related to the shear modulus G of the material by the following equation:

$$c_{\mathrm{T}} = \sqrt{\frac{G}{\rho}}. \qquad (11.2.2)$$

Constants K, E, G and ν above are related to each other by eqs $(6.1.8)_2$ and $(6.2.21)_2$. Thus, the ratio between c_{L} and c_{T} can be expressed as

$$\frac{c_{\mathrm{L}}}{c_{\mathrm{T}}} = \sqrt{\frac{2(1-\nu)}{(1-2\nu)}}. \qquad (11.2.3)$$

These formulae are standard and can be found in many textbooks on continuum mechanics. They are usually applied to solids, on the grounds that in fluids $G = 0$, implying that no shear wave can travel through a fluid. However, such a conclusion is, in fact, an assumption. It is acceptable in many cases but, strictly, it is not supported by experiment. A growing body of experimental evidence has been accumulated in the last few decades, which shows that not only can shear waves propagate in liquids [30] but also, albeit with strong attenuation, in gases [14], [40]. For this reason, formulae (11.2.1)-(11.2.3) are to be considered as applicable, in particular, to fluids.

It should be observed, however, that in fluids—and above all, in gases—the quantities K, E, G, v and ρ depend strongly on pressure and on temperature. This is at variance with what happens in solids, where the above quantities are to a large extent independent of pressure. It should also be remarked that the values of the elastic constants K, E, G, and v that appear in formulae (11.2.1)-(11.2.3) refer to adiabatic conditions, as the frequency of elastic waves is so high that heat conduction has no time to occur as the wave propagates. For this reason, the values of the above constants, as obtained from measurements of c_L and c_T, may differ from the values of the same constants under isothermal conditions.

These details can be ignored as far as the main results of this chapter are concerned. However, the existence of wave propagation in a fluid is crucial to indicating that, in general, the deformation of the fluid may include an elastic component, even when the fluid is not in static equilibrium. No matter how small, this component is essential to the analysis that follows. It enables us to prove a formula, to be presented in Sect. 11.5, which determines the transition to turbulence of the laminar flow of any incompressible linearly viscous fluid. Fortunately, that formula turns out to be independent of the elastic moduli of the fluid, which frees us from needing to know their actual values.

Because the speed of propagation of elastic waves is usually quite high, the values of K, E and G of most materials are comparatively large. This implies that small elastic strains are produced, even for high values of stress. Thus, it is obvious that the total relative displacement of any two adjacent points of a flowing viscous fluid will generally be many orders of magnitude greater than the part of that displacement which is due to any elastic deformation of the fluid in the same flow. For this reason, when studying subsonic motion of viscous fluids, elastic deformation is ignored by the current approach to fluid dynamics. For a vast majority of applications, this is quite sensible. However, small though it may be, elastic deformation can be vital for the flow of a viscous fluid to be compatible with the laws of motion. As discussed below, this applies in particular to the steady simple shearing flow considered in the previous section.

But how is it that a viscous fluid can maintain a constant elastic strain while flowing steadily through space, i.e., in the absence of a rapidly time-varying velocity field such as that created during wave propagation? To answer this question, let us consider the constitutive model made by a spring and a dashpot joined in series, known as Maxwell's model (Fig. 11.2.1).

When a constant force $\boldsymbol{F}$ is applied to the model, the dashpot elongates at the constant rate $\dot{x}_b = F/b$. Here b and x_b are the viscosity coefficient and the length of the dashpot, respectively. The same force also acts on the spring and elongates it by the amount $\Delta x_k = F/k$, where k is the spring constant. Although the dashpot keeps elongating under the action of $\boldsymbol{F}$, the length of the deformed spring remains constant in time, as long as $\boldsymbol{F}$ is applied to the model. The spring elongation may be ignored if k is appropriately large. In any case, the spring elongation is bound to become negligible with respect to the elongation of the dashpot, because the length of the latter grows steadily at the rate $\dot{x}_b$.

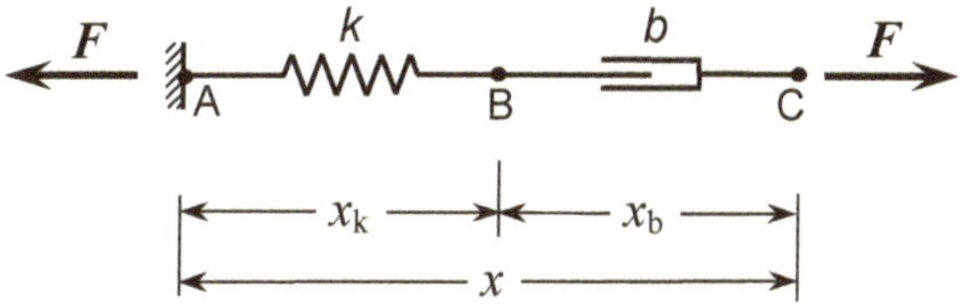

Fig. 11.2.1. Maxwell's constitutive model of viscous fluid.

In steady motion, relation $\dot{x} = \dot{x}_b$ applies. In this condition, the spring does not offer any contribution to the time rate at which the length of the model increases. The spring, however, plays a crucial role in keeping the motion steady. If the spring is removed or breaks, the dashpot accelerates under the applied force and a steady motion is not possible.

A similar situation occurs in the simple shearing flow considered in the previous section. As observed experimentally, that flow is steady. The experiment shows, moreover, that a viscous force F/A is applied per unit area to the faces of a fluid element parallel to the flow, as indicated in Fig. 11.1.1(b). An analogous shearing force must be acting on the element faces that are normal to the flow direction [cf. Fig. 11.1.1(c)], otherwise the element could not meet the angular momentum balance equation and the considered motion would be impossible.

However, viscosity cannot generate that force because there is no shearing flow over an element face that is normal to the flow direction. Where does the shearing force applied to such faces come from?

The answer is obtained by observing that, as with any other force applied to a deformable material, the viscous forces applied to faces AD and BC of the cubic element represented in Fig. 11.1.1 deform the element. The shearing stress σ_{12} that these forces produce is given by eq. (11.1.7). If we do not ignore the fact that the fluid can deform elastically, the elastic strain due to σ_{12} is given by:

$$\varepsilon_{12}^{e} = \frac{\sigma_{12}}{2G} = \frac{1}{2G}\eta\frac{\partial v_1}{\partial x_2}, \qquad (11.2.4)$$

which follows from the general linear elastic stress-strain relation (6.1.3) and from eq. $(6.1.8)_2$. In eq. (11.2.4) and in what follows, the quantity $\boldsymbol{\varepsilon}^{e}$ denotes the elastic part of the total strain tensor. G is the elastic shear modulus of the fluid, which can be determined from eqs (11.2.1) or (11.2.2). Because the strain tensor is symmetric, we have that $\varepsilon_{21}^{e} = \varepsilon_{12}^{e}$. Thus, from the linear elastic stress-strain relation recalled above, we obtain

$$\sigma_{21} = 2\,G\,\varepsilon_{21}^{e} = 2\,G\,\varepsilon_{12}^{e} = \eta\frac{\partial v_1}{\partial x_2} = \sigma_{12}\,. \qquad (11.2.5)$$

This replaces eq. (11.1.8) and eliminates the inconsistency with the angular momentum balance. Analogously to what happens to the spring of Maxwell's model, in the considered flow the angular momentum balance of the viscous forces is only possible because of the elastic deformation of the element. This deformation is bound to be very small because η is small and G is large. Yet, it is this elastic deformation that makes the considered laminar motion physically admissible.

The above analysis refers to homogeneous simple shearing motion. However, any continuous motion can be approximated, in the neighbourhood of any point, by a homogeneous motion. Moreover, any homogeneous strain can be decomposed into a uniform dilatation and three simple shears in mutually orthogonal planes. (The reader may refer to Sects. 46 and 142 of [69] for more details on these classical issues.) By applying eqs (11.1.7) and (11.2.5) to each simple shearing motion that composes a general homogeneous flow, we conclude that in any flow of a linearly viscous fluid the following equation must hold true:

$$\sigma_{ij} = \eta\left(\frac{\partial v_i}{\partial x_j} + \frac{\partial v_j}{\partial x_i}\right) \qquad \textit{for} \;\; i \neq j\,. \qquad (11.2.6)$$

This equation generalizes eq. (11.1.7) and can be taken as the constitutive equation for the shearing components of the viscous stress of a linearly viscous fluid. From eq. (11.2.6) it follows that:

$$\varepsilon^{e}_{ij} = \frac{\sigma_{ij}}{2G} = \frac{1}{2G}\eta\left(\frac{\partial v_i}{\partial x_j} + \frac{\partial v_j}{\partial x_i}\right) \qquad \textit{for} \;\; i \neq j, \qquad (11.2.7)$$

because eq. (11.2.4) applies to each shearing stress component.

Because c_L and c_T must be finite and real, we must have that $v < 0.5$, as is evident from eqs (11.2.1) and (11.2.3). This means that the elastic deformation of the fluid cannot be isochoric (which would require $v = 0.5$). However, because the elastic deformation is several orders of magnitude less than the overall deformation that is produced by the flow, any volume change due to elastic deformation can be neglected when studying the gross motion of the fluid. Thus, no contradiction arises between assuming that the value of v introduced in eqs (11.2.1) - (11.2.3) is less than 0.5 and treating the fluid as incompressible when determining its flow.

In the classical approach to linearly viscous fluids, eq. (11.2.6) is also applied, though with a different coefficient, to the diagonal components of the stress tensor, i.e., for i = j. If the fluid is incompressible, the derivatives $\partial v_i/\partial x_i$ for i = (1, 2, 3) and, thus, the diagonal components of $\boldsymbol{\sigma}$, are restricted by condition (11.1.1). The same components must also meet restriction $p = -\,\mathrm{tr}(\boldsymbol{\sigma})/3$, which can be proven to hold true in any linearly viscous fluid, compressible or not. In the incompressible case, therefore, only one of the three diagonal components of $\boldsymbol{\sigma}$ is independent of the others. However, p can be assigned arbitrarily, because a constant pressure does not affect the motion of incompressible fluids. It is concluded that the stress constitutive equation of an incompressible linearly viscous fluid can always be expressed in the form:

$$\sigma_{ij} = -p\,\delta_{ij} + \eta\left(\frac{\partial v_i}{\partial x_j} + \frac{\partial v_j}{\partial x_i}\right), \qquad (11.2.8)$$

or

$$\boldsymbol{\sigma} = -p\,\boldsymbol{I} + 2\,\eta\,\boldsymbol{D}\,. \qquad (11.2.9)$$

Here $\boldsymbol{D}$ denotes the rate-of-deformation tensor. Its components are defined as:

$$D_{ij} = \frac{1}{2}\left(\frac{\partial v_i}{\partial x_j} + \frac{\partial v_j}{\partial x_i}\right). \tag{11.2.10}$$

Equations (11.2.8) and (11.2.9) are the classical constitutive equations of an incompressible linearly viscous fluid. They are fully compatible with eq. (11.2.6) and provide a relationship between stress and the space derivatives of the velocity of fluid particles. Once these equations are introduced in field equations (11.1.1)-(11.1.3), the velocity field can be determined for given boundary conditions. In the solution of this problem, the elastic deformation (11.2.7) of the fluid plays no role. Accordingly, the interrelation of viscous and elastic response in a viscous fluid flow is ignored in classical fluid dynamics. The consequence is, however, that the theory is incapable of predicting the onset of turbulence.

11.3 Limit to the Laminar Flow

Every force that deforms a material elastically makes the material store internal energy—elastic energy, in this case. As discussed in the previous section, the viscous forces acting on the fluid during laminar flow deform the fluid elastically. The elastic shearing strain that they produce is given by eq. (11.2.7). If that strain is sufficiently small, the associated elastic energy can be calculated using the same expression that applies to any linear elastic material, solid or not. Under isothermal conditions, this energy coincides with the free energy of the material, to within an inessential additive constant. In the notation of Chapter 6, the specific value per unit volume of this energy is ψ. Introducing eq. (6.1.4) into eq. (6.3.4), we can express ψ in the general form:

$$\psi = \frac{1+\nu}{2E}\sigma_{ij}\sigma_{ij} - \frac{\nu}{2E}\sigma_{rr}\sigma_{ss}. \tag{11.3.1}$$

Here and in what follows, we are assuming that $\psi = 0$ when the fluid is at rest.

For linearly viscous fluids we have that $p = -\mathrm{tr}(\boldsymbol{\sigma})/3 = -\sigma_{kk}/3$, as recalled in the previous section. Therefore, for the same fluids, eq. (11.3.1) can be written as

$$\psi = \frac{1+\nu}{2E}\sigma_{ij}\sigma_{ij} - \frac{9\nu}{2E}p^2 . \tag{11.3.2}$$

This equation applies to all linearly viscous fluids, incompressible or not. Every real fluid is elastically compressible. Strictly speaking, elastic incompressibility would require $\nu = 0.5$. This value of ν must be excluded, though, because it would produce an infinite speed of propagation of longitudinal waves in the fluid, as evident from eq. (11.2.1). As previously observed, however, in incompressible fluids any volume change produced by elastic deformation is so small that it can be ignored when solving the motion equations of classical fluid dynamics. For this reason the fluid can be treated as incompressible, even though it is, in fact, elastically compressible.

From eqs (11.3.2), (11.2.9), $(6.1.8)_2$ and $(6.2.21)_2$ we obtain, after simple algebra:

$$\psi = \frac{1}{2K}p^2 - \frac{\eta}{G}pD_{ii} + \frac{\eta^2}{G}D_{ij}D_{ij} . \tag{11.3.3}$$

If the fluid is incompressible (in the usual sense of fluid dynamics, i.e., ignoring the elastic deformation), then from eqs (11.1.1) and (11.2.10) we can write that $D_{ii} = 0$. In this case, eq. (11.3.3) simplifies to:

$$\psi = \frac{1}{2K}p^2 + \frac{\eta^2}{G}D_{ij}\,D_{ij} . \tag{11.3.4}$$

For small values of p, an approximate expression for this equation is the following:

$$\psi = \frac{\eta^2}{G}D_{ij}\,D_{ij} . \tag{11.3.5}$$

This result applies rigorously for $p = 0$, even if the fluid is compressible, cf. eq. (11.3.3).

The maximum amount of free energy that a system can store in isothermal conditions cannot exceed the limit imposed by inequality (3.6.7). This applies, in particular, to the viscous fluids under consideration. When expressed per unit volume, that inequality becomes

$$\psi \le \psi_{max} , \tag{11.3.6}$$

where ψ_{max} is the maximum amount of free energy that, per unit volume, the fluid can store at the considered temperature. ψ_{max} is a property that depends on the fluid and it has to be determined experimentally. Introducing eq. (11.3.4) into inequality (11.3.6), we obtain:

$$\frac{1}{2K}p^2 + \frac{\eta^2}{G}D_{ij}\,D_{ij} \leq \psi_{max}\,. \tag{11.3.7}$$

This relation introduces a limitation to the values that tensor ***D*** can assume in an incompressible linearly viscous fluid. The same relation also restricts the velocity field attainable by the fluid, because ***D*** is related to the velocity gradients by eq. (11.2.10).

An alternative expression of result (11.3.7) can be found by introducing eq. (11.3.2) into inequality (11.3.6). Thus, we obtain:

$$\frac{1+\nu}{2E}\sigma_{ij}\sigma_{ij} - \frac{9\nu}{2E}p^2 \leq \psi_{max}\,. \tag{11.3.8}$$

This shows that the limitation that follows from inequality (11.3.7) is due to the fact that the stress of an incompressible linearly viscous fluid, as the stress of any other material, cannot exceed the values that correspond to the maximum amount of free energy that the material can store.

The restrictions to the fluid motion imposed by inequality (11.3.7) or (11.3.8) are quite general. They apply to any incompressible linearly viscous fluid in any flow. In a turbulent flow, however, the velocity field is so irregular and varies so quickly in time that it is practically impossible to determine the velocity gradient at the points of the fluid with the accuracy that is required for the above inequalities to be of practical use.

The situation is different when the flow is laminar. In this case the velocity field can be predicted theoretically and it is measured quite easily. Inequality (11.3.7) or (11.3.8) can then be used to predict the maximum speed of the flow beyond which laminar flow ceases to be admissible. If that limit is exceeded, the viscous forces applied to the faces of a fluid element cannot be balanced by the elastic stress. As explained in Section 11.2, this means that eq. (11.1.3) and, thus, the angular momentum balance law cannot be met. Unbalanced angular momentum generates vortices and the flow turns to turbulent.

11.4 Free Energy Limit in Fluids

Plane Couette flow is particularly useful for determining the value of ψ_{max} of a fluid. This flow is homogeneous, which makes the transition to turbulence particularly evident because it occurs simultaneously at all points of the fluid. This section shows how the value of ψ_{max} can actually be determined from a plane Couette flow test.

The velocity field in a plane Couette flow is given by eq. (11.1.4). From eq. (11.2.10), we calculate that for such flow

$$\boldsymbol{D} = \begin{vmatrix} 0 & \frac{1}{2}\frac{\overline{v}}{d} & 0 \\ \frac{1}{2}\frac{\overline{v}}{d} & 0 & 0 \\ 0 & 0 & 0 \end{vmatrix} . \tag{11.4.1}$$

Thus, from eq. (11.2.9), we have that

$$\boldsymbol{\sigma} = \begin{vmatrix} -p & \eta\frac{\overline{v}}{d} & 0 \\ \eta\frac{\overline{v}}{d} & -p & 0 \\ 0 & 0 & -p \end{vmatrix} . \tag{11.4.2}$$

For simplicity, we assume that the test is performed at $p = 0$. This is not unrealistic when dealing with liquids. Therefore, from relations (11.4.2) and (11.3.8), we obtain, upon setting $p = 0$:

$$\frac{\eta^2}{2G}\frac{\overline{v}^2}{d^2} \le \psi'_{max} . \tag{11.4.3}$$

Here, ψ'_{max} indicates the value of ψ_{max} for $p = 0$ and use has been made of eq. $(6.1.8)_2$. The main variable entering relation (11.4.3) is the velocity $\overline{v}$ of the upper plate (Fig. 11.1.1).

The test under consideration is usually done to determine the value of the Reynolds number

$$R_e = \frac{\rho\,\overline{v}\,d}{\eta} \tag{11.4.4}$$

at which the laminar to turbulent transition occurs. Let R_e^* be the transition value of R_e. What is actually measured in this experiment is the value, $\overline{v}^*$, of the velocity $\overline{v}$ at that transition. It is related to R_e^* by the relation:

$$\overline{v}^* = \frac{\eta\, R_e^*}{\rho\, d}, \tag{11.4.5}$$

which follows from eq. (11.4.4). This is the limit velocity of the upper plate of the plane Couette flow apparatus, beyond which the flow starts becoming turbulent. Therefore, for $\overline{v} = \overline{v}^*$, the equality sign applies in relation (11.4.3) because the elastic energy of the fluid reaches its maximum value. Thus, if $\overline{v}^*$ is known, ψ'_{max} can be calculated as

$$\psi'_{max} = \frac{\eta^2}{2G} \frac{\overline{v}^{*2}}{d^2}. \tag{11.4.6}$$

Also, by inserting $\overline{v}^*$ for $\overline{v}$ into eq. (11.4.2), we find the maximum value that shear component σ_{12} can attain in the considered fluid. This value will be referred to as τ_{max} and is given by:

$$\tau_{max} = \eta \frac{\overline{v}^*}{d}. \tag{11.4.7}$$

In terms of τ_{max}, eq. (11.4.6) assumes the simpler form:

$$\psi'_{max} = \frac{1}{2G} \tau_{max}^2, \tag{11.4.8}$$

which follows from eqs (11.4.6) and (11.4.7).

As an example of the application of the above formulae, let us take the value of R_e^* from the literature and calculate the value of τ_{max} for water at ambient temperature (20 °C). We refer, in particular, to the experimental results reported in [67], where the laminar to turbulent transition occurs at $R_e^* = 360$. The apparatus used in these experiments and the accuracy of experimental method adopted suggest that this value of R_e^* is likely to be quite correct. However, as observed in the same paper [67], the Reynolds number relevant to the laminar to turbulent transition in a plane Couette flow is variously reported in the literature to fall anywhere between 280 and 750.

The viscosity coefficient of water at 20 °C is $\eta = 10^{-3}$ N sec/m^2, while water density is $\rho = 10^3$ Kg/m^3. The results reported in [67] refer to a distance $d = 5$ mm between the moving plane and the plane where the fluid velocity is zero (Fig. 11.1.1). By introducing these values into eq. (11.4.5) and by taking $R_e^* = 360$, we calculate that:

$$\overline{v}^* = 72 \cdot 10^{-3} \frac{m}{sec}. \tag{11.4.9}$$

If we insert this value of $\overline{v}^*$ into eq. (11.4.7), we find that for water at 20 °C:

$$\tau_{max} = 14.4 \cdot 10^{-3} \text{ Pa}. \tag{11.4.10}$$

This value is a characteristic of water at the considered temperature. It is, therefore, independent of the particular flow that is considered. Such a tiny value of τ_{max} should come as no surprise because it represents the elastic limit in shear of water. Similar tests would show that τ_{max} is generally quite small in all fluids. It is no wonder, then, that the elastic part of the response of a fluid is usually ignored. Yet, small though it is, τ_{max} controls the transition from laminar to turbulent flow, as will be shown in the next section.

11.5 Transition to Turbulence for General Laminar Flows

To extend the results of the previous section to the case $p \neq 0$, we observe that for incompressible fluids the deviatoric (or traceless) part of $\boldsymbol{\sigma}$ is:

$$\boldsymbol{\sigma}' = 2\eta\, \boldsymbol{D} \quad \text{or} \quad \sigma'_{ij} = 2\eta D_{ij} = \eta\left(\frac{\partial v_i}{\partial x_j} + \frac{\partial v_j}{\partial x_i}\right). \tag{11.5.1}$$

This can be verified immediately from eq. (11.2.9) because the incompressibility condition (11.1.1) implies that tr $\boldsymbol{D} = 0$. Consequently, the spherical part of $\boldsymbol{\sigma}$ is:

$$\overline{\boldsymbol{\sigma}} = -p\,\boldsymbol{1} \quad \text{or} \quad \overline{\sigma}_{ij} = -p\,\delta_{ij}. \tag{11.5.2}$$

Because the deviatoric and spherical parts of any symmetric tensor are independent of each other, it is not difficult to verify using eqs (11.3.4) and (11.2.9) that ψ can be decomposed into two independent parts according to the relation:

$$\psi = \overline{\psi} + \psi' . \tag{11.5.3}$$

Here, $\overline{\psi}$ is the part of ψ that depends on $\overline{\boldsymbol{\sigma}}$, while ψ' is the part of ψ that depends on $\boldsymbol{\sigma}'$. More precisely:

$$\overline{\psi} = \frac{1}{2K} p^2 \tag{11.5.4}$$

and

$$\psi' = \frac{1}{4G} \sigma'_{ij} \sigma'_{ij} . \tag{11.5.5}$$

The independence of these two parts of ψ means that the value of ψ_{max} when $p \neq 0$ is simply obtained by adding part (11.5.4) to the maximum value that the deviatoric part (11.5.5) can attain. This value is given by eq. (11.4.8). Thus, for an incompressible linearly viscous fluid at $p \neq 0$, the value of ψ_{max} is given by:

$$\psi_{max} = \overline{\psi} + \psi'_{max} = \frac{1}{2K} p^2 + \frac{1}{2G} \tau_{max}^2 . \tag{11.5.6}$$

By introducing eq. $(11.5.6)_2$ into inequality (11.3.7) we obtain, after a few passages:

$$2 \eta^2 D_{ij} D_{ij} \leq \tau_{max}^2 . \tag{11.5.7}$$

This expresses, quite generally, the condition that must be met by the laminar flow of an incompressible linearly viscous fluid to comply with condition (11.1.3).

By means of eq. (11.5.1), inequality (11.5.7) can also be expressed as

$$\frac{1}{2} \sigma'_{ij} \sigma'_{ij} \leq \tau_{max}^2 \tag{11.5.8}$$

or

$$2\eta^2 \left(\frac{\partial v_i}{\partial x_j} + \frac{\partial v_j}{\partial x_i} \right)^2 \leq \tau_{max}^2 . \tag{11.5.9}$$

Inequalities (11.5.7), (11.5.8) and (11.5.9) are equivalent to each other. A velocity field that does not meet them cannot meet the angular momentum balance law and, therefore, is not physically possible (cf. Section 11.3).

In particular, if the velocity field of a laminar flow is given as a function of a certain characteristic velocity, then by applying any one of these inequalities we can determine the maximum value of the characteristic velocity beyond which the considered laminar flow cannot meet the angular momentum balance law. The unbalanced angular momentum produces vortices in the fluid and the motion becomes turbulent. Thus, the above inequalities govern the transition from laminar to turbulent flow.

None of the said inequalities involve the elastic constants of the fluid. This means that to determine the transition from laminar to turbulent flow the actual values of these constants can be ignored. What matters is that the fluid possesses a limit to the maximum elastic shear stress that it can sustain or, equivalently, to the maximum elastic energy that it can store.

11.6 Turbulence Threshold in Taylor-Couette Flow

Taylor-Couette flow refers to the steady motion that is established in a fluid confined between two coaxial circular cylinders, when the cylinders rotate at different angular velocities about their common axis z (Fig. 11.6.1). Let r_1 and ω_1 be the radius and the angular velocity of the inner cylinder. The analogous quantities relevant to the outer cylinder are r_2 and ω_2. In steady flow conditions, the solution of the field equations (11.1.1)-(11.1.3) shows that the fluid particles move in laminar motion along concentric circles on planes normal to the z-axis. The angular velocity ω of the particles in each circle is given by

$$\omega = \omega(r) = -\frac{A}{2r^2} + B, \tag{11.6.1}$$

where r indicates the circle radius, while A and B are defined by

$$A = \frac{2r_1^2 r_2^2}{r_2^2 - r_1^2}(\omega_2 - \omega_1) \quad \text{and} \quad B = \frac{r_2^2\omega_2 - r_1^2\omega_1}{r_2^2 - r_1^2}, \tag{11.6.2}$$

cf., e.g., [68], [18] or [7].

Denoting by **v** the velocity vector of any fluid particle and by $\mathrm{v} = \mathrm{v}(r)$ the magnitude of **v**, we have that

$$\omega = \omega(r) = \frac{\mathrm{v}(r)}{r} . \tag{11.6.3}$$

From this and from eq. (11.6.1), we obtain

$$\mathrm{v} = \mathrm{v}(r) = -\frac{\mathrm{A}}{2\,r} + \mathrm{B}\,r . \tag{11.6.4}$$

By taking the axes x and y as in Fig. 11.6.1, the components of **v** are given by:

$$\mathrm{v_x} = -r\,\omega \sin \omega t = -\omega y, \quad \mathrm{v_y} = r\,\omega \cos \omega t = \omega x \quad \text{and} \quad \mathrm{v_z} = 0. \tag{11.6.5}$$

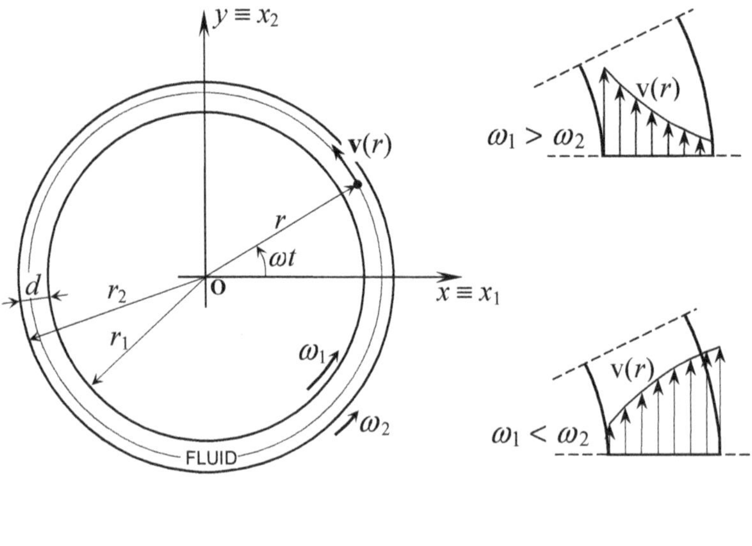

Fig. 11.6.1. Laminar flow of a fluid between two coaxial cylinders in relative rotation about their common axis (*Taylor-Couette flow*). Insets: fluid velocity diagrams in steady conditions.

The symmetry of the flow implies that all fluid particles that belong to the same circle of radius r possess the same specific free energy and, thus, the same value of ψ'. Therefore, in order to calculate the values of ψ' in the fluid, we can refer to the points ($x = r$, $y = 0$) on the x-axis. We identify x_1, x_2 and x_3 with x, y and z, respectively, so that for $y = 0$ eq. (11.6.1) can be written as $\omega = \omega(x) = -\,\mathrm{A}/(2\,x^2) + \mathrm{B}$. Thus, for $y = 0$ we have that $\mathrm{d}\omega/\mathrm{d}x_2 = \mathrm{d}\omega/\mathrm{d}y = 0$ and $x_1\mathrm{d}\omega/\mathrm{d}x_1 = r\,\mathrm{d}\omega/\mathrm{d}r$, because $x_1 \equiv x \equiv r$ if $y = 0$. The components of $\boldsymbol{\sigma}'$ at a point ($x = r$, $y = 0$) can therefore be expressed in the form

$$\boldsymbol{\sigma}'\big|_{y=0} = \boldsymbol{\sigma}'(r)\big|_{y=0} = \begin{vmatrix} 0 & \eta\, r\dfrac{\mathrm{d}\omega}{\mathrm{d}r} & 0 \\ \eta\, r\dfrac{\mathrm{d}\omega}{\mathrm{d}r} & 0 & 0 \\ 0 & 0 & 0 \end{vmatrix}, \tag{11.6.6}$$

as follows from eqs (11.6.5), (11.6.3) and (11.5.1). From eqs (11.6.6), (11.5.5) and (11.6.1), we infer that ψ' attains its maximum value for $r = r_1$, i.e., at the surface of the inner cylinder.

By applying relation (11.5.8) to a fluid element in contact with that surface, we obtain the condition for the considered laminar flow to be compatible with the angular momentum balance equation:

$$\eta\, r_1 \frac{\mathrm{d}\omega}{\mathrm{d}r}\bigg|_{r=r_1} \leq \tau_{\max}\,. \tag{11.6.7}$$

In view of eq. (11.6.1), this relation can also be written as

$$\eta\,\frac{\mathrm{A}}{r_1^2} \leq \tau_{\max}\,, \tag{11.6.8}$$

which, in view of eq. $(11.6.2)_1$, can be used to determine the value of the difference $(\omega_2 - \omega_1)$ at which the considered laminar flow reaches the turbulence threshold.

To exemplify, let us refer to the case in which the outer cylinder is maintained at rest ($\omega_2 = 0$). From eq. $(11.6.2)_1$, it can be inferred that in this case:

$$\mathrm{A} = -\,\frac{2\,r_1^2\,r_2^2}{r_2^2 - r_1^2}\,\omega_1\,. \tag{11.6.9}$$

The threshold value ω_1^* of ω_1 is obtained by inserting this equation into relation (11.6.8) and taking the equality sign in the same relation. Thus, we obtain:

$$\omega_1^* = \frac{r_2^2 - r_1^2}{2\, r_2^2} \frac{\tau_{max}}{\eta} . \tag{11.6.10}$$

Note that we dropped a negative sign in front of the right-hand side term of this equation because the sense of rotation of the cylinder is immaterial.

The Reynolds number associated with the considered flow is usually defined as:

$$\mathrm{R_e} = \frac{\rho\, \omega_1\, r_1\, d}{\eta} , \tag{11.6.11}$$

where $d = r_2 - r_1$ is the gap between the two cylinders. By inserting ω_1^* for ω_1 in this relation, we obtain

$$\mathrm{R_e^*} = \frac{\rho\, \omega_1^*\, r_1\, d}{\eta} . \tag{11.6.12}$$

This is the limit value of the Reynolds number beyond which a full laminar Taylor-Couette flow is not possible if $\omega_2 = 0$.

Let us apply eq. (11.6.12) to predict the value of $\mathrm{R_e^*}$ in the case of the experiments considered in [64]. These experiments refer to an apparatus in which $r_1 = 3.81$ cm, $r_2 = 4.21$ cm, and $d = 0.4$ cm. The fluid is water at room temperature, which means $\rho = 10^3$ Kg/m^3 and $\eta = 10^{-3}$ N sec/m^2. According to our result (11.4.10), we take $\tau_{max} = 14.4\ 10^{-3}$ Pa. With these values, we calculate from eq. (11.6.10) that $\omega_1^* = 1.7$ rad/sec. By inserting this value into eq. (11.6.12), we obtain $\mathrm{R_e^*} = 197$. Compare this with the experimental value $\mathrm{R_e} = 239$, determined in [64] for $\omega_2 = 0$ in the absence of any axial flow. The value of $\mathrm{R_e^*}$ predicted by the present theory appears satisfactory because it refers to the initiation of turbulence, whereas the experimental value refers to full turbulent flow. Obviously, in a non-homogeneous flow, as is the present one, the full turbulent flow is reached for an angular velocity ω_1 and, thus, a value of $\mathrm{R_e}$ that is slightly in excess of the threshold value. Notice also that no precise indication of the temperature at which the experiments were done is given in [64]. This may be a further reason for the discrepancy between the experimental results and the theoretical values obtained here.

Similar results were obtained by the present author for some other data on Taylor-Couette flow taken from the vast literature on the subject. However, the argument will not be pursued further, because a complete assessment of this topic is beyond the scope of this book.

11.7 Turbulence Threshold in Hagen-Poiseuille Flow

Laminar flow of an incompressible linearly viscous fluid through a rectilinear circular pipe is a classic topic of fluid mechanics. Assuming adherence of the fluid to the pipe's walls, the steady flow of fluid, as obtained by integrating eqs (11.1.1) - (11.1.3), is laminar. In this flow, all fluid particles move along rectilinear paths, parallel to the pipe axis. In a cylindrical coordinate system (r, θ, z) in which the inner wall of the pipe is $r = d/2$ and the z-axis is directed as the flow, the velocity field of this laminar flow is given by:

$$\mathrm{v}_r = \mathrm{v}_\theta = 0, \qquad \mathrm{v}_z = \mathrm{v}(r) = \mathrm{v}_{\max} - \frac{4\,\mathrm{v}_{\max}}{d^2}\, r^2 \tag{11.7.1}$$

(cf., e.g., [7]). The notation adopted here is explained in Fig. 11.7.1. The maximum velocity $\mathrm{v}_{\max}$ depends on the viscosity coefficient and on the pressure gradient $\Delta p/L$ along the pipe axis. It is attained for $r = 0$ and it is found to be

$$\mathrm{v}_{\max} = \frac{1}{16\,\eta}\frac{\Delta p}{L} d^2 . \tag{11.7.2}$$

The flow's mean velocity, as calculated from eq. (11.7.1), is

$$\mathrm{v}_{\mathrm{m}} = \frac{1}{2}\mathrm{v}_{\max} . \tag{11.7.3}$$

This kind of flow is known as the Hagen–Poiseuille flow.

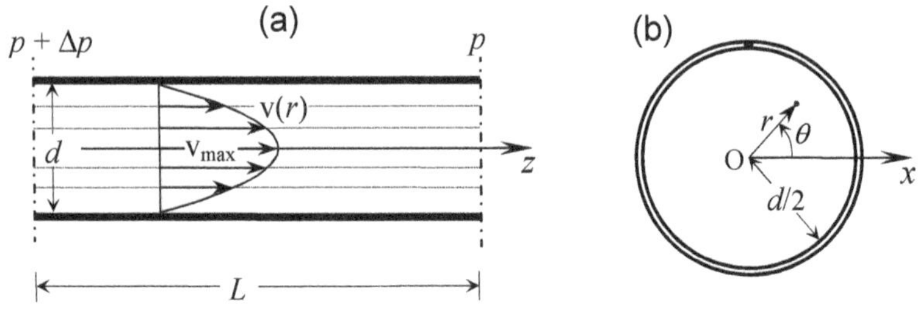

Fig. 11.7.1. (a) Velocity profile in the steady laminar flow of an incompressible linearly viscous fluid in a circular pipe (*Hagen-Poiseuille flow*). (b) Pipe cross section.

The rate-of-deformation tensor $\boldsymbol{D}$ relevant to this flow is obtained by applying definition (11.2.10) to the velocity field (11.7.1). The only non-vanishing components of $\boldsymbol{D}$ in the cylindrical coordinate system (r, θ, z) turn out to be $D_{\mathrm{rz}} = D_{\mathrm{zr}}$. Therefore, by making use of eq. (11.7.3), we obtain:

$$\boldsymbol{D} = \begin{vmatrix} 0 & 0 & -8\dfrac{\mathrm{v_m}}{d^2}r \\ 0 & 0 & 0 \\ -8\dfrac{\mathrm{v_m}}{d^2}r & 0 & 0 \end{vmatrix}. \tag{11.7.4}$$

This shows that $\boldsymbol{D}$ is not uniform throughout the fluid. Apart from an inessential sign, which depends on the orientation of the axes, the largest value of the non-vanishing components of $\boldsymbol{D}$ is attained at the pipe walls and it is equal to

$$D_{\mathrm{rz}} = D_{\mathrm{zr}} = -\frac{4\,\mathrm{v_m}}{d}. \tag{11.7.5}$$

From this equation and from eq. (11.5.1), we infer that the only non-vanishing components of $\boldsymbol{\sigma}'$ are $\sigma'_{\mathrm{rz}} = \sigma'_{\mathrm{zr}}$. They reach their largest absolute value at the pipe wall ($r = d/2$), where $\boldsymbol{\sigma}'$ is given by

$$\sigma' = \begin{vmatrix} 0 & 0 & -8\eta\dfrac{v_m}{d} \\ 0 & 0 & 0 \\ -8\eta\dfrac{v_m}{d} & 0 & 0 \end{vmatrix}. \qquad (11.7.6)$$

As observed in Section 11.5, any laminar flow must meet inequality (11.5.8). Therefore, by introducing eq. (11.7.6) into relation (11.5.8) and taking the equality sign in that relation, we can determine the upper limit of v_m for the considered laminar flow to be compatible with eq. (11.5.8). Denoting this limit as v_m^*, we obtain, after some simple algebra:

$$v_m^* = \frac{d}{8\eta}\,\tau_{max}. \qquad (11.7.7)$$

As the mean velocity of the flow exceeds this value, the shear stress in the fluid layer adjacent to the pipe walls becomes greater than the maximum elastic shear stress that the fluid can exert. Consequently, eq. (11.1.3) cannot be met and some rotational motion must be produced in order to satisfy the angular momentum balance law. The original laminar motion is no longer possible and the flow begins to exhibit some turbulence.

In Hagen-Poiseuille flow, $\boldsymbol{D}$ and σ' are not uniform; thus, as v_m exceeds limit (11.7.7), turbulence is produced initially at a thin layer of fluid near the pipe walls where σ' exceeds the elastic limit. However, as v_m is increased, the depth of that layer grows until full turbulent flow is established throughout the pipe. This explains why, in this flow, the transition to turbulence is comparatively gradual.

For fluid flow in circular pipes, the Reynolds number is usually defined as

$$R_e = \frac{\rho\, v_m\, d}{\eta}. \qquad (11.7.8)$$

The Reynolds number at which the flow ceases to be fully laminar can be calculated from eq. (11.7.8) by substituting for v_m the value v_m^* given by eq. (11.7.7). Thus, denoting this value by R_e^*, we obtain:

$$R_e^* = \frac{\rho\, v_m^*\, d}{\eta} = \frac{\rho\, \tau_{max}\, d^2}{8\eta^2}. \qquad (11.7.9)$$

This gives the threshold Reynolds number above which full laminar flow through the pipe is no longer possible.

To exemplify, for water at 20 °C we have that $\rho = 10^3$ Kg/m^3 and $\eta = 10^{-3}$ N sec/m^2. We also have that $\tau_{max} = 14.4\ 10^{-3}$ Pa, according to result (11.4.10). By inserting these values into eq. (11.7.9), we calculate:

$$R_e^* = 18 \cdot 10^5\, d^2 \quad \text{(with } d \text{ expressed in [m])} \qquad (11.7.10)$$

or

$$R_e^* = 180\, d^2 \quad \text{(with } d \text{ expressed in [cm])}. \qquad (11.7.11)$$

For instance, for $d = 3$ cm, this formula yields $R_e^* = 1620$, which is not unreasonable if compared with experiment. The plot of Fig. 11.7.2 gives the value of R_e^* for other values of d, as obtained from eq. (11.7.11).

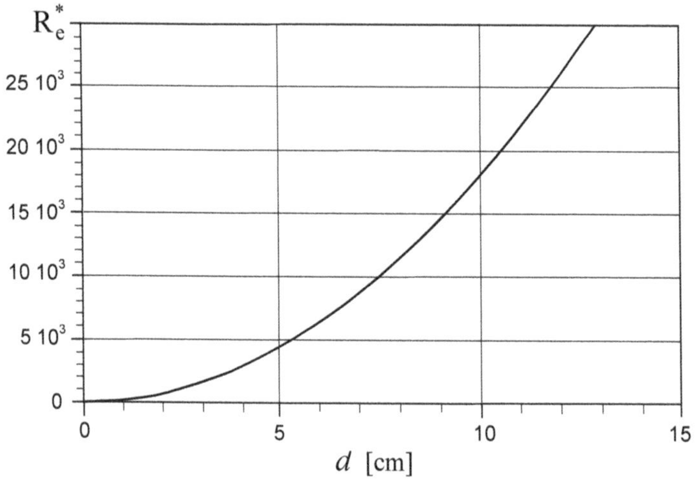

Fig. 11.7.2. Threshold Reynolds number R_e^* as a function of the pipe diameter d, as predicted by eq. (11.7.9) for water at 20 °C.

No limit to laminar flow is predicted by classical fluid dynamics and Hagen-Poiseuille flow is known to be linearly stable for every value of R_e. However, experiments show that for sufficiently large values of R_e the flow is turbulent. The limit Reynolds number above which pipe flow is turbulent is usually considered to be somewhere between 2,000 and 4,000. However, experiments have reported laminar flows in pipes for values of R_e that are orders of magnitude greater than these (cf., e.g., [57], [15]). Moreover, in pipes of very small hydraulic diameter, turbulence has been detected for values of R_e as low as 200–400, cf. [56]. What all this means is that the transition to turbulence is not controlled fully by the Reynolds number.

The present approach shows that R_e^*, i.e., the value of R_e at which the flow ceases to be fully laminar, depends on the elastic limit τ_{max}. This limit is a property of the fluid that does not enter its motion equations. Of course, for any given value of R_e less than R_e^*, all flows in pipes of different diameters are similar for every linear viscous fluid because, when expressed in dimensionless variables, they are all governed by the same dimensionless equations. However, the transition to turbulence depends on the maximum value of the shear stress component reached in the fluid, a quantity that does not enter the flow equations.

For the kind of flow considered in the present section, the shear stress component in question is σ'_{13} and its maximum value is reached at the pipe walls. This value is different for pipes of different diameters, which makes the transition to turbulence depend on d in any fluid and for any value of R_e. The dependence on d of the maximum value of σ'_{13} is explicitly shown by the following expression of the components of $\boldsymbol{\sigma}'$ at the pipe walls:

$$\boldsymbol{\sigma}' = \begin{vmatrix} 0 & 0 & -8\,\eta^2 \dfrac{R_e}{\rho\, d^2} \\ 0 & 0 & 0 \\ -8\,\eta^2 \dfrac{R_e}{\rho\, d^2} & 0 & 0 \end{vmatrix}, \qquad (11.7.12)$$

which follows from eqs (11.7.8) and (11.7.6). When the magnitude of σ'_{13} exceeds τ_{max}, the laminar flow ceases to be admissible because the angular momentum balance condition (11.1.3) can no longer be met. Classical theory ignores that there is a limit to σ'_{13}. As a consequence, it is unable to predict when a laminar solution of the flow equations (11.1.1) - (11.1.3) ceases to be admissible.

12

Intrinsic Heat

Internal energy can be decomposed into two parts: a temperature dependent part and a temperature independent one. In this chapter, the temperature dependent part is referred to as *intrinsic heat* and it is shown to determine the entropy of the system. The consequence is that the entropy of a system cannot be assigned independently of the system's internal energy.

The general relation between entropy and internal energy is derived in Section 12.5. The very fact that this relation turns out to be somewhat involved explains why entropy is such a useful quantity to be used, although the information it conveys is already contained in the internal energy function. Incidentally, intrinsic heat could be a more direct quantity to use than entropy. Historically, however, entropy has been used, and, at this stage of development of thermodynamics, there appears to be little chance for a change.

The present analysis provides a full macroscopic explanation of the nature of entropy and of its physical meaning. Thermodynamics is a macroscopic theory. Therefore, all quantities it uses should be understood and justified in macroscopic terms. However, the physical meaning of entropy is usually sought outside of the macroscopic approach, since a simple macroscopic explanation is lacking. Quite often, reference is made to statistical mechanics. In that case, entropy is related to the state of microscopic disorder of the system, a concept that, in macroscopic terms, is still more obscure than the one it aims to explain. Order and disorder have no citizenship in macroscopic physics, which makes these explanations far from satisfactory even from a logical standpoint. The present chapter should help liberate thermodynamics from such unwarranted dependence on statistical mechanics, which, in principle, is no more general than the macroscopic approach.

The existence of a relationship between entropy and internal energy opens the way to check their reciprocal consistency. In the last sections of this chapter, we check for consistency the expressions of internal energy and entropy that are in use for systems as different as thermoelastic materials, photon gases, and ideal gases. In the case of ideal gases, the expressions turn out to be inconsistent with each other. The origin of the inconsistency is traced back to the fact that the theory of ideal gases ignores thermal radiation. Thermodynamic equilibrium should take thermal radiation into proper account. This cannot be done in the ideal gases theory, because any interaction of the gas particles with radiation is excluded at the outset. The consequences of this may not matter for most applications. However, there are some specific situations in which the interaction between gas particles and radiation is vitally important in order to describe the physical phenomenon correctly. One such situation is when a precise assessment of the temperature change of a low pressure gas that expands in a vacuum is sought [53].

12.1 Simple Heating and Cooling Processes

When two uniform-temperature systems are brought into thermal contact, the hotter system cools down and the colder system heats up. The process comes to an end when the temperatures of the two systems become the same. This phenomenon is a basic pillar of thermodynamics. It justifies the physical definition of heat as that which is transferred from the hotter to the colder system simply by virtue of their difference in temperature [74, p. 73].

If the heat exchange takes place while all of the system's state variables except temperature are constant, we say that the system undergoes a *simple heating* or a *simple cooling* process, depending on whether its temperature increases ($\dot{T} > 0$) or decreases ($\dot{T} < 0$), respectively. When applied to a system that undergoes a simple heating/cooling process, the first law (1.4.8) requires that, at any time during the process, the following equation should hold true:

$$\mathrm{d}U = \mathrm{d}Q \tag{12.1.1}$$

or, equivalently,

$$\dot{U} = \dot{Q}. \tag{12.1.2}$$

This is because $\dot{W}_{\mathrm{in}} = -\dot{W}_{\mathrm{out}} = 0$ in the process, since all of the system's state variables, other than T, are constant.

Of course, no work can be dissipated in a process in which no work is done. Yet a simple heating/cooling process is irreversible if it takes place at a non-vanishing temperature gradient. In this case, irreversibility arises because a finite temperature gradient means a temperature difference between the system's parts. In principle, this could be exploited to turn some of the heat that flows from the hotter to the colder parts of the system into useful work. Simple heating/cooling does not produce such work. For all practical purposes, this is equivalent to producing that work and dissipating it as heat as soon as it is produced. This makes simple heating/cooling a dissipative and hence an irreversible process when it occurs at a non-vanishing temperature gradient.

On the contrary, if a simple heating/cooling process takes place at uniform temperature, no work can be produced from the heat transfer. Thus, no work can be dissipated, and the process is reversible. Though rather ideal, the uniform temperature condition can, in practice, be met to any degree of accuracy by the appropriate reduction of the time rate at which the heat transfer to/from the system takes place.

It is interesting to observe that eqs (12.1.1) and (12.1.2) apply irrespective of whether the simple heating/cooling process takes place reversibly. This is because these equations are a direct consequence of the first law, which holds true irrespective of the reversibility of the process. It follows that any amount of heat, Q, which is transferred to/from the system through a simple heating/cooling process, always produces the internal energy change, ΔU, given by

$$\Delta U = Q, \tag{12.1.3}$$

whether or not the process is reversible.

12.2 Intrinsic Heat

Internal energy is a function of the system's state variables, namely $\boldsymbol{\xi}$ and T. That is, $U = U(\boldsymbol{\xi}, T)$. In a simple heating/cooling process, the variables $\boldsymbol{\xi}$ are constant. Then, from relation (12.1.1) or (12.1.2), it follows that the amount of heat that the system absorbs/releases during a simple heating/cooling process between any two given temperatures is also a state function.

In this chapter, we are interested in the amount of heat that the system will release if cooled at constant $\boldsymbol{\xi}$ from a given temperature T to absolute zero (0 K). This is equal to the amount of heat, Γ, that the system absorbs when heated at constant $\boldsymbol{\xi}$ from 0 K to T, because, as just observed, the heat absorbed/released during a simple heating/cooling process is a state function. Of course, Γ depends on T and the considered values of $\boldsymbol{\xi}$. Thus, we can write:

$$\Gamma = \Gamma(\boldsymbol{\xi},T) = \int_0^T \mathrm{d}Q\Big|_{\boldsymbol{\xi}=\text{const}} . \tag{12.2.1}$$

From eq. (12.1.1), we then have

$$\Gamma = \Gamma(\boldsymbol{\xi},T) = \int_0^T \mathrm{d}U\Big|_{\boldsymbol{\xi}=\text{const}} . \tag{12.2.2}$$

However,

$$\mathrm{d}U = \frac{\partial U}{\partial \boldsymbol{\xi}}\mathrm{d}\boldsymbol{\xi} + \frac{\partial U}{\partial T}\mathrm{d}T . \tag{12.2.3}$$

This means that

$$\mathrm{d}U\Big|_{\boldsymbol{\xi}=\text{const}} = \frac{\partial U}{\partial T}\mathrm{d}T . \tag{12.2.4}$$

Therefore, eq. (12.2.1) can be expressed as

$$\Gamma = \Gamma(\boldsymbol{\xi},T) = \int_0^T \frac{\partial U(\boldsymbol{\xi},T)}{\partial T}\mathrm{d}T . \tag{12.2.5}$$

This enables us to determine state function $\Gamma = \Gamma(\boldsymbol{\xi},T)$ from the internal energy function $U = U(\boldsymbol{\xi},T)$.

Heat Γ, as defined above, will be referred to as the *intrinsic heat* of the system. It is the amount of heat that is needed to increase the temperature of the system from 0 K to T at constant $\boldsymbol{\xi}$. Incidentally, the notion of intrinsic heat introduced here gives a precise meaning to what is sometimes referred to in the literature, although rather vaguely, as the amount of heat that a system must possess to exist at a given state ($\boldsymbol{\xi}$, T).

By differentiating eq. (12.2.5) with respect to T, we obtain

$$\mathrm{d}\Gamma\big|_{\boldsymbol{\xi}=\mathrm{const}} = \frac{\partial U}{\partial T}\,\mathrm{d}T\,, \qquad (12.2.6)$$

or

$$\frac{\partial U}{\partial T} = \frac{\partial \Gamma}{\partial T}\,. \qquad (12.2.7)$$

This implies that

$$U(\boldsymbol{\xi}, T) = \Gamma(\boldsymbol{\xi}, T) + \mathrm{f}(\boldsymbol{\xi}), \qquad (12.2.8)$$

where $\mathrm{f}(\boldsymbol{\xi})$ is a differentiable function of $\boldsymbol{\xi}$. This function can be determined by observing that from the definition of Γ, or, equivalently, from eq. (12.2.5), it follows that

$$\Gamma(\boldsymbol{\xi}, 0) = 0 \qquad \textit{for every}\ \ \boldsymbol{\xi}\,. \qquad (12.2.9)$$

From eq. (12.2.8), this means that $\mathrm{f}(\boldsymbol{\xi})$ is given by

$$\mathrm{f}(\boldsymbol{\xi}) = U(\boldsymbol{\xi}, 0). \qquad (12.2.10)$$

From eqs (12.2.8) and (12.2.10), it can be concluded that

$$\Gamma(\boldsymbol{\xi}, T) = U(\boldsymbol{\xi}, T) - U(\boldsymbol{\xi}, 0), \qquad (12.2.11)$$

which is a useful formula for determining the intrinsic heat after the internal energy function is given.

12.3 The Purely Mechanical Part of Internal Energy

Rewriting eq. (12.2.11) in the form

$$U(\boldsymbol{\xi}, 0) = U(\boldsymbol{\xi}, T) - \Gamma(\boldsymbol{\xi}, T) \qquad (12.3.1)$$

makes it apparent that $U(\boldsymbol{\xi}, 0)$ is the internal energy in excess of Γ. We shall refer to it as U_{w}. That is:

$$U_w = U_w(\boldsymbol{\xi}) = U(\boldsymbol{\xi}, 0). \tag{12.3.2}$$

The internal energy of the system can then be expressed as

$$U(\boldsymbol{\xi}, T) = \Gamma(\boldsymbol{\xi}, T) + U_w(\boldsymbol{\xi}). \tag{12.3.3}$$

By differentiating both sides of this equation we obtain:

$$\mathrm{d}U = \mathrm{d}\Gamma + \mathrm{d}U_w\,, \tag{12.3.4}$$

where

$$\begin{aligned} \mathrm{d}\Gamma &= \frac{\partial \Gamma}{\partial \boldsymbol{\xi}}\cdot \mathrm{d}\boldsymbol{\xi} + \frac{\partial \Gamma}{\partial T}\mathrm{d}T \\ &= \left(\frac{\partial U(\boldsymbol{\xi},T)}{\partial \boldsymbol{\xi}} - \frac{\partial U(\boldsymbol{\xi},0)}{\partial \boldsymbol{\xi}}\right)\cdot \mathrm{d}\boldsymbol{\xi} + \frac{\partial U(\boldsymbol{\xi},T)}{\partial T}\mathrm{d}T, \end{aligned} \tag{12.3.5}$$

and

$$\mathrm{d}U_w = \frac{\partial U(\boldsymbol{\xi},0)}{\partial \boldsymbol{\xi}}\cdot \mathrm{d}\boldsymbol{\xi} = \frac{\mathrm{d}U_w(\boldsymbol{\xi})}{\mathrm{d}\boldsymbol{\xi}}\cdot \mathrm{d}\boldsymbol{\xi}. \tag{12.3.6}$$

$U_w = U_w(\boldsymbol{\xi})$ is a state function that does not depend on temperature. It represents the component of the internal energy of the system that is not affected by the temperature of the system or by the amount of heat that the system absorbs or loses. Of course, U_w, like any non-thermal energy, can be dissipated, entirely or partially, into heat. However, because it is independent of both temperature and absorbed heat, U_w can be regarded as the purely mechanical part of U, and referred to as the *non-thermal part* of the internal energy of the system.

From eq. (12.3.3), it can be concluded that the internal energy of any system can always be decomposed into two parts, i.e., a thermal part or intrinsic heat, Γ, and a non-thermal part, U_w. Both Γ and U_w are constitutive properties of the system. As apparent from eqs (12.2.5) and (12.3.2), both of these constitutive properties are determined by the system's internal energy function, which is a constitutive property itself.

12.4 Maximum Absorbable Heat

As recalled at the beginning of Section 12.1, heat is defined as the physical entity that two uniform temperature systems in thermal contact exchange only by virtue of their temperature difference. Thus, heat is a thermostatic quantity in that both its definition and the calorimetric procedure used to measure it refer to systems in thermal equilibrium. The first law establishes the equivalence between this thermostatic definition of heat and energy. This justifies considering heat as thermal energy.

It must be concluded that, strictly speaking, one cannot speak of heat (or of thermal energy) as an independent entity separated from the system that supplies or absorbs it. In particular, there is no sense in speaking of the transformation of heat into non-thermal energy in the absence of a system that absorbs the heat to be transformed. To be defined and measured, heat must be absorbed or lost by a system. Likewise, to be transformed into non-thermal energy, heat must be absorbed by a system first.

As obvious as these statements may seem, they lie at the very core of thermodynamics. They justify the following proposition, the validity of which is taken for granted without further discussion:

No amount of heat can be converted into non-thermal energy if it is not first absorbed as internal energy by a system.

In a process in which the variables $\boldsymbol{\xi}$ are not constant, some work is done on the system or supplied by it. Part of this work is generally dissipated as heat, which may be absorbed, if only partially, by the system, thus reducing the amount of heat that the system needs to absorb from the surroundings.

Now, let $\mathrm{d}\Gamma$ be the change in Γ in a process from state ($\boldsymbol{\xi}$, T) to state ($\boldsymbol{\xi}+\mathrm{d}\boldsymbol{\xi}$, $T+\mathrm{d}T$). From the definition of Γ, we know that $\mathrm{d}\Gamma$ equals the change in the amount of heat absorbed by the system in a simple heating/cooling process (i.e., a process at constant $\boldsymbol{\xi}$) from 0 K to the final state, as the final state changes from ($\boldsymbol{\xi}$, T) to ($\boldsymbol{\xi}+\mathrm{d}\boldsymbol{\xi}$, $T+\mathrm{d}T$). On the other hand, in a general process from state ($\boldsymbol{\xi}$, T) to state ($\boldsymbol{\xi}+\mathrm{d}\boldsymbol{\xi}$, $T+\mathrm{d}T$), some work is done on the system or absorbed by it, because of the change in $\boldsymbol{\xi}$. If the process is irreversible, then some of this work is dissipated as heat, which, in turn, is absorbed entirely or partially by the system. It follows that the heat $\mathrm{d}Q$ that the system absorbs from the surroundings in any process from ($\boldsymbol{\xi}$, T) to ($\boldsymbol{\xi}+\mathrm{d}\boldsymbol{\xi}$, $T+\mathrm{d}T$) cannot exceed $\mathrm{d}\Gamma$:

$$\mathrm{d}Q \leq \mathrm{d}\Gamma. \tag{12.4.1}$$

The inequality sign in this relation applies when the considered process is irreversible. In that case, dissipation provides an extra source of heat to the system, thus reducing the amount of heat that the system must absorb from the surroundings. Of course, no dissipation takes place if the process occurs reversibly. In this case, relation (12.4.1) applies with the equality sign.

Alternatively, to justify relation (12.4.1) we can also observe that the non-thermal part, U_w, of the internal energy of a system is insensitive to the heat absorbed or lost by the system itself. This means that all of the heat that the system absorbs must be used to increase the intrinsic heat part, Γ, of the internal energy. It follows that relation (12.4.1) must apply with the equality sign whenever the heat absorbed is the sole source of heat available to the system. This occurs when the process is reversible. On the contrary, for an irreversible process, the heat that the system takes from the surroundings will be only a part of the total heat that is supplied to the system, because some of that heat comes to the system directly from the dissipated energy. This makes the heat, $\mathrm{d}Q$, which the system absorbs from the surroundings, less than $\mathrm{d}\Gamma$ and provides another justification of the validity of relation (12.4.1).

When a system absorbs heat, its intrinsic heat, Γ, must increase appropriately, because the change in Γ cannot be less than the heat absorbed, as requested by relation (12.4.1). Since $\Gamma = \Gamma(\boldsymbol{\xi}, T)$, this generally requires that both T and $\boldsymbol{\xi}$ change appropriately. Of course, appropriate changes in the variables $\boldsymbol{\xi}$ may make the system's intrinsic heat vary at a constant T. In this case, if there is no heat exchange, the variation in Γ will be balanced by the work done on the system and the change in U_w, as evident from eqs (12.3.4) and (1.4.7).

In any case, inequality (12.4.1) sets an upper bound to the amount of heat that a system can absorb in any infinitesimal time interval of any process. Such a bound represents a general physical requirement that is independent of the first law of thermodynamics, because that law expresses a balance of energy, quite apart from the kinds of the energies involved. The bound in question is defined by Γ. Since Γ is a state function, the bound will depend on the initial and the final state of the system, but it will not depend on the particular process that joins these states.

Also, since Γ is a constitutive function, each system has its own maximum capacity to absorb heat, which depends on the state of the system. By dividing both sides of inequality (12.4.1) by $\mathrm{d}t$, the same bound can be expressed in terms of the time rates of Q and Γ:

$$\dot{Q} \leq \dot{\Gamma}. \tag{12.4.2}$$

Similarly, by integrating both sides of inequality (12.4.1) along a finite interval, say, from state A to state B, we have

$$Q \leq \Delta\Gamma. \tag{12.4.3}$$

Here, $\Delta\Gamma = \Gamma_B - \Gamma_A$ represents the maximum amount of heat that the system can ever absorb in a process that joins state A to state B.

No analogous limitation exists on the heat that a system can supply. The reason is that there is no limit to the amount of energy that a system can dissipate and thus supply as heat (as long as it has enough energy to dissipate).

12.5 Relationship between Entropy and Intrinsic Heat

In classical thermodynamics, the entropy of a system is the state function defined as:

$$S = S(\boldsymbol{\xi}, T) = \int_{T=0\mathrm{K}}^{(\boldsymbol{\xi}, T)} \left.\frac{\mathrm{d}Q}{T}\right|_{\mathrm{rev}}, \tag{12.5.1}$$

cf. eq. (1.7.2). The subscript *rev* reminds us that the integral is to be calculated along a *reversible* process starting from absolute zero temperature and ending at state ($\boldsymbol{\xi}$, T). Any reversible process will do, since S is a state function. The value assumed by $\boldsymbol{\xi}$ at $T = 0$ K is immaterial, since the so-called third law of thermodynamics ensures that the entropy of any system vanishes at the temperature of absolute zero, irrespective of the value of $\boldsymbol{\xi}$ (see Section 1.7).

To see how S relates to Γ, recall that the second law of thermodynamics can be expressed by the entropy inequality (1.6.7):

$$\mathrm{d}Q \leq T\,\mathrm{d}S. \tag{12.5.2}$$

Like relation (12.4.1), relation (12.5.2) sets an upper bound to the amount of heat that the system can absorb in any infinitesimal interval of any process. In both relations the equality sign applies if the process is reversible. This means that the right-hand sides of these relations must be equal. That is:

$$\mathrm{d}\Gamma = T\,\mathrm{d}S. \tag{12.5.3}$$

By observing that

$$T\,\mathrm{d}S = \mathrm{d}(TS) - S\,\mathrm{d}T, \tag{12.5.4}$$

we can integrate eq. (12.5.3) to obtain

$$\Gamma(\boldsymbol{\xi},T) = T\,S(\boldsymbol{\xi},T) - \int_0^T S(\boldsymbol{\xi},T)\,\mathrm{d}T\,, \tag{12.5.5}$$

which enables us to calculate $\Gamma(\boldsymbol{\xi}, T)$ once $S(\boldsymbol{\xi}, T)$ is given.

Conversely, we can put eq. (12.5.3) in the form:

$$\mathrm{d}S = \frac{\mathrm{d}\Gamma}{T} = \mathrm{d}\left(\frac{\Gamma}{T}\right) + \frac{1}{T^2}\Gamma\,\mathrm{d}T\,, \tag{12.5.6}$$

which can be integrated to yield

$$S = S(\boldsymbol{\xi},T) = \frac{\Gamma(\boldsymbol{\xi},T)}{T} + \int \frac{\Gamma(\boldsymbol{\xi},T)}{T^2}\,\mathrm{d}T + \text{const}\,. \tag{12.5.7}$$

This equation provides a means of calculating $S(\boldsymbol{\xi}, T)$ once $\Gamma(\boldsymbol{\xi}, T)$ is given. Figure 12.5.1 presents a graphic interpretation of eq. (12.5.5), thus giving that equation an immediate geometrical meaning. It shows that the concepts of intrinsic heat and entropy are intimately related.

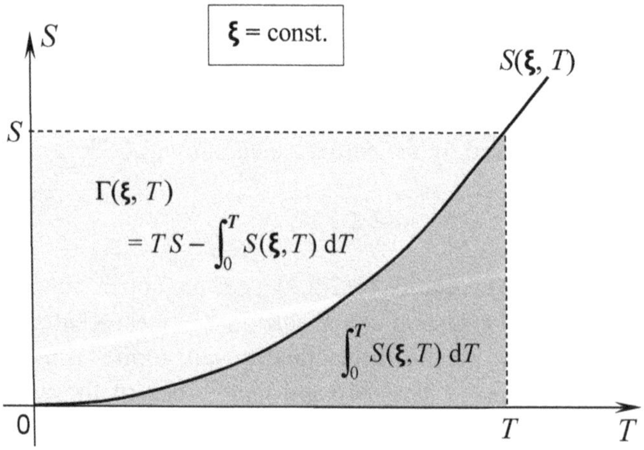

Fig. 12.5.1. Geometrical interpretation of the relationship between S and Γ, as obtained by representing eq. (12.5.5) in the $[S, T]$-plane. With reference to a generic curve $S(\boldsymbol{\xi}, T)$, the product $T\,S$ is the rectangular area between the dashed lines and the coordinate axes. The part of this area above curve $S(\boldsymbol{\xi}, T)$ represents $\Gamma(\boldsymbol{\xi}, T)$.

In view of eq. (12.2.11), we can express eq. (12.5.7) as:

$$S = S(\boldsymbol{\xi}, T) = \frac{U(\boldsymbol{\xi}, T) - U(\boldsymbol{\xi}, 0)}{T} + \int \frac{U(\boldsymbol{\xi}, T)}{T^2}\, \mathrm{d}T - \int \frac{U(\boldsymbol{\xi}, 0)}{T^2}\, \mathrm{d}T + \text{const.} \tag{12.5.8}$$

By means of eq. (12.3.2), this relation can also be written as

$$S = S(\boldsymbol{\xi}, T) = \frac{U(\boldsymbol{\xi}, T) - U_{\mathrm{w}}(\boldsymbol{\xi})}{T} + \int \frac{U(\boldsymbol{\xi}, T)}{T^2}\, \mathrm{d}T - \int \frac{U_{\mathrm{w}}(\boldsymbol{\xi})}{T^2}\, \mathrm{d}T + \text{const.} \tag{12.5.9}$$

Equation (12.5.8), or equivalently eq. (12.5.9), uncovers the relationship between S and U.

In the three sections that follow, we apply this result to check the consistency of the classical expressions of internal energy and entropy that are usually assumed for thermoelastic materials, photon gases, and ideal gases, respectively.

12.6 Linear Thermoelastic Materials

Linear thermoelastic materials are discussed at length in Chapter 6. In the present section, we show how eq. (12.5.8) enables us to obtain the entropy function of these materials directly from the expression of their internal energy function.

As usual in continuum mechanics, we refer to the specific values of U and S (per unit mass) and denote them as u and s, respectively. In addition, we confine our attention to homogeneous systems undergoing homogeneous processes. This means that the state of the system is the same at every point of the system. Since the state variables of these materials do not include spatial gradients, the results that apply to the homogeneous case also apply in non-homogeneous conditions, provided that they are referred to an infinitesimal element of material rather that to the system as a whole.

The specific internal energy of a thermoelastic material was derived in Section 6.3. It is expressed by eq. (6.3.19), which we rewrite as:

$$u = u(\xi_1, \xi_2, T) = \frac{1}{\rho}\left(\frac{K}{2}\xi_1^{\,2} + \beta \xi_1 T + 2\mu \xi_2\right) + c(T - T_0) + u(0, 0, T_0), \tag{12.6.1}$$

where $\beta = 3\alpha K$. The variables ξ_1 and ξ_2 are defined by eqs (6.3.17) and (6.3.18). All the constants that appear in the right-hand side of eq. (12.6.1) are defined in Chapter 6. Because they are constitutive constants, they depend on the particular thermoelastic material under consideration.

Once the expression of the internal energy is given, the entropy function can be calculated from eq. (12.5.8) by substituting u and s for U and S, respectively. This is admissible, because we are considering homogeneous systems. Since u is given by eq. (12.6.1), we obtain:

$$s = s(\xi_1, T) = \frac{1}{\rho}\left(\beta\xi_1 + \beta\int\frac{\xi_1}{T}\mathrm{d}T\right) + c + c\int\frac{1}{T}\mathrm{d}T\,, \qquad (12.6.2)$$

which shows that in a linear thermoelastic material, s does not depend on ξ_2. By calculating the integrals in this equation, we can express the specific entropy of a thermoelastic material as

$$s = s(\xi_1, T) = \frac{1}{\rho}\beta\xi_1 + \left(\frac{\beta}{\rho}\xi_1 + c\right)\ln T + \text{const}\,. \qquad (12.6.3)$$

In practice, one often deals with very small deformations. In this case, the variable ξ_1 also assumes very small values. The term $\beta\xi_1/\rho$ can then be neglected with respect to c, and eq. (12.6.3) simplifies to

$$s = s(\xi_1, T) = \frac{1}{\rho}\beta\xi_1 + c\ln T + \text{const}\,, \qquad (12.6.4)$$

which is nothing but eq. (6.4.5) expressed differently. Though in a different notation, eq. (12.6.4) also coincides with the expression of the entropy of a linear thermoelastic material that is reported in the literature (cf., e.g., [28, p. 536]). The standard procedure for obtaining it, however, is less straightforward than the one given here. The fact that the two procedures lead to the same result supports the validity of the results of the previous section.

12.7 Photon Gas

According to a well-established model in quantum physics, the electromagnetic radiation that fills an otherwise empty cavity consists of a multitude of photons of different wavelengths. These photons move at the speed of light in every direction within the cavity and do not interact with each other. On a macroscopic scale, this system of photons behaves like a gas and thus is endowed with temperature, pressure, internal energy, and entropy. For this reason, it is often referred to as *photon gas* or *radiation gas*.

According to a classical result of quantum mechanics, when such a gas is in thermal equilibrium with the walls of the cavity, its energy density *per unit volume* (referred to as u) is given by

$$u = \frac{4}{c}\sigma T^4, \qquad (12.7.1)$$

where c is the speed of light, σ is the well-known Stefan-Boltzman constant, and T is the temperature of the walls of the cavity (σ= 56.697 nW m^{-2} K^{-4}). By introducing the *radiation density constant*, *a*, defined by

$$a = \frac{4}{c}\sigma \qquad (12.7.2)$$

(a = 75.646 10^{-8} nJ m^{-3} K^{-4}), eq. (12.7.1) can be written as

$$u = aT^4. \qquad (12.7.3)$$

If V denotes the volume of the cavity, the total energy of the photon gas within the cavity is

$$U = uV = aT^4 V. \qquad (12.7.4)$$

A standard analysis shows that the pressure that the photon gas exerts upon the walls of the cavity is given by:

$$p = \frac{1}{3}u = \frac{1}{3}aT^4 \qquad (12.7.5)$$

(see, e.g., [74, Sect. 13.16] and [17, Sect. 3.13]). Therefore, as the volume of the cavity is increased by dV, the work done by the walls on the gas is given by

$$\mathrm{d}W_{\mathrm{in}} = -p\,\mathrm{d}V = -\frac{1}{3}u\,\mathrm{d}V = -\frac{1}{3}aT^4\,\mathrm{d}V. \qquad (12.7.6)$$

From the first law (1.4.8), we then infer that

$$\mathrm{d}U = \mathrm{d}Q - \frac{1}{3}aT^4\,\mathrm{d}V. \qquad (12.7.7)$$

However,

$$\mathrm{d}U = 4\,aT^3\,V\,\mathrm{d}T + aT^4\,\mathrm{d}V, \qquad (12.7.8)$$

as immediately follows from eq. (12.7.4). Therefore, by introducing eq. (12.7.8) into eq. (12.7.7) we obtain

$$\mathrm{d}Q = 4\,aT^3\,V\,\mathrm{d}T + \frac{4}{3}aT^4\,\mathrm{d}V. \qquad (12.7.9)$$

By applying this equation to a reversible process, it is concluded that

$$\mathrm{d}S = 4\,aT^2\,V\,\mathrm{d}T + \frac{4}{3}aT^3\,\mathrm{d}V, \qquad (12.7.10)$$

since $\mathrm{d}Q = T\mathrm{d}S$ for reversible processes. eq. (12.7.10) is a total differential equation. It can easily be integrated to give

$$S = \frac{4}{3}aT^3V, \qquad (12.7.11)$$

the integration constant being equal to zero, since $S = 0$ for $T = 0$.

The same result (12.7.11) follows from eq. (12.5.8) after we introduce the expression (12.7.4) for U and identify ξ with V. This provides further support to the validity of the relationship between entropy and intrinsic heat presented in Section 12.5.

12.8 Ideal Gases

In dealing with gases, it is customary to measure their mass in gram-moles (*moles*). As a consequence, the density, a, per unit mass of any extensive physical quantity, A, represents the amount of A per mole of the substance to which A refers. This is called *molar density* of A. That is

$$a = \frac{A}{n}, \tag{12.8.1}$$

where n is the number of moles of the substance to which the quantity A refers. In what follows, the symbols v, u, and s indicate the molar densities of volume, internal energy, and entropy, as obtained from eq. (12.8.1) by taking $A \equiv V$, $A \equiv U$, and $A \equiv S$, respectively.

Ideal gases, also called *perfect gases*, play a major role in classical thermodynamics. In homogeneous conditions, their state is described by two variables, namely v and T or p and T, where p is the gas pressure. We take molar volume, v, and temperature as the state variables of these gases. In terms of these variables, p is given by the celebrated equation of state of an ideal gas:

$$p\,v = R\,T, \tag{12.8.2}$$

where R is the universal gas constant (R = 8.3143 J/mole K).

The classical approach assumes that the internal energy of an ideal gas does not depend on v. That is,

$$u = u(T) = c_V\,T, \tag{12.8.3}$$

where c_V is the molar specific heat of the gas at constant volume. Although not excluding that $c_V = c_V(T)$, the classical treatment usually refers to the case where c_V is constant. In that case,

$$du = c_v \, dT, \tag{12.8.4}$$

and

$$c_v = \frac{\partial u}{\partial T} = \frac{d u}{d T}. \tag{12.8.5}$$

The rationale behind the rather drastic hypothesis (12.8.3) is the assumption that the free adiabatic expansion of a gas in a vacuum (the so-called Joule-Thomson expansion) does not produce any change in the temperature of the gas. This would rule out any dependence of u on v, or, equivalently, on p. However, the experimental evidence in support of this hypothesis is rather shaky. The best results available actually provide evidence to the contrary, even under rarefied gas conditions (see, e.g., the discussion contained in [74, Sect. 5.2]). Even so, eq. (12.8.3) is currently assumed as the internal energy of an ideal gas.

When the molar volume of a gas undergoes a change dv, the *specific work per mole* that the gas supplies to the surroundings is:

$$dw_{out} = p \, dv. \tag{12.8.6}$$

By using eq. (12.8.2), this also can be written as

$$dw_{out} = \frac{RT}{v} \, dv. \tag{12.8.7}$$

On the other hand, when referred to molar densities, the first law (1.4.8) yields:

$$du = dq - dw_{out}, \tag{12.8.8}$$

where q is the heat absorbed by the system per mole of gas. For reversible processes, we have that $dQ = T\,dS$ and thus $dq = T\,ds$. Therefore, by referring to a reversible process and by introducing eqs (12.8.4), $(12.8.5)_1$ and (12.8.7) into eq. (12.8.8) we obtain, after rearranging:

$$\mathrm{d}s = \frac{1}{T}\frac{\partial u}{\partial T}\mathrm{d}T + \frac{R}{v}\mathrm{d}v = \frac{c_V}{T}\mathrm{d}T + \frac{R}{v}\mathrm{d}v . \tag{12.8.9}$$

At this point the classical arguments assume that s is a function of v and T, which implies that

$$\mathrm{d}s = \frac{\partial s}{\partial T}\mathrm{d}T + \frac{\partial s}{\partial v}\mathrm{d}v . \tag{12.8.10}$$

From this equation and from eq. (12.8.9) it then follows that

$$\frac{\partial s}{\partial T} = \frac{c_V}{T} \tag{12.8.11}$$

and

$$\frac{\partial s}{\partial v} = \frac{R}{v} . \tag{12.8.12}$$

If c_V is constant, these equations can be integrated easily to give the classical expression of the molar entropy of an ideal gas:

$$s = s(v, T) = c_V \ln T + R \ln v + \text{const} . \tag{12.8.13}$$

Up to this point, the analysis is standard. We may want to determine, however, the expression of s that follows from the general eq. (12.5.8) established above, once eq. (12.8.3) is assumed to be valid. In the present case, $\boldsymbol{\xi}$ coincides with v. By introducing eq. (12.8.3) into eq. (12.5.8) we therefore obtain:

$$s = c_V \ln T + \text{const} , \tag{12.8.14}$$

which is clearly different from eq. (12.8.13).

This result is rather surprising, because eq. (12.5.8) – and hence eq. (12.8.14) – is a necessary consequence of general physical principles. We may wonder, therefore, to what extent a gas, which is described by the classical constitutive equations (12.8.2), (12.8.3) and (12.8.13), is admissible physically.

To answer this question we observe that a gas with internal energy (12.8.3) cannot produce work while operating isothermally in thermal equilibrium with the surroundings. This is a direct consequence of the first law (12.8.8). The reason is that, according to eq. (12.8.3), du must vanish in any isothermal process. Moreover, the gas cannot exchange any heat with the surroundings because they are at the same temperature as the gas and two bodies at the same temperature do not exchange heat. In these conditions, therefore, $dq = 0$. From eq. (12.8.8) it then follows that the considered gas cannot exchange work isothermally with the surroundings, if the latter are at the gas temperature. It must be concluded that in a gas with internal energy (12.8.3) the following equations

$$du = 0, \quad dq = 0 \quad \textit{and} \quad dw_{\mathrm{out}} = 0 \tag{12.8.15}$$

apply to every isothermal process in thermal equilibrium with the surroundings.

It follows, in particular, that the only process that an ideal gas can perform isothermally while remaining at the same temperature as its surroundings is a free expansion in a vacuum. At a first sight, this conclusion appears at variance with the large body of experimental evidence according to which a gas can produce work while expanding at a given temperature and absorbing heat from its surroundings at the same temperature. This is true even in the range of temperatures and pressures in which the gas behaves as an ideal gas and, therefore, eq. (12.8.3) applies to it.

The point here, however, is that the process in which an ideal gas produces work and absorbs heat while remaining at the same temperature as its surroundings is not a true isothermal processes. This process is actually a sequence of small adiabatic steps (producing work and temperature changes), each of which is followed by a constant-volume (workless) step that restores the gas temperature to the original value by exchanging heat with the surroundings. For small enough steps performed at a slow enough rate, the gas temperature during the whole process can, for many purposes, be considered as a constant to any desired degree of accuracy. Such a process may be referred as *pseudo-isothermal*. It should be distinguished from a truly isothermal process, because in a pseudo-isothermal process the gas does exchange heat with the surroundings at the gas temperature, while no heat exchange is possible in a truly isothermal process in the same conditions.

From eq. (12.8.7) it follows that the work done by an ideal gas in a pseudo-isothermal expansion from v_o to v is given by:

$$w_{out} = \int_{v_0}^{v} \mathrm{d}w_{out} = \int_{v_0}^{v} \frac{RT}{v}\mathrm{d}v = RT\,(\ln v - \ln v_0)\,, \qquad (12.8.16)$$

since T has the same value at the beginning of each infinitesimal adiabatic step making up the process. On the other hand, from eq. (12.8.8) we infer that in this process

$$\mathrm{d}q = \mathrm{d}w_{out}\,, \qquad (12.8.17)$$

because also u is almost constant in the process. This means that in a pseudo-isothermal process the gas absorbs as much heat as the work that it supplies to the surroundings. This is a very well-known result, although it is usually derived by referring to the incorrect entropy (12.8.13) and to the incorrect process (i.e., isothermal, rather than pseudo isothermal).

Once the correct entropy expression (12.8.14) is adopted, the Helmholtz free energy ($u - Ts$) of the ideal gas reduces to a constant. This is consistent with the fact that, as observed above, a gas cannot produce work by expanding in a strict isothermal way, if its internal energy depends only on temperature. On the contrary, the work produced by the same gas in a pseudo-isothermal process is different than zero and can be calculated from eq. (12.8.16). This work equals the amount of heat that the gas absorbs in the process and it coincides with the isothermal work that is traditionally calculated for ideal gases expansion.

12.8.1 Dependence of c_v on volume

As recalled above, the claim that the adiabatic free expansion of a rarefied gas supports assumption (12.8.3) is not firmly confirmed by the experiments. However, there also are other arguments based on the kinetic theory of gases that are advanced in support of the same assumption. In that theory, an ideal gas is modelled as a multitude of microscopic, rigid mass particles that are unable to interact at a distance and are in continuous thermal motion in every direction within the volume occupied by the gas. The volume itself is mostly empty space, since the gas particles have negligible volume. In this model, the internal energy of the gas coincides with the sum of the kinetic energy of its particles. There is no other way in which these particles can store energy, since they are rigid and long-range forces are ruled out. Thus, the internal energy of such a gas is independent of volume, since the speed of the gas particles is not affected by the volume of the gas.

However, the kinetic theory ignores the fact that the space between the particles is filled by radiation. This is especially true in thermal equilibrium conditions, which the classical theory assumes. From the analysis of Section 12.7 we know that, though devoid of matter, a space filled with radiation has its own internal energy and can absorb or lose heat to reach thermal equilibrium with the walls of the container and with any matter and material particle within that space. At thermal equilibrium, the energy of the radiation is given by eq. (12.7.4) and depends both on T and V. In particular, at a constant volume, any change in temperature will make a radiation-filled space absorb or lose heat from its surroundings.

The same applies to the radiation-filled space between the particles of a volume V of an ideal gas. Thus, there will be two different contributions to the amount of heat that the gas will absorb at constant V as its temperature is increased by $\mathrm{d}T$. The first contribution, $\mathrm{d}Q_\mathrm{k}$, is the heat that is needed to increase the kinetic energy of the gas particles. In an ideal gas, this contribution is independent of the gas volume. The other contribution, $\mathrm{d}Q_\mathrm{r}$, is the heat that is needed for the thermal radiation in the space between the gas particles to reach the new thermal equilibrium. This depends on the volume of the gas and can be calculated by applying eq. (12.7.9) to a constant volume process:

$$\mathrm{d}Q_\mathrm{r} = 4\,aT^3\,V\,\mathrm{d}T\,. \tag{12.8.34}$$

This contribution is ignored both in the classical theory and in the kinetic theory of ideal gases.

Of course, the total heat $\mathrm{d}Q$ absorbed by the gas is the sum of $\mathrm{d}Q_\mathrm{k}$ and $\mathrm{d}Q_\mathrm{r}$. This makes $\mathrm{d}Q$ depend on V since $\mathrm{d}Q_\mathrm{r}$ does. It follows that c_V also depends on V or v, since it is defined as

$$c_V = \left.\frac{dQ}{n\,dT}\right|_V . \tag{12.8.35}$$

In this equation, n indicates the number of gas moles contained in volume V, while subscript V reminds us that dQ and dT refer to a constant volume process. From eqs (12.8.34) and (12.8.35) it follows that the classical assumption that c_V is independent of v is not valid, even if we refer to the rather idealized kinetic gas model. The reason is that it does not take thermal radiation into due account. In nature, every material absorbs and emits thermal radiation. Thus, any model that ignores radiation is, at the outset, unphysical. The interaction between gas particles and thermal radiation was shown in [53] to provide a resolution to the so-called Gibbs paradox concerning the entropy changes of two mixing gases.

APPENDIX A

Processes and Inverse Processes

A process is an orderly sequence of values for the state variables of the system. Time is the natural ordering parameter of the sequence, although other ordering parameters can also be used. Unless otherwise stated, we assume that during the process the state variables are represented by continuous functions of the ordering parameter. If this parameter is time, then the process is defined by the following set of functions:

$$\boldsymbol{\xi}^{(i)} = \boldsymbol{\xi}^{(i)}\ (t, \boldsymbol{X}) \qquad (i = 1, 2, \ldots, n), \tag{A.1}$$

which give the state variables $\boldsymbol{\xi}^{(i)}$ as functions of time t at each point $\boldsymbol{X}$ of the system. In homogeneous systems the dependence on $\boldsymbol{X}$ does not appear in functions (A.1), and the same functions simplify to:

$$\boldsymbol{\xi}^{(i)} = \boldsymbol{\xi}^{(i)}(t) \qquad (i = 1, 2, \ldots, n). \tag{A.2}$$

Any state function, say function

$$u = u(\boldsymbol{\xi}^{(1)}, \boldsymbol{\xi}^{(2)}, \ldots, \boldsymbol{\xi}^{(n)}), \tag{A.3}$$

will generally vary in time during the process. That is,

$$u = u(\boldsymbol{\xi}^{(1)}(t), \boldsymbol{\xi}^{(2)}(t), \ldots, \boldsymbol{\xi}^{(n)}(t)) = \tilde{u}(t), \tag{A.4}$$

where $\tilde{u}(t)$ is an appropriate function of time that depends on the process. When time is the ordering parameter, the time rates of the state variables are obtained by direct time derivation of eq. (A.1) or (A.2):

$$\dot{\boldsymbol{\xi}}^{(i)} = \frac{d\,\boldsymbol{\xi}^{(i)}}{d\,t}. \tag{A.5}$$

The corresponding time derivative of the state function can be calculated by time derivation of its expression $(A.4)_2$:

$$\dot{u} = \frac{d\,\tilde{u}}{d\,t}. \tag{A.6}$$

Equivalently, by applying the derivation chain rule to eq. $(A.4)_1$ the same time derivative can be expressed as

$$\dot{u} = \frac{\partial\, u}{\partial\, \boldsymbol{\xi}^{(1)}}\dot{\boldsymbol{\xi}}^{(1)} + \frac{\partial\, u}{\partial\, \boldsymbol{\xi}^{(2)}}\dot{\boldsymbol{\xi}}^{(2)} + \ldots + \frac{\partial\, u}{\partial\, \boldsymbol{\xi}^{(n)}}\dot{\boldsymbol{\xi}}^{(n)}. \tag{A.7}$$

If a given process is referred to as the *direct* process, the *inverse* process is a process that retraces the same states of the direct process but in the inverse sequence. Thus, the inverse process starts from the final state of the direct process and ends where the latter started. Moreover, it proceeds in such a way that, at any point of the inverse process, the time derivatives of the state variables are the opposite of those at the same point in the direct process. The same is true for the time derivatives of the state functions, as evident from eq. (A.7).

Let $\boldsymbol{\xi} = \boldsymbol{\xi}(t)$ be the function that represents the values of state variable $\boldsymbol{\xi}$ during the direct process. This function is represented as a curve in Fig. A1.1. Let us focus our attention on the part of this process that goes from point A to point B of this curve, corresponding to time t_0 and time t_1, respectively. To determine the function that represents $\boldsymbol{\xi}$ in the inverse process, i.e. from t_1 to t_0, we substitute $-t$ for t in function $\boldsymbol{\xi}(t)$ to obtain curve $\boldsymbol{\xi}'(t) = \boldsymbol{\xi}(-t)$, which represents the mirror image of the original curve with respect to the $\boldsymbol{\xi}$-axis. In the mirrored curve, the image of point B is B′, which represents the value of $\boldsymbol{\xi}$ at time $-t_1$. As t increases from $-t_1$ to $-t_0$, the variable $\boldsymbol{\xi}$ recovers in the reverse order the values that it assumed in the direct process. Moreover, thanks to the relation

$$\frac{\mathrm{d}\,\boldsymbol{\xi}'(t)}{\mathrm{d}\,t} = \frac{\mathrm{d}\,\boldsymbol{\xi}(-t)}{\mathrm{d}\,t} = -\frac{\mathrm{d}\,\boldsymbol{\xi}(t)}{\mathrm{d}\,t}, \tag{A.8}$$

the time derivative of $\boldsymbol{\xi}'$ is the opposite of the time derivative of $\boldsymbol{\xi}$ at the corresponding state of the direct process. In other words, the process $\boldsymbol{\xi}'(t)$ is the particular inverse process of $\boldsymbol{\xi}(t)$ that starts at time $-t_1$.

If we want the inverse process to start at time t rather than $-t_1$, we must displace curve $\boldsymbol{\xi}'(t)$ by the amount $\tau = (t + t_1)$ in the positive direction of the time axis. Analytically, this is achieved by substituting $t-\tau$ for t in the same function:

$$\tilde{\boldsymbol{\xi}}(t) = \boldsymbol{\xi}'(t-\tau) = \boldsymbol{\xi}(-t+\tau). \tag{A.9}$$

An inverse process, as defined above, can be associated with any direct process. However, it may be not admissible from the physical standpoint. An effective way to visualize an inverse process is to film the real process and then run the filmed sequence backward. Although this makes the inverse process visible, such a process is seldom a representation of a physically realistic process. For the inverse process to be physically admissible, the direct process must be thermodynamically reversible. Otherwise, the inverse process cannot comply with the requirements of the second law of thermodynamics—the only physical law that discriminates between the two opposite directions in which a process can link any two neighbouring states of a system.

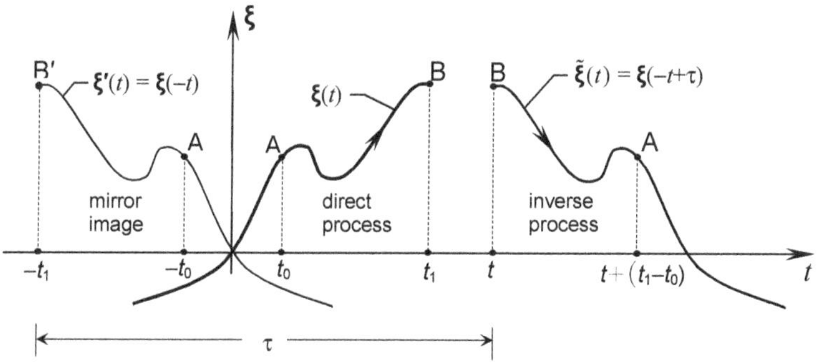

Fig. A1.1. Direct process and its inverse starting at time t

APPENDIX B

Natural Variables and Gibbs Fundamental Equation

One of the basic tenets of classical thermodynamics is the possibility of obtaining all thermodynamically relevant information about a system from a single potential function. The function in question is internal energy, expressed by taking entropy instead of temperature as an independent variable. When such approach is followed, temperature is regarded as a dependent variable, defined by an appropriate constitutive function of entropy and the other independent variables of the system.

The main steps of that approach are reviewed in the next section. In the subsequent section, we discuss why that approach can be applied only to ideal systems and how it should be modified when dealing with real systems.

B.1 Thermodynamics of Ideal Systems

For any given system, let us consider two equilibrium homogeneous states belonging to the same infinitesimal neighbourhood in state space. From all of the possible processes that link these states, let us choose a reversible process. This is a possibility since the two states are equilibrium homogeneous states (cf. Section 1.5). Following common usage in classical thermodynamics, we distinguish the work done by pressure over the change in the system's volume (*volume work*) from the other kinds of work done by or on the system. Equation (1.3.1) can thus be written as:

$$\mathrm{d}W_{\mathrm{in}} = -p\,\mathrm{d}V + \boldsymbol{F}\cdot\mathrm{d}\boldsymbol{\xi}\,. \tag{B.1}$$

From this and from eq. (1.10.1), it follows that $\mathrm{d}W_{\mathrm{in}}^{\mathrm{e}} = \boldsymbol{F}\cdot\mathrm{d}\boldsymbol{\xi}$.

Since the considered process is reversible, eq. (1.9.5) applies. By introducing eqs (1.9.5) and (B.1) into the first law (1.4.7), we obtain:

$$\mathrm{d}U = T\,\mathrm{d}S - p\,\mathrm{d}V + \boldsymbol{F}\cdot\mathrm{d}\boldsymbol{\xi}\,, \tag{B.2}$$

which is often referred to as the *fundamental equation* of (classical) thermodynamics. Equation (B.2) is valid irrespective of whether the infinitesimal increments $\mathrm{d}U$, $\mathrm{d}S$, $\mathrm{d}V$, and $\mathrm{d}\boldsymbol{\xi}$ are the differentials of appropriate functions. The first law requires that U should be a continuous, single-valued function of the state variables of the system. However, it does not impose any restriction on the derivability or differentiability of that function.

Though derived by making reference to a reversible process, eq. (B.2) also applies if the process that joins the two considered states is irreversible. To see why, we observe that the values of the quantities T, p, and $\boldsymbol{F}$ that appear in that equation refer to the initial state of the process (they only undergo negligible changes during the process, because the process is infinitesimal and it joins two equilibrium states belonging to the same infinitesimal neighbourhood). On the other hand, since the quantities U and S are state functions, the values of $\mathrm{d}U$ and $\mathrm{d}S$ depend on the two states that are joined by the process, but not on how the process is performed. The same applies to $\mathrm{d}V$ and $\mathrm{d}\boldsymbol{\xi}$, because V and $\boldsymbol{\xi}$ are state variables or related by appropriate state functions to the state variables of the system. Thus, eq. (B.2) depends on the initial and final states of equilibrium to which it refers, but is independent of whether the process that joins these two states is reversible or irreversible.

The values of p and $\boldsymbol{F}$ during an irreversible process differ from the values of the same quantities in a reversible process, because an irreversible process takes place through non-equilibrium states. As a consequence, the work done on the system in an irreversible process that joins the above two equilibrium states in the same infinitesimal neighbourhood is different from the term $-p\mathrm{d}V+\boldsymbol{F}\cdot\mathrm{d}\boldsymbol{\xi}$ that appears in eq. (B.2), because that term refers to the values of p and $\boldsymbol{F}$ at equilibrium conditions. For this reason eq. (B.2) is fundamentally different than the first law (1.4.7). The first law involves the actual work that is done on the system in the considered process. It applies irrespective of whether the process is reversible or irreversible. By contrast, the fundamental equation (B.2) relates the properties of two equilibrium states in the same infinitesimal neighbourhood quite apart from the process that joins them.

Natural variables

Equation (B.2) can be interpreted as expressing $\mathrm{d}U$ in terms of the infinitesimal increments of the independent variables S, V, and $\boldsymbol{\xi}$. In this interpretation, S, V, and $\boldsymbol{\xi}$ are taken as state variables of the system. Since U is a state function and eq. (B.2) holds true in general, we can conclude that the internal energy of any system can always be expressed as a function of S, V, and $\boldsymbol{\xi}$. That is,

$$U = U(S, V, \boldsymbol{\xi}). \tag{B.3}$$

Thus, because eq. (B.2) follows directly from the basic laws of thermodynamics, the variables S, V, and $\boldsymbol{\xi}$ appears to be the state variables that most naturally follow from the laws of thermodynamics. For this reason, the same variables are often referred to as the *natural* (or *canonical*) state variables of the system.

Of course, other choices of independent variables are possible and, in many cases, more convenient. Observe, for instance, that T is not included among the natural variables, in spite of the fact that temperature is the thermal variable that is directly accessible to the experiment. Consequently, choosing the natural variables as state variables for the system calls for an appropriate state function that gives T as a function of S, V, and $\boldsymbol{\xi}$. Equation (B.6), presented below, can, in fact, provide such an equation. However, the use of eq. (B.6) is confined to theoretical considerations, due to the difficulty of measuring and controlling S during a process (cf. Kestin [31, p. 533]).

Temperature is a far easier quantity to measure and control than entropy. Therefore, in the applications, temperature rather than entropy is chosen as state variable. In this case, the system is described by state variables that are not the natural variables, and entropy must be specified by an appropriate function of T and the other independent state variables of the system. However, no simple relation such as eq. (B.2) exists between $\mathrm{d}U$ and the differentials of the state variables, if these variables are not the natural variables.

The theoretical determination of $U(S, V, \xi)$

Even with the above difficulties, assuming S rather than T as an independent state variable is of some theoretical value in that it leads to the explicit theoretical determination of function (B.3), which reveals some interesting relations between T and p and between $\boldsymbol{F}$ and U.

The theoretical procedure for determining function $U(S, V, \boldsymbol{\xi})$ starts by differentiating eq. (B.3) to obtain:

$$\mathrm{d}U = \frac{\partial U}{\partial S}\,\mathrm{d}S + \frac{\partial U}{\partial V}\,\mathrm{d}V + \frac{\partial U}{\partial \boldsymbol{\xi}}\cdot \mathrm{d}\boldsymbol{\xi}\,. \tag{B.4}$$

In this step, one must assume that $U(S, V, \boldsymbol{\xi})$ is both continuous and differentiable in its domain of definition. From the physical standpoint, this is actually a rather restrictive hypothesis. There is no question that the internal energy should be a continuous function of the state variables of the system. If not, an infinitesimal change in these variables could bring about a finite change in the system's internal energy, which would hardly be compatible with the first law of thermodynamics. However, the differentiability of U is a much more questionable issue. All real systems change their states of aggregation (*phase changes*) for large enough changes in their state variables. This phenomenon is always accompanied by some discontinuity in at least some of the partial derivatives of U, which therefore cannot be differentiable at the points where the discontinuities appear. Thus, the assumption that U should be differentiable in the whole domain of definition of its independent variables is not strictly applicable to real systems.

A system will be referred to as an *ideal* system if its internal energy function is differentiable over the entire range of definition of the state variables from which it depends. In particular, as far as temperature is concerned, that range goes from absolute zero to infinity. Ideal systems are useful both for theoretical purposes and when an axiomatic approach to classical thermodynamics is sought. They may also provide good approximations to real systems within a limited range of values of their state variables. However, ideal systems are not real, which means that the conclusions that apply to them may not apply to *real* systems. This issue is considered in more detail in the next section.

Equation (B.4) can also be written, slightly more explicitly, as

$$\mathrm{d}U = \frac{\partial U}{\partial S}\mathrm{d}S + \frac{\partial U}{\partial V}\mathrm{d}V + \frac{\partial U}{\partial \xi_1}\mathrm{d}\xi_1 + \frac{\partial U}{\partial \xi_2}\mathrm{d}\xi_2 + \ldots + \frac{\partial U}{\partial \xi_m}\mathrm{d}\xi_m, \quad \text{(B.5)}$$

where $\xi_1, \xi_2, \ldots, \xi_m$ are the m components of vector $\boldsymbol{\xi}$, conjugated with the analogous components, $F_1, F_2, \ldots, F_m$, of $\boldsymbol{F}$. By comparing eq. (B.5) with eq. (B.2), we can infer that the choice of S, V, and $\boldsymbol{\xi}$ as independent variables entails the following relations:

$$T = \left(\frac{\partial U}{\partial S}\right)_{V,\boldsymbol{\xi}}, \quad \text{(B.6)}$$

$$p = -\left(\frac{\partial U}{\partial V}\right)_{S,\boldsymbol{\xi}}, \quad \text{(B.7)}$$

and

$$F_i = \left(\frac{\partial U}{\partial \xi_i}\right)_{S,V,\boldsymbol{\xi}_S} \quad (i = 1, 2, \ldots m). \quad \text{(B.8)}$$

In eq. (B.8) and in what follows, the index $\boldsymbol{\xi}_S$ stands for the sequence of $(m-1)$ variables $\xi_1, \xi_2, \ldots, \xi_{i-1}, \xi_{i+1}, \ldots, \xi_m$. The rather redundant notation used in eqs (B.6) through (B.8) is used often in thermodynamics. It helps to remind us that the internal energy U must be understood as a function of S, V, and $\boldsymbol{\xi}$, according to eq. (B.3). Of course, the validity of the same equations is confined to equilibrium states where function U is differentiable, since this is the condition for the validity of eqs (B.2) and (B.5).

Function U and variables S, and V, are extensive quantities. In the chemistry of mixtures and solutions, the variables ξ_i are taken as the number of moles of the components of the mixture (see Appemdix C). In this case, all the independent variables of the system are extensive. For homogeneous systems, this means that any given fraction of the system should contribute an equal fraction of the total values of U, S, V, and $\boldsymbol{\xi}$. For this to be true, function (B.3) should be a homogeneous function of degree 1. According to Euler's theorem on homogeneous functions, this requires that U must satisfy the following equation:

$$U(S,V,\xi_1,\ldots,\xi_m)=S\frac{\partial U}{\partial S}+V\frac{\partial U}{\partial V}+\xi_1\frac{\partial U}{\partial \xi_1}+\xi_2\frac{\partial U}{\partial \xi_2}+\ldots+\xi_m\frac{\partial U}{\partial \xi_m},$$
(B.9)

where, for brevity, we refrained from using the lengthy notation concerning the partial derivatives of U that we used in writing eqs (B.6) to (B.8). By inserting eqs (B.6) to (B.8) into eq. (B.9), we obtain:

$$U(S,V,\xi_1,\ldots,\xi_m)=T\,S-p\,V+\sum_{i=1}^{m}F_i\,\xi_i+\mathrm{C}\,. \qquad \text{(B.10)}$$

Of course, the quantities T, p, and F_i that appear in eq. (B.10) stand for the functions $T(S, V, \boldsymbol{\xi})$, $p(S, V, \boldsymbol{\xi})$, and $F_i(S, V, \boldsymbol{\xi})$, which, in fact, spoils much of the apparent simplicity of this equation.

Constant C that appears in the above equation represents the value that U assumes for $S = 0$, $V = 0$, and $\boldsymbol{\xi} = 0$. This constant should be zero for U to be a homogeneous function of its independent variables. We must observe, however, that the reference value of U cannot be fixed in an absolute way. This justifies the introduction of constant C in the above equation. However, if the function $U = U(S, V, \boldsymbol{\xi})$ is written as in eq. (B.10), one should make reference to the difference $[U(S, V, \boldsymbol{\xi}) - \mathrm{C}]$, rather than the function $U(S, V, \boldsymbol{\xi})$ itself, if an expression of U in the form of a homogeneous function is sought.

Equation (B.10) provides an explicit expression of the internal energy of the system for every possible homogeneous state of equilibrium. It is often referred to as the *Gibbs fundamental equation.* However, the same denomination is often reserved instead to the differential expression (B.16), listed below.

Other potential functions

We consider now the implications of eq. (B.2) on the other classical potential functions of thermodynamics, namely the enthalpy, H, the Helmholtz free energy, Ψ, and the Gibbs free energy, G. These potentials are defined in terms of U as follows:

$$H = U + p\,V, \qquad \text{(B.11)}$$

$$\Psi = U - T\,S, \qquad \text{(B.12)}$$

and

$$G = U + p\,V - TS, \qquad \text{(B.13)}$$

cf. eqs (1.10.3), (1.11.3), and (1.11.10). First, let us determine the infinitesimal increments of these potentials in a reversible process that joins any two equilibrium states of a system that belong to the same infinitesimal neighbourhood. Let U, p, V, T, and S be the values of internal energy, pressure, volume, temperature, and entropy, respectively, at the beginning of the process, and let $U + \mathrm{d}U$, $p + \mathrm{d}p$, $V + \mathrm{d}V$, $T + \mathrm{d}T$, and $S + \mathrm{d}S$ be the analogous values at the end of the process. By using eq. (B.2) and by neglecting all terms that contain the product of infinitesimal increments, we obtain, quite generally:

$$\mathrm{d}H = T\,\mathrm{d}S + V\,\mathrm{d}p + F_1\,\mathrm{d}\xi_1 + F_2\,\mathrm{d}\xi_2 + \ldots + F_{\mathrm{m}}\,\mathrm{d}\xi_{\mathrm{m}}, \qquad \text{(B.14)}$$

$$\mathrm{d}\Psi = -S\,\mathrm{d}T - p\,\mathrm{d}V + F_1\,\mathrm{d}\xi_1 + F_2\,\mathrm{d}\xi_2 + \ldots + F_{\mathrm{m}}\,\mathrm{d}\xi_{\mathrm{m}}, \qquad \text{(B.15)}$$

and

$$\mathrm{d}G = -S\,\mathrm{d}T + V\,\mathrm{d}p + F_1\,\mathrm{d}\xi_1 + F_2\,\mathrm{d}\xi_2 + \ldots + F_{\mathrm{m}}\,\mathrm{d}\xi_{\mathrm{m}} \qquad \text{(B.16)}$$

These equations are valid irrespective of the differentiability of the potential functions H, Ψ, and G. At variance with eq. (B.2), however, eqs (B.14) through (B.16) contain the infinitesimal increments of the variables p and/or T, which are not extensive variables, as was the case for S, V, and $\boldsymbol{\xi}$. As a consequence, the explicit expression of the functions $H = H(S, p, \boldsymbol{\xi})$, $\Psi = \Psi(T, V, \boldsymbol{\xi})$, and $G = G(T, p, \boldsymbol{\xi})$ cannot be obtained by a procedure similar to the one that led to eq. (B.10) above.

Granted that the function U can be expressed in the form (B.10), an explicit expression of the functions H, Ψ, and G can nonetheless be achieved by inserting eq. (B.10) directly into equations (B.11), (B.12), and (B.13), respectively. Thus, we obtain:

$$H = T\,S + F_1\,\xi_1 + F_2\,\xi_2 + \ldots + F_{\mathrm{m}}\,\xi_{\mathrm{m}} + \mathrm{C}, \qquad \text{(B.17)}$$

$$\Psi = -\,pV + F_1\,\xi_1 + F_2\,\xi_2 + \ldots + F_{\mathrm{m}}\,\xi_{\mathrm{m}} + \mathrm{C}, \qquad \text{(B.18)}$$

and

$$G = F_1\,\xi_1 + F_2\,\xi_2 + \ldots + F_{\mathrm{m}}\,\xi_{\mathrm{m}} + \mathrm{C}. \qquad \text{(B.19)}$$

Equation (B.19) is particularly interesting, because it does not involve explicitly the pairs of variables (T, S) or (p, V) that appear in the equations for H and Ψ. By omitting the usually unessential constant C, eq. (B.19) is often written as

$$G = G(T, p, \boldsymbol{\xi}) = F_1\, \xi_1 + F_2\, \xi_2 + ... + F_m\, \xi_m. \qquad \text{(B.20)}$$

Of course, the quantities F_i are given by appropriate constitutive functions of the independent variables.

Eq. (B.16) can be regarded as the total differential of function $G(T, p, \boldsymbol{\xi})$. The same differential can also be written as:

$$dG = \frac{\partial G}{\partial T} dT + \frac{\partial G}{\partial p} dp + \frac{\partial G}{\partial \xi_1} d\xi_1 + \frac{\partial G}{\partial \xi_2} d\xi_2 + ... + \frac{\partial G}{\partial \xi_m} d\xi_m. \qquad \text{(B.21)}$$

By comparing eq. (B.21) with eq. (B.16), we conclude that

$$F_i = \left(\frac{\partial G}{\partial \xi_i}\right)_{p,T,\xi_s}, \quad (i = 1, 2, \dots m) \qquad \text{(B.22)}$$

$$S = -\left(\frac{\partial G}{\partial T}\right)_{p,\boldsymbol{\xi}} = -\left(\frac{\partial \Psi}{\partial T}\right)_{V,\boldsymbol{\xi}}, \qquad \text{(B.23)}$$

and

$$V = \left(\frac{\partial G}{\partial p}\right)_{T,\boldsymbol{\xi}}. \qquad \text{(B.24)}$$

Equations (B.21) to (B.24) apply to any equilibrium state of any ideal system, as do their analogous eqs (B.5) to (B.8). All these equations assume the differentiability of the potential function involved, whereas eqs (B.14) to (B.16) as well as eq. (B.2) do not.

For isothermal processes at constant p, eq. (B.16) gives:

$$dG = F_1\, d\xi_1 + F_2\, d\xi_2 + ... + F_m\, d\xi_m, \qquad \text{(B.25)}$$

Comparing this with eq. (B.20) and considering the Euler theorem on homogeneous functions leads to the conclusion that G must be a homogeneous function of degree 1 in the variable $\boldsymbol{\xi}$. Of course, this result holds true if the arbitrary constant C is ignored or, more precisely, if reference is made to the difference $[G(T, p, \boldsymbol{\xi}) - C]$ rather than to the function $G(T, p, \boldsymbol{\xi})$ itself.

When applied to functions $H = H(S, p, \boldsymbol{\xi})$ and $\Psi = \Psi(T, V, \boldsymbol{\xi})$ and to eqs (B.14) and (B.15), similar arguments to the ones that led to eq. (B.22) enable us to obtain the following further expressions for F_i:

$$F_i = \left(\frac{\partial H}{\partial \xi_i}\right)_{S,\,p,\,\xi_s} \qquad (i = 1, 2, \ldots m) \tag{B.26}$$

and

$$F_i = \left(\frac{\partial \Psi}{\partial \xi_i}\right)_{V,T,\xi_s}, \quad (i=1,2,\ldots m). \tag{B.27}$$

From eqs (B.20) to (B.27) and from eq. (B.8), it can be concluded that

$$\left(\frac{\partial U}{\partial \xi_i}\right)_{S,\,V,\,\xi_s} = \left(\frac{\partial G}{\partial \xi_i}\right)_{p,\,T,\,\xi_s} = \left(\frac{\partial H}{\partial \xi_i}\right)_{S,\,p,\,\xi_s} = \left(\frac{\partial \Psi}{\partial \xi_i}\right)_{V,T,\,\xi_s}. \tag{B.28}$$

As the adopted notation should make it clear, in each of the above derivatives the functions U, G, H and Ψ are understood to be expressed in terms of different sets of independent variables.

The Gibbs-Duhem equation

By differentiating eq. (B.19), we obtain:

$$\mathrm{d}G = F_1\mathrm{d}\xi_1 + F_2\mathrm{d}\xi_2 + \ldots + F_m\mathrm{d}\xi_m + \xi_1\mathrm{d}F_1 + \xi_2\mathrm{d}F_2 + \ldots + \xi_m\mathrm{d}F_m. \tag{B.29}$$

For this value of dG to coincide with that given by eq. (B.16), the following equation must be met:

$$-S\,\mathrm{d}T + V\,\mathrm{d}p = \xi_1\mathrm{d}F_1 + \xi_2\mathrm{d}F_2 + \ldots + \xi_m\mathrm{d}F_m\,. \tag{B.30}$$

This is the so-called *Gibbs-Duhem equation*. It tells us that, in the process that joins two homogeneous equilibrium states, the quantities F_i cannot all be increased independently of the changes in T and p.

To provide an example of application of eq. (B.30), let us refer to a system possessing just one single independent generalized displacement, say ξ_1. Then, equation (B.30) reduces to

$$-S\,\mathrm{d}T + V\,\mathrm{d}p = \xi_1\mathrm{d}F_1\,. \tag{B.31}$$

That is

$$\mathrm{d}F_1 = -\frac{S}{\xi_1}\mathrm{d}T + \frac{V}{\xi_1}\mathrm{d}p\,. \tag{B.32}$$

This means that, in any process that occurs at constant temperature and constant pressure, the increment dF_1 should be equal to zero. In this system, therefore, the generalized force F_1 cannot depend on the variable ξ_1 that is conjugated to it. That is, $F_1 = F_1(p, T)$. From this result and from eq. (B.32), it also follows that, in the considered system, the quantities V and S must be linear functions of ξ_1.

B.2 Thermodynamics of Real Systems

In the previous section, the variables T, p, and $\boldsymbol{F}$ that appear in eq. (B.2) were considered to be functions of the natural state variables, namely S, V, and $\boldsymbol{\xi}$. Accordingly, $T = T(S, V, \boldsymbol{\xi})$, $p = p(S, V, \boldsymbol{\xi})$, and $\boldsymbol{F} = \boldsymbol{F}(S, V, \boldsymbol{\xi})$. In doing so, we assumed that eq. (B.2) was integrable over the entire interval of definition of the variables S, V, and $\boldsymbol{\xi}$. However, there is no guarantee that eq. (B.2) is integrable when it refers to a real system. If it cannot be integrated, it cannot be used to determine the function $U = U(S, V, \boldsymbol{\xi})$. In that case, T can no longer be related to S by eq. (B.6), and the natural variable approach loses much of its use.

The integrability conditions of a total differential form, such as eq. (B.2), are well known. These conditions restrict the coefficients of the infinitesimal increments of the independent variables that appear on the right side of eq. (B.2) by requiring that the quantities T, p, and $\boldsymbol{F}$ be appropriate partial derivative of one single scalar function of the variables S, V, and $\boldsymbol{\xi}$. In the present case, the scalar function is $U = U(S, V, \boldsymbol{\xi})$, and the relations that link the coefficients T, p, and $\boldsymbol{F}$ to it are eqs (B.6) to (B.8). In the particular case of two independent variables, say, S and V, we have that $U = U(S, V)$. In this case, the integrability conditions reduce to the single condition:

$$\frac{\partial T}{\partial V} = -\frac{\partial p}{\partial S}, \tag{B.33}$$

as immediately follows from eqs (B.6) and (B.7) and from the fact that

$$\frac{\partial^2 U}{\partial S \partial V} = \frac{\partial^2 U}{\partial V \partial S}, \tag{B.34}$$

In the case of more than two independent variables, additional integrability conditions analogous to eq. (B.33) must be added. The argument is a well-known topic in the theory of differential equations and needs no further explanation here.

However, there is no compelling physical reason for a real system to comply with the integrability conditions above. If anything, the opposite is true; i.e., a real system will not generally comply with such conditions. As discussed in the previous section, an almost universal feature of any real system is that it exhibits changes of phase if the state variables are allowed to span large enough intervals of values. A phase change invariably means a discontinuity in one or more partial derivatives of the internal energy function of the system. This results in loss of integrability for eq. (B.2), if the equation is considered in the whole range of states that the system can attain. As far as the relations presented in the previous sections, the consequence of this is profound. Many of the equations lose their validity, and many of the conclusions reached must be modified or re-interpreted.

The internal energy surface of a real system

To substantiate the above claim, let us refer to a system whose internal energy partial derivatives become discontinuous for certain values of the system's state variables. For simplicity, we shall assume that the state of the system is defined by S and V. In this case, the system's internal energy, $U = U(S, V)$, represents a surface in the space of S, V, and U. Figure B.1 portrays a portion of that surface. It is actually made of two sub-surfaces, U_1 and U_2, which intersect along a line d that projects as a curve $\varphi = \varphi(S, V)$ on plane $[S, V]$. The region of plane $[S, V]$ on one side of curve φ is denoted as $\mathscr{D}_1$, while the region on the other side of the same curve is denoted as $\mathscr{D}_2$. The internal energy function, $U = U(S, V)$, is represented by surface U_1 in domain $\mathscr{D}_1$ and by surface U_2 in domain $\mathscr{D}_2$. It is a continuous function over the entire domain $\mathscr{D} = \mathscr{D}_1 + \mathscr{D}_2$, and it possesses continuous partial derivatives inside the domains $\mathscr{D}_1$ and $\mathscr{D}_2$. However, it is not differentiable at the common boundary φ of these domains, because, at that boundary, some of the partial derivatives of U become discontinuous, for instance, due to a phase change taking place at that boundary.

In Fig. B.1 the discontinuity in the derivatives of U reveals itself as an abrupt change in the tangent plane to surface U as the independent variables S and V cross curve φ. The lack of differentiability at the points of φ means that function $U = U(S, V)$ cannot be obtained by integrating eq. (B.2) over the domain $\mathscr{D}$, in spite of the fact that eq. (B.2) holds true throughout that domain.

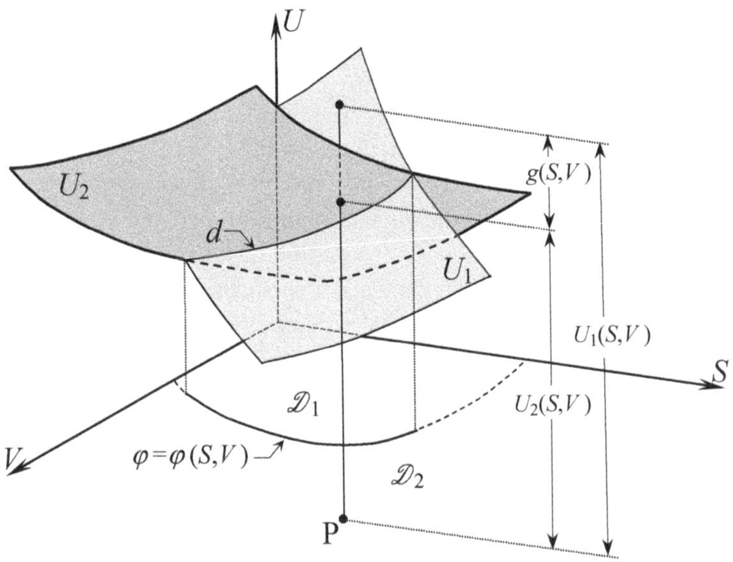

Fig. B.1. The internal energy surface $U = U(S, V)$ of a system that exhibits phase changes within the domain $\mathcal{D} = \mathcal{D}_1 + \mathcal{D}_2$. The surface is composed of two sub-surfaces, U_1 and U_2, which intersect along a curve d where a phase change occurs.

To better illustrate this point, let us refer to the difference $g(S, V) = U_2(S, V) - U_1(S, V)$. With the help of function $g(S, V)$, the internal energy of the system outside region $\mathcal{D}_1$ can be expressed as $U = U_2(S, V) = U_1(S, V) - g(S, V)$. This shows that $g = g(S, V)$ accounts for the discontinuity in the partial derivatives of U that takes place at the points where a phase change takes place. However, eq. (B.2) contains no information whatsoever concerning these points, let alone the extent of discontinuity that is produced there. This makes the same equation incapable of determining the function $g(S, V)$ and therefore incapable of determining the internal energy surface over its entire domain of definition.

Though very important from the conceptual standpoint, the above issue is rarely considered in the literature. Consider still the example of Fig. B.1 from another perspective. As observed, some of the partial derivatives of U are not continuous at the intersection of the two sub-surfaces, U_1 and U_2. From eqs (B.6) and (B.7), it then follows that the pair of functions $T = T(S, V)$ and $p = p(S, V)$ are different in the two sub-domains $\mathcal{D}_1$ and $\mathcal{D}_2$. Because it involves the functions T and p, eq. (B.2) will assume a different form in each sub-domain. Of course, that equation can be integrated separately in sub-domain $\mathcal{D}_1$ and sub-domain $\mathcal{D}_2$, provided that it meets the appropriate integrability conditions in these sub-domains. The integration, however, will produce a different function in each sub-domain. In Fig. B.1, the functions that result from the integration of eq. (B.2) in sub-domains $\mathcal{D}_1$ and $\mathcal{D}_2$ are represented by surfaces $U_1(S, V)$ and $U_2(S, V)$, respectively.

Both $U_1(S, V)$ and $U_2(S, V)$ are defined within an arbitrary integration constant. Thus, to determine $U(S, V)$ we must assign two arbitrary constants. We can take care of one of these constants by fixing the value of U at an arbitrarily chosen reference point. To determine the other constant, we may impose that $U_1 = U_2$ at a point where the phase change occurs. However, the locus of all such points must belong both to surface U_1 and to surface U_2, because the internal energy is a continuous function of the state variables of the system. This imposes a rather stringent condition on the intersection of the two surfaces, which is extraneous to eq. (B.2). As a consequence, at variance with what is true for ideal systems, eq. (B.2) alone cannot determine the internal energy, $U = U(S, V)$, of a system that exhibits phase changes.

Application of ideal system formulae to real systems

The presence of discontinuities in the partial derivatives of U means in particular that relations (B.6) and (B.7)—and relation (B.8) in the case of more than two independent variables—cannot hold true at the points where phase changes occurs. This means that eq. (B.10) cannot represent the internal energy function of a real system for all values of S, V, and $\boldsymbol{\xi}$. Moreover, relations (B.17) to (B.19) cannot be applied to real systems, since these relations derive directly from relation (B.10).

However, relation (B.10) and relations (B.17) to (B.20) can also be of use when dealing with real systems, provided that those relations are employed to determine the change in U, H, Ψ, or G rather than the actual values of these quantities. In this case, the following restrictions must be met. First, the application of the relations (B.10) and (B.17) to (B.20) must be confined to a sub-domain in which eq. (B.2) is integrable. We denote $\mathcal{D}_\mathsf{s}$ as the generic of such sub-domains. The index s introduced here ranges from 1 to the number n of sub-domains in which $\mathcal{D}$ is integrable. Second, relations (B.10) and (B.17) to (B.20) must refer to the particular function U_s that results from the integration of eq. (B.2) in domain $\mathscr{D}_s$. This means that the functions H, Ψ, and G in these sub-domains are obtained by inserting U_s in their respective definitions (B.11)-(B.13).

Let us refer, for instance, to the Gibbs free energy, G, of a real system. Similar arguments apply to the other thermodynamic potentials as well. Suppose that we are interested in the changes in G due to a process that occurs in a sub-domain $\mathscr{D}_s$ in which eq. (B.2) is integrable. Let $U_s = U_s(S, V, \boldsymbol{\xi})$ be the function that is obtained by integrating eq. (B.2) in the considered sub-domain. By introducing U_s into eq. (B.13), we define the corresponding potential function G_s. Then, by the same arguments that we followed to establish eq. (B.19), it is easy to verify that eq. (B.19) applies to G_s. That is,

$$G_s = F_1\,\xi_1 + F_2\,\xi_2 + \ldots + F_m\,\xi_m + C_s\,, \tag{B.35}$$

where C_s is an appropriate constant. Of course, G_s is far from being the Gibbs free energy function of the system over the entire domain $\mathscr{D}$. The reason is that U_s does not coincide with U if eq. (B.2) is not integrable in $\mathscr{D}$, as Fig. B.1 shows. However, in each sub-domain $\mathscr{D}_s$. we have

$$\Delta G_s = \Delta G\,, \tag{B.36}$$

because $\Delta U_s = \Delta U$ in that sub-domain. Equation (B.36) means that, in $\mathscr{D}_s$, the two functions, G_s and G, differ by a constant. Therefore, as far as the changes in G are concerned, it makes no difference whether we refer to G_s or G. This is true, of course, only for states that fall within the sub-domain $\mathscr{D}_s$.

An important consequence of eq. (B.36) follows from the fact that the operation of partial derivation commutes with the increment operator "Δ". Thus, by applying that operator to both sides of eq. (B.23), we obtain, with the help of eq. (B.36):

$$\Delta S = -\left(\frac{\partial \Delta G}{\partial T}\right)_{p,\xi} = -\,\Delta\left(\frac{\partial G_s}{\partial T}\right)_{p,\xi}\,. \tag{B.37}$$

Similarly, the following relations

$$\Delta V = \left(\frac{\partial \Delta G}{\partial p}\right)_{T,\xi} = \Delta\left(\frac{\partial G_s}{\partial p}\right)_{T,\xi} \tag{B.38}$$

and

$$\Delta F_{\mathrm{k}} = \left(\frac{\partial \Delta G}{\partial \xi_{\mathrm{k}}}\right)_{p,T,\xi_{\mathrm{s}}} = \Delta\left(\frac{\partial G_{\mathrm{s}}}{\partial \xi_{\mathrm{k}}}\right)_{p,T,\xi_{\mathrm{s}}} \qquad (k = 1, 2, \dots m) \tag{B.39}$$

can be obtained from eqs (B.24) and (B.22), respectively.

The consequence is that, for a real system, the increments ΔS, ΔV, and ΔF_{k}, rather than the values of S, V and F_{k}, are related to the partial derivatives of G_{s} in each sub-domain $\mathscr{D}_{\mathrm{s}}$ in which eq. (B.2) is integrable. In each of these sub-domains, we therefore have

$$S = -\left(\frac{\partial G_{\mathrm{s}}}{\partial T}\right)_{p,\xi} + S^{\Delta}, \tag{B.40}$$

$$V = \left(\frac{\partial G_{\mathrm{s}}}{\partial p}\right)_{T,\xi} + V^{\Delta}, \tag{B.41}$$

and

$$F_{\mathrm{k}} = \left(\frac{\partial G_{\mathrm{s}}}{\partial \xi_{\mathrm{k}}}\right)_{p,T,\xi_{\mathrm{s}}} + F_{\mathrm{k}}^{\Delta}, \quad (k = 1, 2, \dots m), \tag{B.42}$$

where S^{Δ}, V^{Δ}, and F_{k}^{Δ} are constants. The values of these constants are not arbitrary; they depend on the extent of discontinuity suffered by the partial derivatives of the system's free energy due to the phase changes that took place outside and at the boundary of the considered sub-domain $\mathscr{D}_{\mathrm{s}}$. Equations (B.40) to (B.42) extend to real systems the results (B.23), (B.24), and (B.22) that apply to ideal systems.

The internal energy function of real systems

Since eq. (B.10) does not apply to real systems, how should the internal energy function of such systems be determined? Because eq. (B.6) generally does not apply to a real system, we must use T instead of S as the independent thermal variable of the system. Suppose then that we chose $(T, V, \boldsymbol{\xi})$ rather than $(S, V, \boldsymbol{\xi})$ as variables of state of the system. The system's entropy becomes a function, $S = S(T, V, \boldsymbol{\xi})$, that can be determined experimentally (e.g., by the calorimetric method discussed in Section 1.7), while the system's internal energy, $U = U(T, V, \boldsymbol{\xi})$, or, better, its difference with respect to the reference value, is obtained from the first law of thermodynamics, as specified by eq. (1.4.6).

More precisely, eq. (1.4.6) enables us to calculate the change in internal energy when the system goes from equilibrium state A to equilibrium state B. The result is independent of whether the process that joins the two states is reversible or not, because U is a state function. If the function $S = S(T, V, \boldsymbol{\xi})$ is known, it may be convenient to consider the case in which the process from A to B is reversible. In this case, the heat absorbed in the process can be calculated as

$$Q = \int_A^B T\,\mathrm{d}S = \int_A^B T\frac{\partial S}{\partial T}\mathrm{d}T + \int_A^B T\frac{\partial S}{\partial V}\mathrm{d}V + \sum_i \int_A^B T\frac{\partial S}{\partial \xi_i}\mathrm{d}\xi_i\,, \qquad \text{(B.43)}$$

which immediately follows from eq. (1.6.4). Likewise, if the equilibrium values of p and $\boldsymbol{F}$ are known, the work done on the reversible process is given by

$$W_{in} = \int_A^B \mathrm{d}W_{in} = -\int_A^B p\,\mathrm{d}V + \sum_i \int_A^B F_i\,\mathrm{d}\xi_i\,, \qquad \text{(B.44)}$$

The change in internal energy is then obtained from eq. (1.4.6):

$$U_B - U_A = \int_A^B T\,\mathrm{d}S + \int_A^B \mathrm{d}W_{in}\,, \qquad \text{(B.45)}$$

where the integrals are given by eqs (B.43) and (B.44), respectively.

By taking A as the reference state and B as any equilibrium state, eq. (B.45) can be written in the following form:

$$U = U(T, V, \boldsymbol{\xi}) = \int T\,\mathrm{d}S + \int \mathrm{d}W_{in} + \mathrm{C}\,, \qquad \text{(B.46)}$$

where $\mathrm{C} = U_A$ is an appropriate constant. The internal energy function of a real system is thus determined.

Isothermal processes

For isothermal processes at any given temperature T, eq. (B.46) can be simplified by taking the reference state for the internal energy as a state at temperature T. In that case, eq. (B.46) can be reduced to

$$U = U(V, \boldsymbol{\xi}; T) = T\,\Delta S + W_{in} + \mathrm{C}\,, \qquad \text{(B.47)}$$

where ΔS is the difference between the entropy of the actual state and the entropy of the reference state. In eq. (B.47), as in eqs (B.45) and

(B.46), the quantity W_{in} is the work done when the process that joins the reference state to the actual state of the system is performed reversibly. As follows from relation (1.8.2), the quantity $T\Delta S$ that appears in formula (B.47) is the reversible heat absorbed by the system in the same process.

Equation (B.47) is often used when dealing with isothermal processes of a real system. It should carefully be distinguished from the formally similar eq. (B.10), which takes S as the thermal state variable, thus making T, p, and $\boldsymbol{F}$ functions of entropy. This contrasts with eq. (B.47) which takes T as a thermal state variable, thus making $S = S(T, V, \boldsymbol{\xi})$ a dependent quantity.

Once internal energy and entropy are expressed as functions of the variables T, V, and $\boldsymbol{\xi}$, the potentials H, Ψ, and G can be expressed as functions of the same variables by applying definitions (B.11), (B.12), and (B.13), respectively. The potentials thus obtained have the same physical meaning as those considered in the previous section, although they are now expressed in terms of a different set of independent variables. For this reason, special care must be exercised in using eqs (B.26)-(B.28), because these equations, as they stand, are valid only for a particular set of independent variables, different for different equations and for different potentials.

APPENDIX C

Thermochemical Systems

For easier reference, we recall here a few basic notions of thermochemistry. Much of the material considered in this appendix is standard. The reader may refer to any of the many excellent textbooks on physical chemistry (e.g., [23] and [2]) for a complete exposition of the matter treated here.

C.1 Partial Molar Gibbs Free Energies

Thermochemistry deals with mixtures of different chemical substances that may or may not react chemically with each other. A homogeneous mixture is also referred to as a solution. In the case of solutions, the state variables that define the generalized displacement vector $\boldsymbol{\xi}$ introduced in eq. (1.3.1) are the number of moles n_i ($i = 1, 2, \ldots, r$) of the r substances that make up the solution. Therefore, eq. (B.1) becomes:

$$\mathrm{d}W_{\mathrm{in}} = -p\,\mathrm{d}V + \sum_{\mathrm{i}=1}^{\mathrm{r}} \mu_{\mathrm{i}}\,\mathrm{d}n_{\mathrm{i}}\,. \tag{C.1}$$

Here, the quantities μ_i are the generalized forces conjugated to the variables n_i. They are functions of the state variables of the system and are also referred to as the *chemical potentials* of the substances to which they are conjugated. This terminology dates back to Gibbs, although it is not the only terminology in use today.

In terms of μ_i and n_i, the fundamental equation (B.2) can be written as

$$dU = T\,dS - p\,dV + \sum_{i=1}^{r} \mu_i \, dn_i \,. \tag{C.2}$$

From this equation, through arguments similar to those leading to eqs (B.16) and (B.20), it can be concluded that

$$dG = -S\,dT + V\,dp + \sum_{i=1}^{r} \mu_i \, dn_i \tag{C.3}$$

and that

$$G = G(p,T,n_1,n_2,...,n_r) = \mu_1 n_1 + \mu_2 n_2 + ... + \mu_r n_r \,. \tag{C.4}$$

The latter equation holds true to within an additional arbitrary constant. It shows that μ_i represents the contribution to the total Gibbs free energy of the system that comes from 1 mol of substance *i*. For this reason, the chemical potential μ_i is also referred to as the *partial molar Gibbs free energy* of substance *i*.

Equation (C.4) applies to systems that behave as ideal systems in the sense specified in Appendix B. It shows that, in an ideal system, the dependence of *G* on *p* and *T* occurs through functions μ_i. For processes that take place at constant *p* and *T*, equation (C.3) yields

$$dG = \mu_1 \, dn_1 + \mu_2 \, dn_2 + ... + \mu_r \, dn_r \,. \tag{C.5}$$

This equation is a particular case of eq. (B.25), which refers to thermochemical systems in general. From eqs (C.4) and (C.5) it can be inferred that *G* must be a homogeneous function of degree 1 in the variables n_i.

An immediate consequence of eq. (C.3) and of the fact that $G = G(p,T,n_1,n_2,\ldots,n_r)$ is

$$S = -\left(\frac{\partial G}{\partial T}\right)_{n_i,\,p} , \tag{C.6}$$

$$V = \left(\frac{\partial G}{\partial p}\right)_{n_i, T}, \tag{C.7}$$

and

$$\mu_i = \left(\frac{\partial G}{\partial n_i}\right)_{p, T, n'}. \tag{C.8}$$

The last equation is by far the most important to the present discussion. The subscript n' stands for all variables n_k, with the exception of n_i.

Note that for real system eqs (C.6) through (C.8) should be written in a form that is analogous to eqs (B.40) through (B.42). To make this point clear, we introduced the notation G_s in Section B.2 to refer to the function G in each sub-domain $\mathcal{D}_s$ where eq. (B.2) is integrable. This issue has been ignored in the literature; no distinction is usually made between ideal and real systems when dealing with the above equations. However, the analysis in Appendix B should make it clear that, for real systems, eqs (B.40) to (B.42) provide the correct expressions of the functions S, V, and G to be introduced in eqs (C.6) to (C.8).

C.2 Chemical Reactions

Up to this point, we have not distinguished between non-reacting mixtures and mixtures of chemically-reacting substances. The difference between the two kinds of mixtures relates to how the amounts n_i can be changed. In a non-reacting mixture, any change in n_i involves adding or subtracting the appropriate component to the mixture. Therefore, such a change can only take place in an *open system* (i.e., a system that exchanges matter with its surroundings). The same does not necessarily apply to systems that contain chemically-reacting substances. In this case, the quantities n_i may change even in a *closed system* because of the chemical reactions that take place within the system.

A typical chemical reaction is represented by a balanced chemical equation such as

$$\nu_A \, \mathrm{A} + \nu_B \, \mathrm{B} = \nu_C \, \mathrm{C} + \nu_D \, \mathrm{D}. \tag{C.9}$$

Here, A, B, C, and D stand for a fixed number of molecules of the homologous chemical species A, B, C, and D that appear in the reaction. The actual number of molecules is immaterial, but it is the same for all species. Quite often, this number is expressed in moles.

(One mole of a species indicates an Avogadro's number, N = 6.0221 10^{23}, of particles of that species). The quantities ν_A, ν_B, ν_C, and ν_D are the so-called *stoichiometric coefficients*. They are pure numbers that specify the relative amounts of reactants (the species on the left side of the equation) and products of reaction (the species on the right side of the same equation). Thus, if A, B, C, and D represent 1 mol of each species, then ν_A moles of A react with ν_B moles of B to produce ν_C moles of C and ν_D moles of D.

It is sometimes convenient to write a chemical equation such as (C.9) in the following form:

$$0 = \sum_{i=1}^{r} \nu_i A_i , \tag{C.10}$$

where the each quantity A_i (i=1,2,..., r) indicates a fixed number, usually one mole, of molecules of the homologous species A_i involved in the reaction. When a chemical reaction is written in this way, the stoichiometric coefficients of reactants are given a negative sign, while those of products of reaction are positive.

We use the term 'reactants' to refer to the species on the left side of a chemical reaction equation expressed in the form (C.9) and the term 'products of reaction' to refer to the species on the right side of the equation. This presumes that the chemical process proceeds from the left to the right. However, the reaction may also proceed in the opposite direction if it is properly driven (e.g., by electrochemical means) or if the initial conditions are appropriate. This poses no problem with the adopted definitions of reactants and products of reaction as long as we agree that, when the reaction proceeds from the right to the left, the 'products' are in fact the species that react, whereas the 'reactants' are in fact the products of the reaction.

C.3 Extent of Reaction

A chemical reaction equation describes the way in which a chemical process occurs. Suppose that, at the beginning of the reaction, the system contains only the reactants. We assume that the system is closed, so that no new material can be added to or extracted from it. The reaction process will go on transforming reactants into products of reaction in the proportions that are specified by the reaction equation. Eventually, a state of chemical equilibrium is attained in which reactants and products have reached definite proportions that remain constant over time.

In certain cases, the final state of the process will contain products of reactions only. In that case the process has gone to completion. A necessary (but by no means sufficient) condition for this to occur is that various reactants that are put together at the beginning of the reaction are in stoichiometric proportions (i.e., their relative amounts are proportional to their stoichiometric coefficients in the reaction equation). In general, however, the process will not go to completion. The final state of the system may contain both reactants and products of reaction even if the reactants were initially in the correct stoichiometric proportions. The amount of each species at this final, equilibrium state can be predicted from thermodynamics, and this is a major success of thermochemistry.

During a chemical reaction, the amount of the species changes proportionally to the stoichiometric coefficients as specified by the relevant chemical equation. Thus, by taking ν_i negative when referred to a reactant, the change in the number of moles of species A_i at any stage of the reaction can be expressed as:

$$n_i - n_i^o = \nu_i \xi \qquad (i = 1, 2, \ldots, r), \qquad (C.11)$$

where n_i is the number of moles of A_i at the considered stage of the process, and n_i^o denotes the amount of the same species at the beginning of the process. The variable ξ introduced above is the *extent of reaction*. It is an extensive, non-negative, scalar variable that has the same physical dimensions as n_i, since the stoichiometric coefficients are dimensionless. Thus, ξ is usually measured in moles. As eqs (C.11) should make it apparent, ξ can be taken as a state variable of the reacting system, since it suffices to determine the quantities n_i once the constants n_i^o are specified.

At the beginning of the process the variable ξ is equal to zero, since $n_i = n_i^o$. As ξ increases, the amount of the reactants decreases, and the amount of the products increases. This follows immediately from eq. (C.11), since the values of ν_i are negative for the reactants and positive for the products. If the reaction is described by an equation such as (C.9), an increase in ξ will make the process proceed from left to right. Starting from any intermediate stage of the process, we can apply appropriate actions on the system that make ξ increase or decrease. If ξ decreases, the reaction will proceed from right to left.

The largest value that ξ can attain depends on the initial amount of the reactants. However, it can never exceed the value that makes a reactant disappear, since, at that value, the reaction is brought to a halt by the depletion of a reactant. The other reactants may or may not be used up, depending on their proportions at the beginning of the reaction.

For any given reaction, a *chemical transformation* is defined as the reaction of the smallest number of atoms or molecules that are needed to perform the reaction itself. Thus, a reaction is generally made of an enormous number of chemical transformations. From eq. (C.10), we see that each of these transformations involves just ν_i atoms/molecules of each component A_i. However, from eq. (C.11), we know that $\nu_i\,\xi$ is the number of moles of A_i that are transformed by the reaction as its extent of reaction reaches the value ξ. Therefore, for closed systems, ξ coincides with the number expressed in moles of chemical transformations brought about by the reaction as the system's composition changes from n_i^0 to n_i. In particular, for a closed, homogeneous system of volume V, the ratio ξ/V coincides with the number of chemical transformations per unit volume that must take place in the system as the extent of reaction goes from zero to ξ.

If the reaction occurs at constant pressure and temperature, chemical equilibrium must correspond to the state at which the Gibbs free energy of the system is at its minimum value [cf. eq. (1.11.18) and the remarks following it]. As observed, that state will generally contain non-vanishing amounts of both reactants and products of reaction. Their actual proportions can be calculated from eq. (C.11) once the value of the extent of reaction at equilibrium is given. This value, denoted by ξ_{eq}, depends on pressure and temperature.

C.4 Reaction Gibbs Free Energy

In a chemically-reacting system, the quantities n_i depend on ξ, according to eq. (C.11). Since $G = G(p,T,n_1,n_2,,\ldots,n_r)$, it follows that the Gibbs free energy of a closed system undergoing a chemical reaction process can be expressed as:

$$G = G(p, T, \xi) \tag{C.12}$$

Here, we assume that just one single chemical reaction is taking place in the system. The generalization of the above formula to cover several simultaneous reactions can be made by introducing a different variable ξ for each reaction. If constant pressure and constant volume conditions are considered, then ξ is the only independent variable that changes during the process. In this case, we have

$$G = G(\xi)_{p,T} \,. \tag{C.13}$$

The derivative with respect to ξ of function (C.13) plays a fundamental role in the progress of the reaction. To illustrate, let us first differentiate both sides of eq. (C.11) to obtain:

$$\mathrm{d}n_\mathrm{i} = \nu_\mathrm{i}\,\mathrm{d}\xi \,. \tag{C.14}$$

When this expression is introduced into eq. (C.3), we have:

$$\mathrm{dG}\big|_{p,T} = \sum_{\mathrm{i=1}}^{\mathrm{r}} \nu_\mathrm{i}\,\mu_\mathrm{i}\,\mathrm{d}\xi \tag{C.15}$$

The summation in this equation can be split into two separate sums, one for reactants and one for products. Thus, the equation can be rewritten as

$$\mathrm{dG}\big|_{p,T} = \left[\left(\sum\nolimits_\mathrm{h} \nu_\mathrm{h}\,\mu_\mathrm{h}\right)_{\text{products}} - \left(\sum\nolimits_\mathrm{k} |\nu_\mathrm{k}|\,\mu_\mathrm{k}\right)_{\text{reactants}}\right]\mathrm{d}\xi \tag{C.16}$$

or

$$\mathrm{dG}\big|_{p,T} = \Delta_\mathrm{r}G\,\mathrm{d}\xi \,, \tag{C.17}$$

where

$$\Delta_\mathrm{r}G = \left[\left(\sum\nolimits_\mathrm{h} \nu_\mathrm{h}\,\mu_\mathrm{h}\right)_{\text{products}} - \left(\sum\nolimits_\mathrm{k} |\nu_\mathrm{k}|\,\mu_\mathrm{k}\right)_{\text{reactants}}\right]. \tag{C.18}$$

The quantity $\Delta_\mathrm{r}G$ defined by this equation is usually called *reaction Gibbs energy*. This terminology is not free of some ambiguity, because it does not make it clear that $\Delta_\mathrm{r}G$ is actually energy per unit extent of reaction. In addition, the same terminology may give the false impression that $\Delta_\mathrm{r}G$ assumes a definite value for any definite reaction. Clearly, this is not true. $\Delta_\mathrm{r}G$ is a function of p, T, and n_1, n_2, …, n_r, because so are the quantities μ_i (see the next section). Thus, $\Delta_\mathrm{r}G = \Delta_\mathrm{r}G(p, T, n_1, n_2, \ldots, n_\mathrm{r})$. However, in standard conditions, $\Delta_\mathrm{r}G$ assumes a definite value for any given reaction. This value is denoted by $\Delta_\mathrm{r}G°$ and is discussed at the end of this section.

For closed systems, $\Delta_r G$ also can be expressed as $\Delta_r G = \Delta_r G(p, T, \xi)$, because, in this case, eq. (C.11) applies. From eqs (C.3) and (C.14), we infer that, for a chemical reaction that occurs in a closed system, $\Delta_r G$ is the energy that the system liberates (if $\Delta_r G < 0$) or absorbs (if $\Delta_r G > 0$) per mole of ξ at any stage ξ of the reaction. Thus, $\Delta_r G$ is an intensive quantity that has units of energy per mole.

Another important interpretation of $\Delta_r G$ can be obtained as follows. By differentiating eq. (C.13), we get:

$$\mathrm{d}G|_{p,T} = \left(\frac{\partial G}{\partial \xi}\right)_{p,T} \mathrm{d}\xi \,. \tag{C.19}$$

From this and from eq. (C.17), we conclude that

$$\Delta_r G = \left(\frac{\partial G}{\partial \xi}\right)_{p,T} . \tag{C.20}$$

Because function (C.13) represents a curve on a plane [G, ξ], eq. (C.19) shows that $\Delta_r G$ can also be interpreted as a measure of the slope of that curve (Fig. C.1). The greater the absolute value of $\Delta_r G$, the steeper the curve.

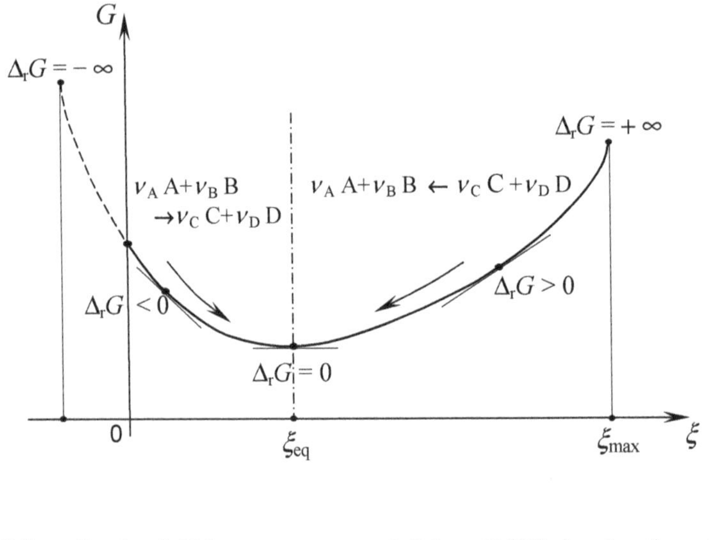

Fig. C.1. Graph of $G(\xi)$ at constant p and T [eq. (C.13)] showing the point of chemical equilibrium, ξ_{eq}, and the spontaneous directions of reaction. The plot refers to a system containing a non-vanishing amount of products of reaction for $\xi = 0$, while for $\xi = \xi_{max}$ at least one of the reactants vanishes (cf. Section C.7).

If function (C.13) is given, the condition $\Delta_r G = 0$ locates the point of zero slope of that function. Accordingly, the corresponding value of ξ is the equilibrium value, ξ_{eq}. Under constant temperature conditions, the system moves toward that point spontaneously, provided that no external forces other than pressure act upon the system. The reason is that, under the same conditions, an increase in the system's Gibbs free energy would be incompatible with the second law of thermodynamics (see Section 1.11 for more details on this point). Once the value of ξ_{eq} is known, the composition of the system at chemical equilibrium follows directly from eq. (C.11).

The situation is different if there are other forces besides p that do work on the system. In this case, the (generalized) forces applied to the system can drive it to a value of ξ far from ξ_{eq}, and, if the forces are not removed, the system can be kept in that state for any length of time. A typical example is when a reaction takes place in an electrochemical cell. By controlling the difference in the electric potential of the cell's electrodes, the extent of reaction ξ can be changed at will, both above and below the value ξ_{eq}, and kept at the non-equilibrium value indefinitely. As soon as the electric field is removed, the spontaneous reaction process starts again. According to eq. (C.20), the value of $\Delta_r G$ at the starting of the process may be positive or negative, depending on the value of the tangent to curve (C.20) when the electric field is removed (Fig. C.1). In both cases, the process that follows the electric field removal will make G decrease, because the reaction heads toward the point of minimum of G, at $\xi = \xi_{eq}$. Accordingly, the extent of reaction ξ increases or decreases depending on whether its initial value was less than or greater than ξ_{eq}.

An increase in ξ decreases the amount of reactants and increases the amount of products. This immediately follows from eqs (C.11). In this case, if the reaction is represented as in eq. (C.9), it will proceed from the left to right of that equation. Conversely, if the process makes ξ decrease, the direction of the reaction is reversed; i.e., the 'products' on the right side of eq. (C.9) will transform into the 'reactants' on the left side of the same equation.

The quantity $\Delta_r G$ introduced by eq. (C.18) should be carefully distinguished from the similar quantity, $\Delta_r G°$, defined by:

$$\Delta_r G^o = \left[\left(\sum_h \nu_h \mu_h^o \right)_{\text{products}} - \left(\sum_k |\nu_k| \mu_k^o \right)_{\text{reactants}} \right], \qquad \text{(C.21)}$$

where μ_i^o indicates the values that μ_i assumes in the standard state, assumed to be that of the pure substance at the pressure of 1 bar and at the considered temperature. Often, $\Delta_r G°$ is referred to as the *standard molar reaction Gibbs energy*. It represents the difference between the standard Gibbs energy of the products of reaction and that of the reactants, when the number of moles of each component is equal to its respective stoichiometric coefficient. Thus, $\Delta_r G°$ need not coincide with any of the values that $\Delta_r G$ attains during the reaction. The reason is that, in general, there may not be any stage of the reaction in which reactants and products are simultaneously at their standard state and at the stoichiometric proportions prescribed by the reaction equation.

The quantity $\Delta_r G°$ can be calculated directly from the tabulated standard values (usually referred to 25 °C and 1 bar) of the energy of formation of the species involved in the reaction. Clearly, in terms of these energies, eq. (C.21) can be written as:

$$\Delta_r G^o = \left[\left(\sum_h \nu_h \, \Delta G_h^F \right)_{\text{products}} - \left(\sum_k |\nu_k| \Delta G_k^F \right)_{\text{reactants}} \right], \tag{C.22}$$

where ΔG_i^F is the standard molar energy of formation of species i. Remember that ΔG_i^F vanishes for any pure element in its most stable state. A negative value of $\Delta_r G°$ usually indicates that the reaction tends to occur spontaneously, although this may not always be true.

C.5 Activities and Activity Coefficients

The chemical potentials μ_i introduced in Section C.1 are functions of the state variables of the system. It is well known that they can be expressed in the form:

$$\mu_i = \mu_i^o + RT \ln a_i \,, \tag{C.23}$$

cf., e.g., [23], [2], [13], and [21]. The quantities R, a_i, and μ_i^o introduced in this formula have the following meaning. R is the universal gas constant ($R = 8.3145$ J K^{-1} mol^{-1}). The quantity a_i is the *activity* of the substance to which μ_i refers. It is a dimensionless function of the state variables of the system that is different for different systems and for different substances within the system. μ_i^o is the so-called *standard chemical potential*. It coincides with the value that μ_i assumes at a reference state in which pressure and composition assume some specified values. This reference state is usually referred to as the standard state.

It should be noted that the term "standard state" has nothing to do with the analogous terms "standard pressure," "standard temperature," and "standard solution" in use in analytical chemistry. The standard state to which μ_i^0 refers can vary for different mixtures and for different components in the same mixture or solution. Moreover, it does not refer to any fixed values of temperature and pressure. Consequently, there are different standard states, and different values of μ_i^0, depending on the considered values of T and p; i.e., $\mu_i^0 = \mu_i^0(T, p)$. In some important cases, such as liquid solutions, the dependence of μ_i^0 on p is negligible. In these cases, $\mu_i^0 = \mu_i^0(T)$. Quite often, the standard state is taken as the state of the pure substance at the standard pressure p^0 and at the considered temperature. The latter need not coincide with the standard temperature, although it frequently does.

Being a reference state, the standard state can, in principle, be fixed arbitrarily. Because μ_i^0 depends on that state, eq. (C.23) does not suffice to define a_i uniquely. Therefore, to specify the activity of a substance in a mixture or in a solution, the conditions that define the standard state of that substance must be assigned. For ideal mixtures and solutions, these conditions are obtained directly from theory, as we shall see in a moment. However, for non-ideal mixtures or solutions, the additional conditions required to fully define the activity of a component or a solute are conventional.

Ideal gas mixtures and ideal liquid solutions

For ideal gas mixtures and ideal liquid solutions, the expression of the chemical potentials μ_i is obtained on theoretical grounds. These potentials have the form (C.23), which fact actually motivated the use of a similar form for the non-ideal case. For each species in an ideal mixture or solution, the expression of the activity a_i and the definition of the standard state are obtained by comparing the theoretically-derived expression of μ_i with eq. (C.23).

More specifically, in the case of mixtures of ideal gases, the theoretical expression of the chemical potential of gas component A_i has the well-known form:

$$\mu_i = \mu_i^0 + RT \ln \frac{p_i}{p^0}. \qquad (C.24)$$

Here, p_i is the partial pressure of the considered gas component, p^0 indicates the pressure of the mixture as a whole, and μ_i^0 denotes the value of μ_i for $p_i = p^0$ at the considered temperature T (the particular case in which $p^{\mathrm{o}} = 1$ bar is considered quite frequently). By comparing eq. (C.23) with eq. (C.24), it follows that, for ideal gas mixtures:

$$a_{\mathrm{i}} = \frac{p_{\mathrm{i}}}{p^{\mathrm{o}}}. \tag{C.25}$$

As for the standard state, we observe that eqs (C.23) through (C.25) indicate that $\mu_{\mathrm{i}}^{\mathrm{o}}$ coincides with the value μ_{i} when $a_{\mathrm{i}} = 1$, since $\ln 1 = 0$. According to eq. (C.25), the value $a_{\mathrm{i}} = 1$ is attained at the particular state at which the partial pressure p_{i} coincides with pressure p^{o}. This occurs when gas *i* is the only component of the mixture. Thus, for ideal gas mixtures, the standard state to be considered in expression (C.23) is that of the pure gas *i* at the considered values of *T* and p^{o}.

The case of ideal liquid solutions is treated similarly. In this case, we know that the theoretical expression of μ_{i} is:

$$\mu_{\mathrm{i}} = \mu_{\mathrm{i}}^{\mathrm{o}} + RT \ln \frac{p_{\mathrm{i}}}{p_{\mathrm{i}}^{*}}, \tag{C.26}$$

where p_{i} is the vapour pressure of the liquid component A_{i}, and p_{i}^{*} is the equilibrium vapour pressure of pure liquid A_{i} at the considered temperature. By comparing eq. (C.26) with eq. (C.23), we conclude that, for ideal solutions:

$$a_{\mathrm{i}} = \frac{p_{\mathrm{i}}}{p_{\mathrm{i}}^{*}}. \tag{C.27}$$

The value $a_{\mathrm{i}} = 1$ is attained when the concentrations of all other species vanish, and the solution reduces to pure liquid A_{i}, because then p_{i} coincides with p_{i}^{*}. If $a_{\mathrm{i}} = 1$, then $\mu_{\mathrm{i}} = \mu_{\mathrm{i}}^{\mathrm{o}}$, as follows from comparing eq. (C.26) with eq. (C.23). It follows that the standard state of a liquid component of an ideal solution is the pure liquid state of that component at the considered values of temperature and pressure. In practice, the standard state of the pure liquid is usually referred to 1 atm pressure and considered as independent of pressure. Unless large pressures above 1 atm are considered, this makes little difference because the chemical potential of a pure liquid is practically independent of pressure. More details on this point are given at the end of this section.

Alternatively, the activities of the components of a mixture of ideal gases or a liquid solution can be expressed in terms of the mole fraction of that component. The mole fraction of species A_{i} in a mixture or a solution is denoted by x_{i} and is defined by:

$$x_i = \frac{n_i}{\sum_k n_k}. \tag{C.28}$$

The summation in eq. (C.28) is supposed to be extended to all components of the mixture or solution. In the case of liquid solutions, the moles of solvent, n_{solv}, are often distinguished from the moles of solute. Accordingly, the above formula is written as

$$x_i = \frac{n_i}{n_{solv} + \sum_j n_j}, \tag{C.29}$$

where the summation is now understood to be restricted to the solutes.

From *Dalton's law*, we know that the partial pressure p_i of gas A_i in an ideal gas mixture is related to the total pressure p of the mixture by the formula:

$$p_i = x_i \, p. \tag{C.30}$$

This applies, in particular, when $p = p^\circ$, in which case the above formula reads:

$$p_i = x_i \, p^o. \tag{C.31}$$

An analogous relation applies to the vapour pressures of the components of an ideal solution at a given temperature T. That relation is the *Rault's law*:

$$p_i = x_i \, p_i^*, \tag{C.32}$$

where p_i is the partial vapour pressure of component A_i of the solution, and p_i^* is the vapour pressure of the same component in the pure state at the considered temperature.

By introducing eqs (C.31) and (C.32) into eqs (C.24) and (C.26), respectively, we can condense formulae (C.24) and (C.26) into one single equation:

$$\mu_i = \mu_i^o + RT \ln x_i. \tag{C.33}$$

In this form, eq. (C.23) applies to both ideal gas mixtures and ideal liquid solutions. In either case, the activities of the species involved can be expressed as:

$$a_i = x_i. \tag{C.34}$$

However, the two cases differ in the definition of the standard state, which is the state of pure gas at temperature T and pressure $p°$ in the case of an ideal gas component and the state of pure liquid at temperature T for a liquid component of an ideal solution.

If the values of $p°$ and T are chosen to be the standard pressure and standard temperature, the corresponding molar Gibbs free energy, μ_i^o, of the considered component can be read directly from the thermodynamic tables. From this value, the value of μ_i^o at any pressure and temperature can be extrapolated through routine calculations of chemical thermodynamics.

Non-ideal solutions

In non-ideal solutions, it is convenient to express the activities, a_i, in the following form:

$$a_i = \gamma_i \, x_i \,. \tag{C.35}$$

The quantity γ_i introduced here is known as the *activity coefficient* of the considered species A_i. It is a function of the state variables of the solution, like a_i. By introducing eq. (C.35) into eq. (C.23), we obtain:

$$\mu_i = \mu_i^o + RT \ln(\gamma_i x_i) \,. \tag{C.36}$$

In the case of ideal solutions, all activity coefficients are equal to unity, as can be verified by comparing eqs (C.34) and (C.35). In the non-ideal case, the same coefficients differ from unity. Their actual expression, $\gamma_i = \gamma_i(p, T, n_1, n_2, ..., n_r)$, depends on a number of factors and is best determined experimentally. In certain cases, especially for dilute solutions, a theoretical determination of the activity coefficients is possible. Semi-empirical formulae, which are valid for electrolyte solutions in a wide range of concentrations, are also available for the same purpose (see Section 7.4).

The functions γ_i introduced above suffer from the same indeterminacy that affects the activities a_i, since their values depend on the choice of the standard state and since eq. (C.36) cannot determine both μ_i^o and γ_i. Consequently, appropriate conventions are needed to determine these quantities. Convenience dictates which convention should be used and when. Different conventions can be used for different solutions and for different components in the same solution.

In order to better understand the various conventions in use, we decompose μ_i into the sum of two components according to the relation

$$\mu_i = \mu_i^{id} + \mu_i^{E} . \tag{C.37}$$

The first component, namely μ_i^{id}, denotes the value that μ_i would assume if the solution behaved ideally. Accordingly, this part must have the form of (C.33):

$$\mu_i^{id} = \mu_i^{o\,id} + RT \ln x_i . \tag{C.38}$$

For $x_i = 1$, the solution reduces to pure species A_i. For this value of x_i, therefore, μ_i and μ_i^{id} coincide. From eqs (C.36) and (C.38) it then follows that:

$$\mu_i^{o\,id} = \mu_i^{o} + RT \ln(\gamma_i)_{x_i=1} . \tag{C.39}$$

The other component of μ_i, i.e., μ_i^{E}, represents the excess over μ_i^{id} due to the non-ideal behaviour of the solution. Accordingly, it is referred to as the *excess chemical potential* of component A_i. From eqs (C.36), (C.37) and (C.38) we have that

$$\mu_i^{E} = \mu_i^{o} - \mu_i^{o\,id} + RT \ln \gamma_i . \tag{C.40}$$

For ideal solutions, $\gamma_i = 1$. This makes $\mu_i^{E} = 0$, because $\mu_i^{o} \equiv \mu_i^{o\,id}$ in this case. For non-ideal solutions, the value of μ_i^{E} can be either positive or negative. We have then positive or negative deviations from ideality.

The conventions most frequently used to fix the standard state and activity coefficients of non-ideal solutions are described below.

Convention I

This convention is appropriate for liquid solutions of species that, in pure form, are liquid at the considered pressure and temperature. In this case, it is not unusual for a species A_i of a non-ideal solution to exhibit ideal-solution behaviour as the concentrations of all other species approaches zero. This means that eq. (C.33) applies to species A_i in the limit as $x_i \rightarrow 1$. Therefore, in the same limit, the quantity μ_i^{E} must tend to zero. A way to make eq. (C.36) consistent with this is to stipulate that the function γ_i should comply with the condition that

$\gamma_i \rightarrow 1$ as $x_i \rightarrow 1$, and, at the same time, to choose the standard state of species A_i as the state of the pure liquid A_i at the considered temperature. Because both μ_i^o and $\mu_i^{o\,id}$ are independent of x_i, these choices make $\mu_i^o = \mu_i^{o\,id}$, as immediately follows from eq. (C.39). Under these conditions, the excess potential μ_i^E will tend to zero as $x_i \rightarrow 1$, as is apparent from eq. (C.40).

Of course, the condition that μ_i^E should tend to zero as $x_i \rightarrow 1$ could also be fulfilled by different choices of the standard state and different restrictions in the function γ_i. However, the above convention is the most natural one for liquid species that, when in solution, behave in the ideal way as their mole ratio tends to unity (that is, as the solution tends to coincide with the pure liquid species under consideration).

Convention II

To decide if we can apply Convention I to solute A_i, we must verify that the solute tends to the ideal behaviour as $x_i \rightarrow 1$, i.e., as the solution tends to coincide with pure liquid A_i. Obviously, this cannot the case if A_i is solid or gaseous when pure, if A_i is not soluble at high concentrations, or if A_i does not exist in the pure form. The latter case applies to ionic species in particular. In all of these cases, we must resort to considering, instead, the behaviour of the solute at low concentration (i.e., as $x_i \rightarrow 0$).

Because any solute tends to exhibit ideal behaviour as its concentration decreases, the quantity μ_i^E should go to zero as $x_i \rightarrow 0$. A way to make eq. (C.40) consistent with this fact is to prescribe that $\gamma_i \rightarrow 1$ as $x_i \rightarrow 0$, and take the standard state of the solute so that $\mu_i^o = \mu_i^{o\,id}$. The catch is, however, that the state that the solute reaches as $x_i \rightarrow 0$ cannot be taken as its standard state, because there is no solute in solution at zero concentration.

To circumvent this difficulty, we consider a hypothetical ideal liquid component whose behaviour in the solution approximates the behaviour of the non-ideal component A_i in the limit as $x_i \rightarrow 0$. As recalled above, both μ_i^o and $\mu_i^{o\,id}$ are independent of x_i. This means that, to define the standard state of solute A_i, we may refer instead to the standard state of the above hypothetical liquid component that approximates the behaviour of non-ideal solute A_i as $x_i \rightarrow 0$. Because that hypothetical component is ideal, its standard state is the pure liquid state and, therefore, corresponds to $x_i = 1$. The same state is taken as the standard state of non-ideal component A_i of the solution. In this way, we meet the condition that $\gamma_i \rightarrow 1$ as $x_i \rightarrow 0$ and that $\mu_i^o = \mu_i^{o\,id}$ for $x_i = 1$.

Here, we have an instance of a standard state that is not a real state for solute A_i in question. This does not pose any problem, since the value of μ_i^o thus obtained is uniquely determined. This value coincides with $\mu_i^{o\,id}$, which is also equal to the standard state chemical potential of the ideal liquid that approximates the real liquid as $x_i \to 0$. It can be calculated from the thermochemical properties of the non-ideal component A_i, or it can be determined from experiments on real solutions at small values of x_i.

In practice, one first defines the hypothetical ideal liquid in question by calculating its Henry's law constant, K^H. This can be done once the vapour-pressure/concentration curve of solute A_i is known, for instance, because it is determined experimentally. The slope of the tangent line to that curve at $x_i=0$ coincides with K^H. Once K^H is determined, the value of μ_i^o can be shown to be given by $\mu_i^o = \mu_i^* + RT\ln(K^H / p_i^*)$, where μ_i^* and p_i^* are the chemical potential and the vapour pressure of pure A_i at the considered values of temperature and pressure [2, p.184]. Because it is based on Henry's law, the standard state of the present convention is often referred to as the *Henry's law standard state*.

Convention III

This convention uses Convention I for the solvent and Convention II for the solutes. Moreover, it takes concentration in mole fraction for the solvent, and molality for the solutes. This is a sensible thing to do, because the concentrations of solvents and solutes are usually measured this way in the laboratory and because the modern thermodynamic tables refer to molality as a measure of concentration.

Independently of the adopted convention, the standard state and the activity of a substance depend on the measure of concentration that is used. This issue is discussed briefly below.

Molality vs. mole fraction

Mole fraction, x_i, as defined by eq. (C.28) or eq. (C.29), is often the most convenient definition of concentration to use when dealing with theoretical arguments. However, in both theoretical and practical chemistry, many other alternative ways are in use to define the concentration of a substance.

For instance, when dealing with solutions, the concentration of a solute is most conveniently defined by its *molal concentration* or *molality*. This is usually denoted as b_i, and it is defined as the moles of solute per unit mass of solvent. That is:

$$b_i = \frac{n_i}{m_{solv}}, \tag{C.41}$$

where n_i are the moles of the considered solute, and m_{solv} is the mass of the solvent in which they are dissolved. Unlike x_i, molality is not dimensionless. In the SI system of units, it is measured in mol/kg. If the *molar mass* of the solvent is indicated by M_{solv}, then, in consistent units:

$$m_{solv} = \mathsf{M}_{solv}\, n_{solv}, \tag{C.42}$$

where n_{solv} is the number of moles of the solvent. The relation between b_i and x_i follows directly from eqs (C.29), (C.41), and (C.42):

$$b_i = \frac{x_i}{\mathsf{M}_{solv}\, x_{solv}}. \tag{C.43}$$

Here, x_{solv} is the mole fraction of the solvent:

$$x_{solv} = \frac{n_{solv}}{n_{solv} + \sum_k n_k}. \tag{C.44}$$

For very dilute solutions, x_{solv} tends to 1, which makes $b_i \cong x_i/\mathsf{M}_{solv}$.

Activities and activity coefficients relevant to different measures of concentration

The activity a_i of a component of a mixture or solution is usually regarded as the effective concentration of that component. However, if this standpoint is adopted, which is quite a sensible thing to do, we have to accept that the definition of activity depends on how we measure concentration. In particular, if concentrations are measured in molalities rather than mole fractions, the definition of activity given by eq. (C.35) should be replaced by

$$a_i = \gamma_i \frac{b_i}{b_i^o}. \tag{C.45}$$

The quantity b_i^o introduced in this formula is the molality of the considered component at some fixed reference state. The latter may not coincide with the standard state to which μ_i^o refers, although it usually does. The values of the activity coefficient, γ_i, depend on the reference state adopted.

Analogous considerations apply when the concentration is measured in *molarities*, i.e., moles of solute per unit volume of solution. The reader should refer to any textbook of classical chemistry for the formula that expresses the relation between molarity and molality (see, e.g., [3] pp. 328-330). For aqueous solutions at room temperature, the measures of concentration in molality and in molarity nearly coincide. In this case, an *r* molal solution of solute A_i is essentially the same as an *r* molar solution of the same solute, although molality and molarity have different physical dimensions. In any case, it is important to emphasize that the activity coefficients that appear in definitions (C.35) and (C.45) are different and are expressed by different functions of the variables that are taken to describe the state of the solution. However, the current literature denotes them with the same symbol.

When concentration is measured in molalities, Convention II is more conveniently applied by taking the standard state of the solutes as the state of unit molality rather than the state of unit molar fraction. In this case, we have that $b_i^o = 1$ molal and eq. (C.23) is usually written as:

$$\mu_i = \mu_i^o + RT \ln(\gamma_i b_i), \tag{C.46}$$

where division by 1 molal is understood in the argument of the logarithm. Eq. (C.46) is formally identical to eq. (C.36), but μ_i^o and γ_i refer to two different measures of concentration.

Biological standard state

As an application of the above observations, let us consider the definition of a *biological standard state*, which is used extensively in biochemistry to deal with aqueous solutions at room temperature. When Convention III is used, the concentration at the standard state of an ionic species in a solution—like that of any other solute—should be 1 molal in the considered ions. For hydrogen ions (H^+), this concentration means pH $\cong 0$, because for aqueous solutions at room temperature 1 molal concentration is about the same as 1 mole per litre concentration. (Remember that the pH index is defined as the negative of the common logarithm of the number that measures the concentration of H^+ ions in moles per litre. At room temperature, therefore, a 1 molal aqueous solution of H^+ has a concentration of about 1 mole/litre, which makes its pH vanish, since log 1 = 0.)

A solution with a pH = 0, however, is very acidic. It is completely different from the nearly neutral solutions (pH = 7, i.e., 10^{-7} mole/litre of H^+) in which biological reactions usually occur. In biochemistry therefore, when dealing with an aqueous solution that contains H^+ ions, a special standard state is taken for these ions. This is the so-called *biological standard state.* It only applies to hydrogen ions and corresponds to a concentration of 10^{-7} mole/litre, rather than the 1 mole/litre prescribed by Convention III. A solution is said to be in the biological standard state if the concentration of its H^+ ions corresponds to pH = 7, while all other solutes are at unity molality.

Let $b^{\oplus}_{H^+}$ denote the molality of H^+ in the biological standard state of an aqueous solution at 25 °C. This means that $b^{\oplus}_{H^+} = 10^{-7}$ molal. If $\mu^{\oplus}_{H^+}$ denotes the molar Gibbs energy of the hydrogen ions at that state, it is easy to show that $\mu^{\oplus}_{H^+}$ is related to the 1-molal-standard-state $\mu^{o}_{H^+}$ of Convention III as follows:

$$\mu^{\oplus}_{H^+} = \mu^{o}_{H^+} - 7RT\ln 10 . \tag{C.47}$$

Accordingly, at 298.1 K, the biological standard state of the solution has a molar Gibbs energy that is 7 RT ln 10 = 39.96 KJ mol^{-1} less than that of the standard state of Convention III.

In agreement with what we observed when writing eq. (C.23), when reference is made to the biological standard state, the molar Gibbs energy of the hydrogen ions can be expressed as:

$$\mu_{H^+} = \mu^{\oplus}_{H^+} + RT\ln a_{H^+} . \tag{C.48}$$

However, the activity, a_{H^+}, of the hydrogen ions that appears in eq. (C.48) must be understood to be given by

$$a_{H^+} = \gamma^{\oplus}_{H^+} b_{H^+} / b^{\oplus}_{H^+} , \tag{C.49}$$

where $\gamma^{\oplus}_{H^+}$ is the appropriate activity coefficient for eq. (C.48) to hold true. For this reason, when the biological standard state is used, the activity of the hydrogen ions refers to the *relative concentration* with respect to the biological standard state, rather than to the absolute concentration of the same ions in the solution.

Activities of solid phases and pure liquid phases

The contribution to G from solid phases and pure liquid (immiscible) phases should finally be considered. The expression of G given by eq. (C.4) may include these contributions. The partial molar Gibbs free energies, μ_i, of these phases are almost independent of pressure. Thus, they practically coincide with their respective standard values μ_i^o. A formal proof can be obtained by observing that, under the usual hypotheses of continuity and derivability, the order of derivation of function $G = G(p,T,n_1,n_2,\ldots,n_r)$ with respect to its independent variables can be interchanged. Thus:

$$\frac{\partial^2 G}{\partial n_i\, \partial p} = \frac{\partial^2 G}{\partial p\, \partial n_i}. \qquad \text{(C.50)}$$

From eqs (C.7) and (C.8), this means that:

$$\left(\frac{\partial \mu_i}{\partial p}\right)_{T,n_i} = \left(\frac{\partial V}{\partial n_i}\right)_{p,T,n'}. \qquad \text{(C.51)}$$

The right side of this equation is the *partial molar volume* of component *i*. It coincides with the *molar volume* of the component if the latter is a solid or a pure liquid. The molar volumes of solids and liquids have comparatively small values, which makes the corresponding value of μ_i almost independent of pressure, according to eq. (C.51). Therefore, their chemical potentials can be considered constant and equal to μ_i^o over the usual range of pressure. Accordingly, the activities of pure liquids and solids can be considered as unity, since this is what is needed to make eq. (C.23) consistent with the fact that $\mu_i \equiv \mu_i^o$ for these components.

C.6 Reaction Quotient

By introducing eq. (C.23) into eq. (C.18) and by recalling eq. (C.21), we can express $\Delta_r G$ as follows:

$$\Delta_r G = \Delta_r G^o + RT\left[\left(\sum\nolimits_h \ln a_h^{\nu_h}\right)_{\text{products}} - \left(\sum\nolimits_k \ln a_k^{|\nu_h|}\right)_{\text{reactants}}\right] \qquad \text{(C.52)}$$

or

$$\Delta_r G = \Delta_r G^o + RT \ln \frac{\left(\prod_h a_h^{\nu_h}\right)_{\text{products}}}{\left(\prod_k a_k^{|\nu_k|}\right)_{\text{reactants}}}, \tag{C.53}$$

since the sum of the logarithms equals the logarithm of the product of their arguments. The symbol Π introduced here indicates the product of a sequence. That is:

$$\prod_{i=1}^{n} y_i = y_1 \cdot y_2 \cdot \ldots \cdot y_n \, . \tag{C.54}$$

Equation (C.53) is usually expressed in the form:

$$\Delta_r G = \Delta_r G^o + RT \ln Q , \tag{C.55}$$

where Q is the so-called *reaction quotient*, given by:

$$Q = \frac{\left(\prod_h a_h^{\nu_h}\right)_{\text{products}}}{\left(\prod_k a_k^{|\nu_k|}\right)_{\text{reactants}}} . \tag{C.56}$$

As an example of application of eq. (C.56), let us calculate the reaction quotient relevant to reaction (C.9). By assuming that the activities can be expressed as in eq. (C.34), from eq. (C.56) we obtain:

$$Q = \frac{x_C^{\nu_C} \, x_D^{\nu_D}}{x_A^{\nu_A} \, x_B^{\nu_B}} . \tag{C.57}$$

We already observed that $\Delta_r G = 0$ at equilibrium. From eq. (C.55), we then infer that the following relation must hold true once chemical equilibrium is reached:

$$\Delta_r G^o = -RT \ln K . \tag{C.58}$$

The quantity K that appears here is the value that Q attains in chemical equilibrium conditions at temperature T. K is commonly referred to as the *equilibrium constant* of the considered reaction at the considered temperature. Its value can be calculated from the standard molar energies of formation of the substances involved in the reaction, once eq. (C.22) is inserted into eq. (C.58) and the latter is solved for K. Since $K = Q$ at equilibrium and the activities are directly related to the concentrations of the species involved in the reaction, we can calculate the equilibrium concentration of one of the species in solution from the value of K if the concentrations of the other species are given.

Also, if the equilibrium concentrations (i.e., the value of K) before and after the temperature change are known, then equation (C.28) can be used to determine how a change in temperature will affect $\Delta_r G$.

C.7 Gibbs Free Energy as a Function of the Extent of Reaction

As shown by eq. (C.13), the Gibbs free energy of a system that is undergoing a chemical process at isothermal and isobaric conditions can be expressed as a function of the variable ξ. In this section, when referring to such a function, we shall for brevity omit the indices p and T that appear in eq. (C.13) and simply write the Gibbs free energy as $G = G(\xi)$. Moreover, we shall focus our attention on mixtures of ideal gases or ideal liquid solutions, since their chemical potentials, μ_i, have comparatively simple forms (C.33). Once the expressions of the activity coefficients, γ_i, are given as functions of p, T, and n_i, eq. (C.35) makes it possible to treat the case of non-ideal systems in a similar way.

In order to obtain the explicit expression of $G = G(\xi)$, the quantities x_i that appear in eq. (C.33) must first be expressed in terms of ξ. This is done by introducing eqs (C.11) into eq. (C.29). The resulting expressions $x_i = x_i(\xi)$ are then introduced into eqs (C.33) to give the function $\mu_i = \mu_i(\xi)$. The latter function, together with eqs (C.11), are substituted into eq. (C.4) to yield $G = G(\xi)$. The procedure involves routine algebraic manipulations and does not present conceptual difficulties. The resulting free energy function will depend on the reaction and on the values of the constants n_i^0. An example of these calculations for an electrolytic system involving the reaction, $Zn + CuSO_4 = Cu + SnSO_4$, is presented in Chapter 7.

The values that ξ can attain range from $\xi = 0$ at the beginning of the reaction, to the value $\xi = \xi_{max}$. The latter value is reached as one or more reactants are exhausted. No further increase in ξ is possible beyond ξ_{max}, simply because of the lack reactants. From eq. (C.11), it can be inferred that ξ_{max} is equal to the smallest of the $n_k^0 / |\nu_k|$ ratios of the reactants. That is:

$$\xi_{\max} = \min\left(\frac{n_k^o}{|\nu_k|}\right)_{\text{reactants}} . \qquad \text{(C.59)}$$

An important feature of the function $G = G(\xi)$ concerns the value of its tangent, $dG/d\xi$, at the ends of the reaction interval $[0, \xi_{\max}]$. From eqs (C.18) and (C.20), we have:

$$\frac{dG}{d\xi} = \left[\left(\sum_h \nu_h \mu_h\right)_{\text{products}} - \left(\sum_k |\nu_k| \mu_k\right)_{\text{reactants}}\right] . \qquad \text{(C.60)}$$

For $\xi = \xi_{\max}$, there will be at least one reactant, j, that disappears from the system. Thus, it follows that the molar fraction of this reactant will vanish for this value of ξ:

$$x_j\Big|_{\xi=\xi_{\max}} = 0 . \qquad \text{(C.61)}$$

From eq. (C.33), it follows that the chemical potential will diverge negatively at that stage of the reaction:

$$\mu_j\Big|_{\xi=\xi_{\max}} = -\infty . \qquad \text{(C.62)}$$

From the fact that j refers to a reactant and from eq. (C.60), we can conclude that

$$\left.\frac{dG}{d\xi}\right|_{\xi=\xi_{\max}} = \infty . \qquad \text{(C.63)}$$

That is, the tangent to the curve $G = G(\xi)$ must be vertical at the right end of the reaction interval $[0, \xi_{\max}]$.

An analogous situation occurs at the beginning of the process, provided that, for $\xi = 0$, at least one of the products of the reaction, r, vanishes. In that case, we have

$$x_r\Big|_{\xi=0} = 0 , \qquad \text{(C.64)}$$

which, in view of eq. (C.33), implies that

$$\mu_r\Big|_{\xi=0} = \infty . \qquad \text{(C.65)}$$

From eq. (C.60), we then infer that

$$\left.\frac{dG}{d\xi}\right|_{\xi=0} = -\infty . \tag{C.66}$$

Under the considered conditions, therefore, curve $G = G(\xi)$ will exhibit a vertical tangent at $\xi = 0$.

The assumption that at least one product of the reaction is missing from the system for $\xi = 0$ is not really restrictive. Starting from any initial composition, we can always drive (or imagine that we can drive) the reaction backwards and force the products to react and transform themselves into the reactants (cf. the dashed part of curve $G = G(\xi)$ in Fig. C.1). Of course, such an inverse reaction cannot proceed beyond the point at which one or more products of the reaction are exhausted. If we take the state of the system at that point as a new initial state, the old initial state becomes an intermediate state of a new, extended range. Then, the new initial point of the latter range corresponds to a composition of the system in which at least one of the products of reaction is missing. Eq. (C.66) must then apply at that point.

By referring to the reaction interval $[0, \xi_{max}]$, as defined above, Fig. C.2 represents two typical shapes for curve $G = G(\xi)$. Since the derivatives of this curve assume opposite values at the ends of the considered range, the curve itself must possess at least one minimum point within that interval. (We assume that the derivative of $G(\xi)$ is continuous, which usually is true in practice.) The minimum point of this curve is the point of chemical equilibrium. The system spontaneously tends to it when set free from external actions.

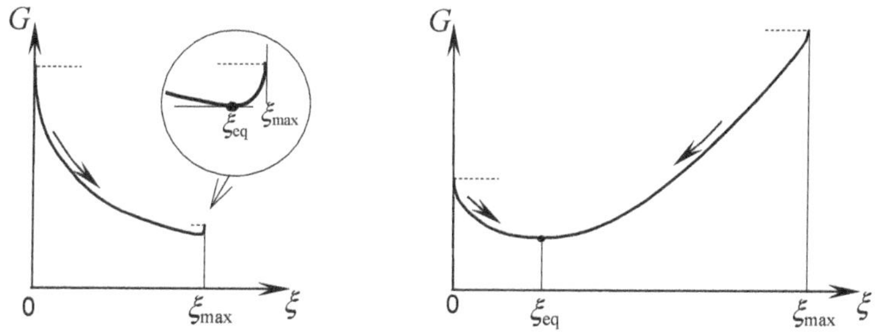

Fig. C.2. Typical curves of $G = G(\xi)$. The left curve refers to a reaction that proceeds almost to the complete exhaustion of a reactant. The horizontal, dashed lines indicate the points at which the tangent to the curve is vertical.

The derivative of the Gibbs free energy is often interpreted as the *driving force* of a chemical reaction. The greater the absolute value of this derivative, the stronger the tendency of the system to head spontaneously in the direction of decreasing G. This derivative tends to infinity at both ends of the reaction interval $[0, \xi_{max}]$, meaning that the driving force for the reaction becomes greater and greater as these ends are approached. This is consistent with the well-known experimental fact that it becomes increasingly difficult to remove a reacting component from a solution as the concentration of that component becomes increasingly smaller.

APPENDIX D

Effect of Pressure on Free Energy of Liquid Solutions

Constant pressure processes are commonplace. However, there are some practical situations in which even small pressure differences play a dominant role in the system's response. Pressure changes are important, in particular, when two liquid solutions of different concentrations are separated by a semi-permeable membrane. In this case, the selective migration (*osmosis*) of solvent or solute molecules through the membrane depends on the difference in the concentrations of the two solutions and on the difference in their pressures. In similar situations the dependence of the solution's Gibbs free energy on pressure cannot be ignored. The argument is especially important in the case of living systems, due to the central role of semi-permeable membranes in keeping these systems alive.

D.1 Partial Molar Volumes

The volume, V, of a liquid solution is a function of pressure, temperature, and composition. The solution's composition is measured by the number of moles, n_i, of each species i in the solution, including the solvent. Thus, we have:

$$V = V(p,T,n_1,n_2,...,n_r). \tag{D.1}$$

The partial derivative of V with respect to n_i (i = 1, 2, ..., r) is the *partial molar volume* of component i of the solution. This derivative is denoted by $\overline{V}_i$:

$$\overline{V}_i = \overline{V}_i(p,T,n_1,n_2,...,n_r) = \frac{\partial V}{\partial n_i}. \tag{D.2}$$

Geometrically speaking, $\overline{V}_i$ represents the slope of the tangent to the curve $V = V(p, T, n_1, n_2, ..., n_r)|_{p,T,n'}$, which is obtained from eq. (D.1) by keeping constant all independent variables except n_i. (According to the notation introduced in Appendix C, the subscript n' stands for all variables n_k, with the exception of n_i.)

By differentiating function (D.1) at constant p and T and by using definition (D.2), we obtain:

$$\mathrm{d}V\big|_{p,T} = \overline{V}_1\,\mathrm{d}n_1 + \overline{V}_2\,\mathrm{d}n_2 + ... + \overline{V}_r\,\mathrm{d}n_r. \tag{D.3}$$

Since the volume of a solution is an extensive property, the following equation must be valid for every positive scalar α:

$$V(p,T,\alpha n_1,\alpha n_2,...,\alpha n_r) = \alpha V(p,T,n_1,n_2,...,n_r). \tag{D.4}$$

This means that V is a homogeneous function of degree 1 in the variables n_i. From the well-known Euler's theorem on homogeneous functions, it then follows that function (D.1) can be expressed in the following form:

$$V = n_1\overline{V}_1 + n_2\overline{V}_2 + ... + n_r\overline{V}_r. \tag{D.5}$$

This shows that the partial molar volumes, $\overline{V}_i$, can also be regarded as the contribution to V from a unit amount of component i; this justifies their name. In eq. (D.5), the solvent is considered to be one of the r components of the solution. Of course, the quantities $\overline{V}_i$ are different for different solutions, even if they refer to the same component. In other words, the partial molar volumes are a property of the solution and not of the component per se.

By taking the partial derivative of both sides of eq. (C.7) with respect to n_i and by making use of definition (D.2), we obtain:

$$\frac{\partial^2 G}{\partial n_i \partial p} = \overline{V}_i \,. \tag{D.6}$$

On the other hand, by taking the derivative of eq. (C.8) with respect to p, we obtain:

$$\frac{\partial^2 G}{\partial p \partial n_i} = \frac{\partial \mu_i}{\partial p} \,. \tag{D.7}$$

Since the order of derivation is immaterial, from the last two equations we conclude that

$$\overline{V}_i = \frac{\partial \mu_i}{\partial p} \,. \tag{D.8}$$

When compared to eq. (C.7), this equation shows that the partial molar volume of each component relates to the partial molar free energy, μ_i, of the same component as the total volume V of the solution relates to the solution's free energy G.

D.2 Pressure Dependence of Gibbs Energy

In determining the explicit relationship between G and p for liquid solutions, we first refer to ideal solutions. The relationship between G and p for real solutions is then obtained from that of ideal solutions by introducing appropriate activity coefficients.

Ideal solutions

According to eq. (C.33), the molar free energy of any solute, i, in an ideal solution can be expressed as

$$\mu_i = \mu_i(p, T, n_{solv}, n_1, n_2, \ldots, n_s) = \mu_i^o(p, T) + RT \ln x_i \,. \tag{D.9}$$

In this equation the variable n_{solv} singles out the moles of solvent from the moles of the solutes. This means that the quantities n_i ($i = 1, 2, \ldots, s$)

refer to solutes, as do their mole fractions x_i. An equation similar to eq. (D.9) applies to the solvent:

$$\mu_{solv} = \mu_{solv}(p, T, n_{solv}, n_1, n_2, \ldots, n_s) = \mu_{solv}^{o}(p, T) + RT \ln x_{solv}, \quad \text{(D.10)}$$

where, according to eq. (C.29), the quantity x_{solv} is expressed as

$$x_{solv} = \frac{n_{solv}}{n_{solv} + \sum_{j=1}^{s} n_j}. \quad \text{(D.11)}$$

The quantities μ_i^o and μ_{solv}^o that appear in eqs (D.9) and (D.10) represent the molar Gibbs free energy of the pure solutes and of the pure solvent, respectively, at the considered values of pressure and temperature. Often, these values are the standard pressure, $p°$, and standard temperature, 298 K. However, there is no compelling need to do so; the quantities μ_i^o and μ_{solv}^o can be defined for any value of p and T.

In the following, the symbol, n_{os}, denotes the total number of moles of all the species of solutes contained in the solution. That is:

$$n_{os} = \sum_{j=1}^{s} n_j. \quad \text{(D.12)}$$

Moreover, every ionic species in the solution is considered distinct from the substance from which it dissociated. Thus, a partially-soluble salt in a solution will generally result in different solutes, namely, the various ionic species that arise from the dissociation of the salt and the undissociated salt itself. In other words, the quantity n_{os} introduced above represents the total number of dissolved particles. This number controls the so-called colligative properties of the solution, particularly its osmotic pressure, that do not depend on the nature of the dissolved particles. As in the case of n_{os}, the index "os" is appended to the symbol of a quantity to stress the pre-eminent role of that quantity in the osmotic phenomenon.

Similarly, the mole fraction of all particles dissolved in the solution is indicated as x_{os}. Therefore, in view of eqs (C.29) and (D.12), we have:

$$x_{\mathrm{os}} = \frac{\sum_{\mathrm{j=1}}^{\mathrm{s}} n_{\mathrm{j}}}{n_{\mathrm{solv}} + \sum_{\mathrm{j=1}}^{\mathrm{s}} n_{\mathrm{j}}} = \frac{n_{\mathrm{os}}}{n_{\mathrm{solv}} + n_{\mathrm{os}}} . \tag{D.13}$$

From eqs (D.11) and (D.12), it can then be verified that

$$x_{\mathrm{solv}} + x_{\mathrm{os}} = 1 . \tag{D.14}$$

Consequently, eq. (D.10) also can be written as

$$\mu_{\mathrm{solv}} = \mu_{\mathrm{solv}}^{\mathrm{o}}(p,T) + RT \ln(1 - x_{\mathrm{os}}) . \tag{D.15}$$

From eqs (D.9), (D.10), and (D.15), we see that, for an ideal solution at constant composition, the dependence of Gibbs free energy on pressure comes through the functions $\mu_{\mathrm{i}}^{\mathrm{o}}$ and $\mu_{\mathrm{solv}}^{\mathrm{o}}$. This means that if we take the partial derivative with respect to p of both sides of eq. (D.9) and keep eq. (D.8) into account we obtain

$$\frac{\partial \mu_{\mathrm{i}}^{\mathrm{o}}}{\partial p} = \frac{\partial \mu_{\mathrm{i}}}{\partial p} = \overline{V}_{\mathrm{i}}(p,T,n_{\mathrm{solv}},n_1,n_2,\ldots,n_{\mathrm{s}}) . \tag{D.16}$$

By integrating this equation along a process in which pressure varies from p° to $p^{\circ}+\Delta p$ while all other variables are kept constant, we quite generally obtain

$$\begin{aligned} \mu_{\mathrm{i}}^{\mathrm{o}}(p^{\mathrm{o}}+\Delta p, T) &= \mu_{\mathrm{i}}^{\mathrm{o}}(p^{\mathrm{o}}, T) + \int_{p^{\mathrm{o}}}^{p^{\mathrm{o}}+\Delta p} \frac{\partial \mu_{\mathrm{i}}^{\mathrm{o}}}{\partial p} \mathrm{d}p \\ &= \mu_{\mathrm{i}}^{\mathrm{o}}(p^{\mathrm{o}},T) + \int_{p^{\mathrm{o}}}^{p^{\mathrm{o}}+\Delta p} \overline{V}_{\mathrm{i}} \mathrm{d}p . \end{aligned} \tag{D.17}$$

The last of these equations follows from eq. (D.8).

In solids and liquids, the dependence of $\overline{V}$ and $\overline{V}_{\mathrm{i}}$ on pressure can usually be neglected. The quantity $\overline{V}_{\mathrm{i}}$ can then be considered constant, and eq. (D.17) yields:

$$\mu_{\mathrm{i}}^{\mathrm{o}}(p^{\mathrm{o}} + \Delta p, T) = \mu_{\mathrm{i}}^{\mathrm{o}}(p^{\mathrm{o}},T) + \overline{V}_{\mathrm{i}} \Delta p . \tag{D.18}$$

In this case, the dependence of μ_i on pressure can be made explicit by inserting eq. (D.18) into eq. (D.9) to obtain:

$$\mu_i(p^o+\Delta p,T)=\mu_i^o(p^o,T)+\bar{V}_i\,\Delta p+RT\ln x_i. \qquad \text{(D.19)}$$

Of course, a similar equation also applies to the solvent, since the solvent is one of the components of the solution. When referred to the solvent, the above equation becomes:

$$\mu_{solv}(p^o+\Delta p,T)=\mu_{solv}^o(p^o,T)+\bar{V}_{solv}\,\Delta p+RT\ln(1-x_{os}), \qquad \text{(D.20)}$$

thanks to eq. (D.14).

In liquids or in liquid solutions, the contribution of pressure changes to the molar Gibbs energy is usually modest, unless the change is very large. For liquid water at standard pressure and temperature, for instance, $\bar{V}_{H_2O}$ = 18 10^{-6} m^3mol^{-1}. For a pressure change, Δp, of 1 atm = 10^5 N m^{-2} we have that $\bar{V}_{H_2O}\,\Delta p$ is 1.8 J mol^{-1}, which is rather small compared to $\mu^o_{H_2O}$ = 237.13 kJ mol^{-1}. Though small in comparison to μ_i^o and μ_{solv}^o, the term $\bar{V}_{solv}\,\Delta p$ has an essential role in the osmotic pressure phenomenon, as shown in the next section.

By introducing eqs (D.19) and (D.20) into eq. (C.4), the Gibbs free energy of the solution can be expressed as

$$\begin{aligned}G=G(p,T,n_{solv},n_1,n_2,\ldots,n_s)&=n_{solv}\mu_{solv}^o(p^o,T)+\sum_{j=1}^{s}n_j\mu_j^o(p^o,T)\\&+V\Delta p+RT\left[n_{solv}\ln(1-x_{os})+\sum_{j=1}^{s}n_j\ln x_j\right]+C,\end{aligned} \qquad \text{(D.21)}$$

where V is a function of the state variables of the solution, according to eq. (D.1). In writing eq. (D.21), we added the arbitrary constant C, which depends on the reference value used for internal energy. We also used eq. (D.5) in the form:

$$n_{solv}\bar{V}_{solv}+\sum_{j=1}^{s}n_j\bar{V}_j=V. \qquad \text{(D.22)}$$

Eq. (D.21) is the relation we were seeking. It expresses the dependence of G on Δp or, equivalently, of G on $p = p^o + \Delta p$. For diluted solutions, we have $x_{os} \ll 1$, and hence $\ln(1-x_{os})\cong -x_{os}$. In that case, eq. (D.21) simplifies to:

$$G = n_{solv}\,\mu^{o}_{solv} + \sum_{j=1}^{s} n_j\,\mu^{o}_j + V\Delta p + RT\left[-n_{solv}\,x_{os} + \sum_{j=1}^{s} n_j \ln x_j\right] + C. \quad (D.23)$$

For more concentrated solutions, a better approximation can be obtained from eq. (D.21) by replacing the term $\ln(1-x_{os})$ with its so-called *virial expansion:*

$$\ln(1-x_{os}) = -x_{os} - \frac{x_{os}^{\;2}}{2} - \frac{x_{os}^{\;3}}{3} - \ldots \quad (D.24)$$

Non-ideal solutions

One way to apply the above formulae to non-ideal solutions, such as high concentration solutions or ionic solutions, is to replace the concentrations x_{os} and x_{solv} with the corresponding effective concentrations (or activities). Thus, eq. (D.21) assumes the following form:

$$\begin{aligned} G = {} & n_{solv}\,\mu^{o}_{solv}(p^{o},T) + \sum_{j=1}^{s} n_j\,\mu^{o}_j(p^{o},T) \\ & + V\Delta p + RT\left[n_{solv} \ln a_{solv} + \sum_{j=1}^{s} n_j \ln a_j\right] + C, \end{aligned} \quad (D.25)$$

where the quantities a_i are the solute activities, while a_{solv} is the solvent activity. The latter replaces the solvent molar concentration, x_{solv}, which, according to eq. (D.14), equals $(1- x_{os})$.

The definition of activity is recalled in Appendix C.5. As observed there, the activity depends on the measure of concentration that is used. In the laboratory, we often measure solute concentrations in molalities (moles of solute per unit mass of solvent). These are denoted by the symbol b in this book. Thus, the symbol b_i indicates the molality of component i of the solution. In particular, when concentrations are measured in molalities, the reference state of a solute is assumed to be the state (whether ideal or hypothetical) of unit molality in that solute, i.e., $b_i^{o} = 1$ mol kg^{-1}. Then, the activity of the same solute at any concentration can be expressed by:

$$a_i = \gamma_i\, b_i\,, \quad (D.26)$$

where γ_i is an appropriate activity coefficient, cf. eq. (C.45). In eq. (D.26), we followed the common practice of leaving the division by 1 mol kg^{-1} understood in the right side of the same equation. Generally speaking, the activity coefficients introduced above are functions of p, T, n_i, and n_{solv}. They must be determined experimentally or from appropriate semi-empirical formulae.

Since molality is a measure of concentration relative to the amount of solvent, it does not make sense to measure the concentration of the solvent in molalities. For this reason, another measure of concentration must be used to specify the concentration of the solvent. If we measure the concentration of the solvent in mole fraction, the solvent activity will be related to its concentration by eq. (C.35). That is:

$$a_{solv} = \gamma_{solv}\, x_{solv} \,, \tag{D.27}$$

where x_{solv} is the mole fraction of the solvent. If the concentrations of the solutes are measured in molalities and the concentration of the solvent is measured in mole fractions, we can express eq. (D.25) in the following form:

$$\begin{aligned} G = {} & n_{solv}\, \mu^{o}_{solv}(p^{o},T) + \sum_{j=1}^{s} n_j\, \mu^{o}_{j}(p^{o},T) \; + V\Delta p \\ & + RT\left[n_{solv} \ln \gamma_{solv}\, x_{solv} + \sum_{j=1}^{s} n_j \ln \gamma_j\, b_j \right] + C \,, \end{aligned} \tag{D.28}$$

thanks to eqs (C.41), (D.26), and (D.27). By means of eqs (D.11), (C.41), and (C.42), we can express eq. (D.28) more explicitly as:

$$\begin{aligned} G = {} & n_{solv}\, \mu^{o}_{solv}(p^{o},T) + \sum_{j=1}^{s} n_j\, \mu^{o}_{j}(p^{o},T) + V\Delta p \\ & + RT\left[n_{solv} \ln \frac{\gamma_{solv}\, n_{solv}}{n_{solv} + \sum_{j=1}^{s} n_j} + \sum_{j=1}^{s} n_j \ln \frac{\gamma_j\, n_j}{m_{solv}} \right] + C \,. \end{aligned} \tag{D.29}$$

Remember that, in formulae (D.28) and (D.29), the argument of the logarithms that refer to the solutes are understood to be divided by 1 mol kg^{-1}.

D.3 Osmotic Pressure

Suppose now that the solution is enclosed in a container that has semi-permeable walls. More specifically, the walls are assumed to be permeable to the solvent but impervious to the solutes. Moreover, the container is immersed in another solution of the same solvent with a different concentration of solutes. We append the index "in" to a quantity that refers to the solution inside the container and append the index "out" to a quantity that refers to the solution outside the container. No index is appended to quantities that assume the same value in both solutions or when no distinction must be made between the solutions. Isothermal conditions are assumed throughout.

Initially, the two solutions are at the same pressure, but they have different concentrations. That is:

$$x_{\mathrm{os}}\big|_{\mathrm{in}} \neq x_{\mathrm{os}}\big|_{\mathrm{out}} . \tag{D.30}$$

Since the container is permeable to the solvent but not to the solutes, a net flow of solvent will ensue from the more diluted to the more concentrated solution. For ideal solutions, this is a colligative effect; i.e., it depends on the relative number of particles of solute and solvent, not on their nature. Since the two solutions have different concentrations, their solvents posses different partial molar free energies. As a consequence, a flow of solvent will take place through the semi-permeable walls of the container from the solution at higher free energy to the solution at lower free energy. This is a spontaneous process, because it reduces the total free energy of the two-solution system. The process ends when the equilibrium condition

$$\mu_{\mathrm{solv}}\big|_{\mathrm{in}} = \mu_{\mathrm{solv}}\big|_{\mathrm{out}} \tag{D.31}$$

is reached.

If the walls of the container are fixed and no other fluid enters or leaves the container, the flow of solvent through the walls of the container also produces a change in the pressure of the solution inside the container. Let Δp be the pressure difference that is reached at equilibrium between the inside and outside solutions as the flow of solvent comes to an end. That is:

$$\Delta p = p|_{\text{in}} - p|_{\text{out}} . \tag{D.32}$$

This quantity can be positive or negative, and it represents the extra pressure that must be applied to the internal solution to stop the flow of solvent through the semi-permeable partition. If $p°$ is the pressure of the outside solution and $p°+\Delta p$ that of the solution in the semi-permeable container, the values of $\mu_{\text{solv}}|_{\text{out}}$ and $\mu_{\text{solv}}|_{\text{in}}$ can be calculated from eq. (D.20) as:

$$\mu_{\text{solv}}|_{\text{out}} = \mu^{\text{o}}_{\text{solv}} + RT\ \ln(1 - x_{\text{os}}|_{\text{out}}) \tag{D.33}$$

and

$$\mu_{\text{solv}}|_{\text{in}} = \mu^{\text{o}}_{\text{solv}} + \bar{V}_{\text{solv}}|_{\text{in}}\ \Delta p + RT\ \ln(1 - x_{\text{os}}|_{\text{in}}) , \tag{D.34}$$

respectively. The extra pressure, Δp, that must be applied to the internal solution to stop the flow of solvent through the semi-permeable partition is obtained by inserting eqs (D.33) and (D.34) into eq. (D.31). Thus, we obtain:

$$\Delta p = \frac{1}{\bar{V}_{\text{solv}}|_{\text{in}}} RT \left[\ln(1 - x_{\text{os}}|_{\text{out}}) - \ln(1 - x_{\text{os}}|_{\text{in}}) \right] . \tag{D.35}$$

In the case where the outside solution is pure solvent, we have $x_{\text{os}}|_{\text{out}} = 0$. Denoting as Π the value of Δp relevant to this case, we deduce from eq. (D.35) that

$$\Pi = -\frac{1}{\bar{V}_{\text{solv}}|_{\text{in}}} RT\ \ln(1 - x_{\text{os}}|_{\text{in}}) . \tag{D.36}$$

Pressure Π is known as the *osmotic pressure* of the solution. It represents the maximum value that Δp can ever attain in the considered setup. This follows directly from eq. (D.35) once it is observed that x_{os} cannot be greater than 1. The osmotic pressure is thus a property of the solution to which it refers, because it is entirely defined in terms of it.

For dilute solutions ($x_{\text{os}} \ll 1$), eq. (D.35) approximates to:

$$\Delta p = \frac{1}{\bar{V}_{\text{solv}}\big|_{\text{in}}} RT\, \Delta x_{\text{os}}\ , \tag{D.37}$$

where

$$\Delta x_{\text{os}} = x_{\text{os}}\big|_{\text{in}} - x_{\text{os}}\big|_{\text{out}}\ . \tag{D.38}$$

For more concentrated solutions, better approximations can be obtained by evaluating the term in brackets in eq. (D.35) by the following expansion:

$$\ln(1 - x_{\text{os}}\big|_{\text{out}}) - \ln(1 - x_{\text{os}}\big|_{\text{in}}) = -\Delta x_{\text{os}} - \frac{\Delta x_{\text{os}}^{\ 2}}{2} - \frac{\Delta x_{\text{os}}^{\ 3}}{3} - \ldots \tag{D.39}$$

Similarly, for solutions that are sufficiently dilute, eq. (D.36) approximates to:

$$\Pi = \frac{1}{\bar{V}_{\text{solv}}\big|_{\text{in}}} RT\, x_{\text{os}}\big|_{\text{in}}\ . \tag{D.40}$$

In this case, the quantities n_{j} are negligible with respect to n_{solv}, and from eq. (D.22) we obtain:

$$n_{\text{solv}}\, \bar{V}_{\text{solv}} \cong V\ . \tag{D.41}$$

From this and from eq. (D.13), we can also express eq. (D.40) in the following form:

$$\Pi = \frac{1}{V} RT\, n_{\text{os}}\ , \tag{D.42}$$

which is the well-known van't Hoff approximate formula for the osmotic pressure of a solution.

For non-ideal solutions, eq. (D.35) and the above formulae of the osmotic pressure must be multiplied by an appropriate *osmotic coefficient*, $\phi = \phi(p, T, n_1, n_2, \ldots, n_{\text{r}})$, which characterizes the deviation from ideal behaviour. In particular, when applied to a non-ideal solvent, eq. (C.30) can be written as:

$$\Delta p = \frac{\phi}{\bar{V}_{\text{solv}}\big|_{\text{in}}} RT \left[\ln(1 - x_{\text{os}}\big|_{\text{out}}) - \ln(1 - x_{\text{os}}\big|_{\text{in}}) \right]. \tag{D.43}$$

In view of eqs (D.11), (D.13), and (D.14), eq. (D.43) can also be written in the following form:

$$\begin{aligned}\Delta p &= \Delta p(p, T, n_{\text{solv}}, n_1, n_2, \ldots, n_s) \\ &= \frac{\phi}{\overline{V}_{\text{solv}}} RT \left[\ln(n_{\text{solv}} + \sum_{j=1}^{s} n_j) - \ln n_{\text{solv}} + \ln(x_{\text{solv}}|_{\text{out}}) \right],\end{aligned} \tag{D.44}$$

where all quantities except $x_{\text{solv}}|_{\text{out}}$ refer to the solution inside the semi-permeable container.

For both ideal and non-ideal solutions, osmotic pressure can be quite large even for low to moderate values of x_{os}. According to eq. (D.42), for instance, 1 mole of an ideal solute dissolved in 1 Kg of water at room temperature (298 K) produces an osmotic pressure of 24.5 atm. For human blood, considered as a solution of 0.295 mol of dissolved particles per litre of water (average normal value), eq. (D.42) gives an osmotic pressure of 7.5 atm at 310 K (body temperature).

D.4 The Helmholtz Free Energy of a Solution

For systems undergoing isothermal processes at constant pressure and volume, the change in the Gibbs free energy, G, coincides with the change in the Helmholtz free energy, Ψ, cf. eqs (1.11.3) and (1.11.4). Under these conditions, there is no need to distinguish between the two kinds of free energies when calculating the change in the free energy of the system.

There are situations, however, in which one should distinguish between the two kinds of free energies of a solution. This happens, in particular, when studying the admissible range of a living cell. As for any other system, the admissible range of a cell is the set of all states that the cell can reach isothermally while abiding by its constitutive equations for internal energy and entropy and hence for Ψ. In the case of living cells, however, changes in pressure and volume are the rule rather than the exception. Pressure changes are brought about by osmotic effects and are fundamental to any living system, while volume changes are the immediate consequence of the fact that the living cell is an open system that exchanges matter with its surroundings. In these conditions, the change in the free energy of the cell is different if we refer to G or to Ψ, and the two free energies must be distinguished from each other.

The relation between G and Ψ is given by definition $(1.11.10)_3$, which we rewrite here as:

$$\Psi = G - p\,V. \tag{D.45}$$

Since

$$p = p^\circ + \Delta p, \tag{D.46}$$

from eq. (D.45) it follows that

$$\Psi = G - p^\circ\,V - V\,\Delta p. \tag{D.47}$$

From this and from eq. (D.29), we obtain:

$$\begin{aligned}\Psi &= n_{\text{solv}}\,\mu^{\text{o}}_{\text{solv}}(p^{\text{o}},T) + \sum_{\text{j=1}}^{\text{s}} n_{\text{j}}\,\mu^{\text{o}}_{\text{j}}(p^{\text{o}},T) - p^{\text{o}}\,V \\ &+ RT\left[n_{\text{solv}} \ln \frac{\gamma_{\text{solv}}\,n_{\text{solv}}}{n_{\text{solv}} + \sum_{\text{j=1}}^{\text{s}} n_{\text{j}}} + \sum_{\text{j=1}}^{\text{s}} n_{\text{j}} \ln \frac{\gamma_{\text{j}}\,n_{\text{j}}}{m_{\text{solv}}} \right] + \text{C}.\end{aligned} \tag{D.48}$$

This is the general expression for the Helmholtz free energy of any liquid solution.

When using the above expression, it should be noted that V is a function of the state variables of the solution, as specified by eq. (D.5). If needed, the change in volume relative to the initial volume can be calculated as

$$\Delta V = V - V_{\text{o}}, \tag{D.49}$$

where V_{o} is the initial volume. From eqs (D.49) and (D.5), we can also express ΔV as

$$\Delta V = \sum_{\text{i=1}}^{\text{r}} \overline{V}_{\text{i}}\,\Delta n_{\text{i}}. \tag{D.50}$$

The molar volumes, $\overline{V}_{\text{i}}$, that appear in this equation are defined by eq. (D.2), while the quantities Δn_{i} are given by

$$\Delta n_{\text{i}} = n_{\text{i}} - n_{\text{i}}^{\text{o}}, \tag{D.51}$$

with n_{i}^{o} being the initial value of n_{i}.

D.5 Gas Solutes in Liquid Solutions

Quite often in practice, liquid solutions are in contact with a gaseous atmosphere. The latter may actually be a liquid medium containing dissolved gases. In either case, the gas molecules from the surrounding medium will diffuse into the solution. Likewise, any gas contained in the solution will spread out of it into the surrounding medium. A dynamic equilibrium is reached when the partial pressure of each gaseous species in the solution is the same as the partial pressure of the same species in the surrounding medium.

This equilibrium condition applies whether or not the gas reacts with other species in the solution. The gas may even originate from components of the solution that are undergoing chemical reactions. Consider the case of a gaseous solute, A, that resulted from the chemical reaction of other species in the solution. Assume that the surrounding atmosphere initially did not contain any A and that it was large enough for the partial pressure of A to remain equal to zero as the gas from the solution diffuses into it. Under these conditions, all of the gas that is produced will leave the solution, because, at equilibrium, its partial pressure in the solution must be zero since its partial pressure in the surrounding medium is zero.

As is true for any solute, the molar Gibbs energy of a gas in solution can be expressed in the form (C.33) or (C.36), depending on whether the gas behaves as an ideal gas. The activity coefficient, γ_i, of a gas is usually referred to as the *fugacity coefficient*. Its marked dependence on pressure distinguishes it from the activity coefficients of non-gaseous species.

In the following, we confine our attention to the case of gases in solution that exhibit ideal gas behaviour. In this case, if the mole ratio of the gas is taken as a measure of its concentration, the fugacity coefficient of the gas will be equal to unity, because the activity of an ideal gas coincides with its mole ratio [as evident from eqs (C.27) and (C.32)]. The following analysis, however, can be almost immediately extended to include the non-ideal gas case, once the fugacity coefficients of the involved gases are specified. We shall not pursue this generalization, because, at room temperature, ideal gas behaviour provides a fair approximation of the behaviour of a majority of real gases, up to pressures of several tens of bar—which is quite enough for the purposes of this book.

In the case of gas solutes, the dependence of μ_i^0 on pressure must be calculated directly from formula (D.17), because the partial molar volume of a gas solute is strongly dependent on pressure. We can relate $\overline{V}_i$ to the gas pressure by remembering that the ideal gas equation, $pV = n\,RT$, also applies to mixtures of ideal gases. Therefore, for an ideal gas mixture, we have:

$$V = n_{gas}\frac{RT}{p}, \tag{D.52}$$

where p denotes the total pressure of the mixture, and n_{gas} indicates the total number of moles of gases in the mixture. That is:

$$n_{gas} = n_a + n_b + ... + n_g, \tag{D. 53}$$

where n_a, n_b, …, n_g are the number of moles of each gas in the mixture. From eqs (D.52) and (D.2), it then follows that the relation

$$\overline{V}_i = \frac{RT}{p} \tag{D. 54}$$

applies to every gas component. This shows that the partial molar volume of every gas in an ideal gas mixture is the same and depends only on p and T (cf. e.g., [63], Sect 1.10).

When referring to an ideal gas in solution, we can insert eq. (D.54) into the integral that appears in eq. (D.17) to obtain, upon integration:

$$\mu_i^o(p^o + \Delta p, T) = \mu_i^o(p^o, T) + RT\ \ln\frac{p^o + \Delta p}{p^o}. \tag{D.55}$$

Once introduced into eq. (D.9), this result enables us to conclude that:

$$\mu_i = \mu_i^o(p^o, T) + RT\ \ln x_i + RT\ \ln\frac{p^o + \Delta p}{p^o}, \tag{D.56}$$

which applies instead of (D.19) to gas solutes. Therefore, for every solute included in eq. (C.4), the quantities μ_i are either expressed by eq. (D.56) or by eq. (D.9), depending on whether the solute is a gas or a liquid. From the same equations, we thus obtain:

$$G = G(p, T, n_{\text{solv}}, n_1, n_2, \ldots, n_s)$$

$$= n_{\text{solv}}\, \mu^{\text{o}}_{\text{solv}}(p^{\text{o}}, T) + \sum_{\text{j}=1}^{\text{s}} n_{\text{j}}\, \mu^{\text{o}}_{\text{j}}(p^{\text{o}}, T) + V\Delta p \qquad \text{(D.57)}$$

$$+ RT\left[n_{\text{solv}} \ln(1 - x_{\text{os}}) + \sum_{\text{j}=1}^{\text{s}} n_{\text{j}} \ln x_{\text{i}} + n_{\text{gas}} \ln \frac{p^{\text{o}} + \Delta p}{p^{\text{o}}} \right] + \text{C}.$$

This is the expression of the Gibbs energy of an ideal solution that contains both non-gaseous and gaseous solutes, the total number of moles of gaseous solutes is n_{gas}, according to eq. (D.53).

For non-ideal solutions, the activities a_{i} should be substituted for x_{i} in the above equation:

$$G = n_{\text{solv}}\, \mu^{\text{o}}_{\text{solv}}(p^{\text{o}}, T) + \sum_{\text{j}=1}^{\text{s}} n_{\text{j}}\, \mu^{\text{o}}_{\text{j}}(p^{\text{o}}, T) + V\Delta p \qquad \text{(D.58)}$$

$$+ RT\left[n_{\text{solv}} \ln a_{\text{solv}} + \sum_{\text{j}=1}^{\text{s}} n_{\text{j}} \ln a_{\text{j}} + n_{\text{gas}} \ln \frac{p^{\text{o}} + \Delta p}{p^{\text{o}}} \right] + \text{C}.$$

In applying this equation, the activity a_{i} relevant to a gas solute should simply be set equal to the gas mole fraction, x_{i}, because we are assuming that the gas in solution behaves as if it were an ideal gas.

More explicitly, let the indices "1, 2, …, g" refer to the gas solutes and the remaining indices, "g+1, g+2, …, s" to non-gaseous solutes. By calculations similar to the ones leading to eq. (D.29), we obtain:

$$G = n_{\rm solv}\,\mu^{\rm o}_{\rm solv}(p^{\rm o},T) + \sum_{\rm j=1}^{\rm s} n_{\rm j}\,\mu^{\rm o}_{\rm j}(p^{\rm o},T) + V\Delta p$$

$$+\ RT\left[\,n_{\rm solv}\ln\frac{\gamma_{\rm solv}\,n_{\rm solv}}{n_{\rm solv}+\sum_{\rm j=1}^{\rm s} n_{\rm j}} + \sum_{\rm j=1}^{\rm g} n_{\rm j}\ln x_{\rm j}\right. \tag{D.59}$$

$$\left. + \sum_{\rm j=g+1}^{\rm s} n_{\rm j}\ln\gamma_{\rm j}\frac{n_{\rm j}}{m_{\rm solv}} + n_{\rm gas}\ln\frac{p^{\rm o}+\Delta p}{p^{\rm o}}\right] + {\rm C}\,,$$

which generalizes eq. (D.29) to the case in which some of the solutes are gases. The Helmholtz free energy relevant to this case can be obtained from eqs (D.47) and (D.58):

$$\Psi = n_{\rm solv}\,\mu^{\rm o}_{\rm solv}(p^{\rm o},T) + \sum_{\rm j=1}^{\rm s} n_{\rm j}\,\mu^{\rm o}_{\rm j}(p^{\rm o},T) - p^{\rm o}\,V$$

$$+\ RT\left[\,n_{\rm solv}\ln\frac{\gamma_{\rm solv}\,n_{\rm solv}}{n_{\rm solv}+\sum_{\rm j=1}^{\rm s} n_{\rm j}} + \sum_{\rm j=1}^{\rm g} n_{\rm j}\ln x_{\rm j}\right. \tag{D.60}$$

$$\left. + \sum_{\rm j=g+1}^{\rm s} n_{\rm j}\ln\gamma_{\rm j}\frac{n_{\rm j}}{m_{\rm solv}} + n_{\rm gas}\ln\frac{p^{\rm o}+\Delta p}{p^{\rm o}}\right] + {\rm C}.$$

D.6 Solvent of Variable Composition

To some extent, the distinction between solutes and solvent is arbitrary. It is commonly used to indicate that the solvent is the largest component of a solution or to signify that we are considering a particular component of the solution as the medium in which the other components (i.e., the solutes) are dissolved. In some applications, however, it may be appropriate to think of the solvent as a mixture of some of the components of the solution. The components of the solvent mixture are then referred to as co-solvents. In that case, the composition of the solvent may change independently of the other components of the solution.

In the formulae of the previous sections, we considered the solvent as consisting of a single component. Here, we consider how the same formulae change when the solvent itself is a mixture of variable composition. We focus on the formulae that concern the Helmholtz free energy of the solution. For simplicity, we consider the case of two co-solvents, which can be extended easily to include any number of co-solvents.

Let A and B be the two co-solvents that compose the solvent and n_A and n_B be their respective mole numbers. The total number of moles of co-solvents that at any time make up the solvent is:

$$n_{solv}\big|_{mix} = n_A + n_B \,. \tag{D.61}$$

We refer to this quantity as the *overall mole number* of solvent. By applying eq. (D.48) to the present case, we obtain:

$$\begin{aligned}\Psi &= n_A\,\mu_A^o(p^o,T) + n_B\,\mu_B^o(p^o,T) + \sum_{j=1}^{S} n_j\,\mu_j^o(p^o,T) - p^o V \\ &+ RT\left[\,n_A \ln\frac{\gamma_A\, n_A}{n_A + n_B + \sum_{j=1}^{S} n_j} + n_B \ln\frac{\gamma_B\, n_B}{n_A + n_B + \sum_{j=1}^{S} n_j}\right. \\ &\left. + \sum_{j=1}^{S} n_j \ln\frac{\gamma_j\, n_j}{m_{solv}}\right] + C,\end{aligned} \tag{D.62}$$

where the summations are understood to run over the components of the solution other than the co-solvents. The total mass of solvent, m_{solv}, entering the last summation in eq. (D.62) can be expressed as:

$$m_{solv} = m_A + m_B = n_A \mathrm{M}_A + n_B \mathrm{M}_B \,. \tag{D.63}$$

Here, m_A and m_B are actual the masses of A and B, while M_A and M_B are their respective molar masses.

In order to calculate the free energy of the solution per overall mole number of solvent, it is convenient to introduce the following densities (per overall mole number of solvent):

$$\bar{n}_{\mathrm{A}} = \frac{n_{\mathrm{A}}}{n_{\mathrm{solv}}\big|_{\mathrm{mix}}}, \tag{D.64}$$

$$\bar{n}_{\mathrm{B}} = \frac{n_{\mathrm{B}}}{n_{\mathrm{solv}}\big|_{\mathrm{mix}}}, \tag{D.65}$$

$$\bar{n}_{\mathrm{j}} = \frac{n_{\mathrm{j}}}{n_{\mathrm{solv}}\big|_{\mathrm{mix}}}, \tag{D.66}$$

and

$$\bar{\mathrm{V}} = \frac{V}{n_{\mathrm{solv}}\big|_{\mathrm{mix}}}. \tag{D.67}$$

In terms of these densities, the free energy of the solution can be expressed in the following form:

$$\Psi = n_{\mathrm{solv}}\big|_{\mathrm{mix}}\ \bar{\Psi}(p, T, \bar{n}_{\mathrm{A}}, \bar{n}_{\mathrm{B}}, \bar{n}_1, \ldots, \bar{n}_{\mathrm{s}}) + \mathrm{C}, \tag{D.68}$$

where the function $\bar{\Psi}$ is the free energy per overall mole number of solvent, obtained by dividing Ψ by $n_{\mathrm{solv}}\big|_{\mathrm{mix}}$. After some simple mathematics, from eqs (D.61) to (D.68), we finally obtain:

$$\begin{aligned}\bar{\Psi} = {} & \bar{n}_{\mathrm{A}}\, \mu_{\mathrm{A}}^{\mathrm{o}}(p^{\mathrm{o}}, T) + \bar{n}_{\mathrm{B}}\, \mu_{\mathrm{B}}^{\mathrm{o}}(p^{\mathrm{o}}, T) + \sum_{\mathrm{j=1}}^{\mathrm{s}} \bar{n}_{\mathrm{j}}\, \mu_{\mathrm{j}}^{\mathrm{o}}(p^{\mathrm{o}}, T) - p^{\mathrm{o}}\, \bar{\mathrm{V}} \\ & + RT\left[\ \bar{n}_{\mathrm{A}} \ln \frac{\gamma_{\mathrm{A}}\, \bar{n}_{\mathrm{A}}}{1 + \sum_{\mathrm{j=1}}^{\mathrm{s}} \bar{n}_{\mathrm{j}}} + \bar{n}_{\mathrm{B}} \ln \frac{\gamma_{\mathrm{B}}\, \bar{n}_{\mathrm{B}}}{1 + \sum_{\mathrm{j=1}}^{\mathrm{s}} \bar{n}_{\mathrm{j}}} \right. \\ & \left. + \sum_{\mathrm{j=1}}^{\mathrm{s}} \bar{n}_{\mathrm{j}} \ln \frac{\gamma_{\mathrm{j}}\, \bar{n}_{\mathrm{j}}}{\bar{n}_{\mathrm{A}}\, \mathrm{M_A} + \bar{n}_{\mathrm{B}}\, \mathrm{M_B}} \right] + \mathrm{C},\end{aligned} \tag{D.69}$$

which is the Helmholtz free energy of the solution per unit overall mole number of solvent.

References

[1] Arampatzis A., Knicker A., Metzler V., Brüggemann G.-P.: Mechanical power in running: a comparison of different approaches. *J. Biomechanics* **33**, 457-463 (2000).

[2] Atkins P., De Paula J.: *Physical Chemistry*, 8th ed., OUP, Oxford, 2006.

[3] Atkins P., Jones L.: *Chemical Principles: The Quest for Insight*, 4th ed., Freeman, New York, 2008.

[4] Atkins P., De Paula J.: *Physical Chemistry for the life sciences*, 2nd ed., OUP, Oxford, 2011.

[5] Bangsbo J., Graham T. E., Kiens B., Saltin B.: Elevated muscle glycogen and anaerobic energy. *J. Physiology* **451**, 205-227 (1992).

[6] Bartlett D.H.: Pressure effects on in vivo microbial processes. *Bioch. Biophys. Acta* **1595**, 367-381 (2002).

[7] Batchelor G.K.: *An Introduction to Fluid Dynamics*, CUP, Cambridge, UK, 2000.

[8] Battley E.H.: Calculation of entropy change accompanying growth of Escherichia coli K-12 on succinic acid. *Biotechnol. Bioeng.* **41**, 422-428 (1993).

[9] Buono M. J., Kolkhorstm F. W.: Estimating ATP resynthesis during a marathon run: a method to introduce metabolism. *Adv. Physiol. Educ.* **25**, 70-71 (2001).

[10] Butler J.N.: *Ionic Equilibrium: Solubility and pH Calculations*, Wiley, Canada, 1998.

[11] Cox M.M. Nelson D.L.: *Leheninger Principles of Biochemistry*, 5th ed., Freeman, N.Y., 2008.

[12] Crowe J.H., Crowe L.M., Oliver A.E. Tsvetkova N., Wolkers W., Tablin F.: The trehalose myth revisited: Introduction to a symposium on stabilization of cells in dry state. *Cryobiology* **43**, 89-105 (2001).

[13] Denbigh K.: *The Principles of Chemical Equilibrium*, 4th ed., CUP, Cambridge, 1981.

[14] Dudko V.V., Yushkanov A.A., Yalamov Yu. I.: Generation of shear waves in gas by a vibrating surface. *High Temperature* **47**, 243-249 (2009).

[15] Eckhardt B.: Introduction. Turbulence transition in pipe flow: 125th anniversary of the publication of Reynolds' paper. *Phil. Trans. R. Soc. A* **367**, 449–455 (2009).

[16] Ellis R.J.: Macromolecular crowding: obvious but underappreciated. *Trends ii Bioch. Sc.* **26**, 597-604 (2001).

[17] Fast J.D.: *Entropy*, 2nd ed., Macmillan, London 1970.

[18] Feynman R.P., Leighton R.B., Sands M.: *The Feynman Lectures on Physics*, vol. **2**, Addison-Wesley, Reading, Ma.1975.

[19] Fulton A.B.: How crowded is the cytoplasm? *Cell* **30**, 345-347 (1982).

[20] Fung Y.C., Tong P.: *Classical and Computational Solid Mechanics*, World Scientific, Singapore 2001.

[21] Gaetzel M., Infelta P.: *The Bases of Chemical Thermodynamics*, vol. 2, Universal Publishers, Parkland, Florida, 2000.

[22] García A.H.: Anhydrobiosis in bacteria: From physiology to applications. *J. Biosc*. **36**, 939-950 (2011).

[23] Glasstone S., Lewis D.: *Elements of Physical Chemistry*, 2nd ed., Macmillan, London, 1966.

[24] Guendouzi M.El., Mounir A., Dinane A.: Water activity, osmotic and activity coefficients of aqueous solutions of Li_2SO_4, Na_2SO_4, K_2SO_4,

$(NH_4)_2SO_4$, $MgSO_4$, $NiSO_4$, $CuSO_4$ and $ZnSO_4$ at T=298.15 K. *J. Chem. Thermodynamics* **35**, 209-220 (2003).

[25] Hamer W.J., Wu Y.-C.: Osmotic coefficients and mean activity coefficients of uni-univalent electrolytes in water. *J. Phys. Chem. Ref. Data*, **1**, 1047-1099, (1972).

[26] Hansen L.D., Criddle R.S., Battley E.H.: Biological calorimetry and the thermodynamics of the origination and evolution of life. *Pure Appl. Chem.* **81**, 1843-1855 (2009).

[27] Hoekstra F.A., Golovina E.A., Buitink J.: Mechanisms of plant desiccation tolerance. *Trends Plant Sc.* **6**, 431-438 (2001).

[28] Hunter S.C.: *Mechanics of Continuous Media*, 2nd ed., Ellis Horwood, Chichester, UK, 1983.

[29] Jain N.K, Roy I.: Effect of trehalose on protein structure. *Protein Sc.* **18**, 24-36 (2009)

[30] Joseph D.D., Riccius O., Arney M.: Shear–wave speeds and elastic moduli for different liquids. Part 2. Experiments. *J. Fluid Mech.* **171**, 309-338 (1986).

[31] Kestin J.: *A course in Thermodynamics*, vol. 1, McGraw-Hilll, New York, 1979.

[32] Kreyszig E.: *Advanced Engineering Mathematics*, 9th ed., Wiley, U.K., 2006.

[33] Lang F.: Mechanism and significance of cell volume regulation. *J. Am. College Nutrition* **26**, 613S-623S (2007).

[34] Leigh D.C.: *Nonlinear Continuum Mechanics*, McGraw-Hill, New York, 1968.

[35] Lewis G.N., Randall M.: *Thermodynamics and the Free Energy of Chemical Substances*, McGraw-Hill, New York, 1923.

[36] Lide D.E.: *Handbook of Chemistry and Physics*, 89th ed., CRC Press, Boca Raton, Fl, 2008.

[37] Lourens A., Molenaar R., van den Brand H., Heetkamp M.J.W., Meijerhof R., Kemp B.: Effect of egg size on heat production and the transition of energy from egg to hatchling. *Poultry Science* **85**, 770-776 (2006).

[38] Lourens A., van den Brand H., Heetkamp M.J.W., Meijerhof R., Kemp B.: Effect of eggshell temperature and oxygen concentration on embrio growth and metabolism during incubation. *Poultry Science* **86**, 2194-2199 (2007).

[39] Lubliner J.: *Plasticity Theory*, Macmillan, New York, 1990.

[40] Luque De Castro M.D., Priego Capote F.: *Analytical Applications of Ultrasound*, Elsevier, Amsterdam, 2007.

[41] Malvern L.E.: *Introduction to the Mechanics of a Continuous Medium*, Prentice-Hall, Englewood Cliffs, New Jersey, 1969.

[42] Mathews C.K., van Holde K.E., Ahern K.G.: *Biochemistry*, 3th ed., Addison Wesley Longman, San Francisco, 2000.

[43] Mendez J., Keys A.: Density and composition of mammalian muscle. *Metabolism* **9**, 184-188 (1960).

[44] Minton A.P., Colclasure G.C., Parker J.C.: Model for the role of molecular crowding in regulation of cellular volume. *Proc. Nat. Acad. Sc. USA* **89**, 10504-10506 (1992).

[45] Minton A.P.: How can biochemical reactions within the cell differ from those in test tubes? *J.Cell Sc.* **119**, 2863-2869 (2006).

[46] Mises, R. von: *Mechanik der festen Körper im plastisch deformablen Zustand.* In: Nachrichten der Gesellschaft der Wissensschaften. Math. Phys. Phys. Klasse **1**, 582-592, Göttingen, 1913.

[47] Monk P.M.S.: *Physical chemistry: understanding our chemical world.* Wiley, Chichester, UK, 2004.

[48] Paglietti A.: The mathematical formulation of the local form of the second principle of thermodynamics. *Ann. Inst. Henri Poincaré* **27**, 207-219 (1977).

[49] Paglietti A.: On the mathematical formulation of the first principle of thermodynamics for non-uniform temperature processes. *Ann. Inst. Henri Poincaré* **30**, 61-82 (1979).

[50] Paglietti A.: Temperature dependence of heat energy in non-equilibrium thermodynamics. *Transport. Theory. Stat. Phys.* **13**, 521-526 (1984).

[51] Paglietti A.: Thermodynamic nature and control of the elastic limit in solids. *Int. J. Non-Linear Mech.* **24**, 571-583 (1989).

[52] Paglietti A.: *Plasticity of Cold Worked Metals. A Deductive Approach.* WIT Press, Southampton and Boston 2007.

[53] Paglietti A.: Ideal gas interaction with thermal radiation in classical thermodynamics and Gibbs' paradox. *Continuum Mech. Thermodyn.* **24**, 201-210 (2012).

[54] Paul B.: Macroscopic criteria for plastic flow and brittle fracture. In: H. Liebowitz (ed.), *Fracture*, vol. **2**, 313-469, Academic Press, New York, 1968.

[55] Paunovic M., Schlesinger M.: *Fundamentals of Electrochemical Deposition*, 2nd ed., Wiley, Hoboken, N.J., 2006.

[56] Peng X.F., Peterson G.P., Wang B.X.: Heat transfer characteristics of water flowing through microchannels. *Experimental heat transfer.* 7, 265-283 (1994).

[57] Pfenniger W.: Transition in the inlet length of tubes at high Reynolds numbers. In: *Boundary layer and flow control*, Lachman G. (ed), 970-980, Pergamon, New York, 1961

[58] Qiao Y., Wang R., Bai Y., Hansen L.D.: Characterizing critical phases of germination in winterfat and malting barley with isothermal calorimetry. *Seed Science Research* **15**, 229-238 (2005).

[59] Romijn C. and Lokhorst W.: Foetal heat production in the fowl. *J. Physiol.* **150**, 239-249 (1960).

[60] Schmoltz E., Lamprecht I.: Calorimetric investigation on activity states and development of holometabolic insects. *Thermochim. Acta* **349**, 61-68 (2000).

[61] Schmoltz E., Lamprecht I.: Thermal investigation on social insects. In: *The Nature of Biological Systems as Revealed by Calorimetric Investigations*, D. Lorinczy (ed), Ch. 10, 251-283, Kluwer Acad. Publ., Dordrecht, (2004).

[62] Schmoltz E., Kösece F., Lamprecht I.: Energetics of honeybee development. Isoperibol and combustion calorimetric investigations. *Thermochim. Acta* **437**, 39-47 (2005).

[63] Silbey R.J., Alberty R.A., Bawendi M.G.: *Physical Chemistry*, 4th ed., Wiley, New York, 2005.

[64] Sinha M., Kevrekidis I.G., Smits A.J.: Experimental study of a Neimark-Sacker bifurcation in axially forced Taylor-Couette flow. *J. Fluid Mech.* **558**, 1-32 (2006).

[65] Strange K.: Cellular volume homeostasis. *Adv. Physiol. Edu.* **28**, 155-159 (2004).

[66] Sun M.S., Harriss D.K., Magnuson V.R.: Activity corrections for ionic equilibria in aqueous solution. *Can. J. Chem.* **58**, 1253-1257 (1980).

[67] Tillmark N., Alfredson P.H.: Experiments in plane Couette flow. *J. Fluid Mech.* **235**, 89-102 (1992).

[68] Tritton D.J.: *Physical Fluid Dynamics*, 2nd ed., OUP, N.Y., 1988.

[69] Truesdell C., Toupin R.A.: The Classical Field Theories, in S. Flügge (ed.), *Encyclopedia of Physics*, vol.**3**, pt.1, 226-793, Springer-Verlag OHG, Berlin, 1960.

[70] Truesdell C., Rajagopal K.R.: *An Introduction to the Mechanics of Fluids*, Birkhäuser, Boston, 2000.

[71] Wagman D.D., Evans W.H., Parker V.B., Schumm R.H., Halow I., Bailey S.M., Churney K.L., Nuttal R.L.: The NBS tables of chemical thermodynamic properties. *J. Phys. Chem. Ref. Data*, **11**, Suppl. 2, (1982).

[72] Wharton D.A.: *Life at the limits*, Cambridge University Press, U.K., 2002.

[73] Zemaitis J.F., Clark D.M., Rafal M., Scrivner N.C., Zemaitis J.F. Jr.: *Handbook of Aqueous Electrolyte Thermodynamics: Theory & Application*, Wiley-AIChE, New York, 1986.

[74] Zemansky M.W.: *Heat and Thermodynamics*, 5th ed., McGraw-Hill, Tokyo, 1968.

Index

D

E

M

N

O

P

T

U

V

W

Y

www.ingramcontent.com/pod-product-compliance
Lightning Source LLC
Chambersburg PA
CBHW030236230326
41458CB00091B/311
* 9 7 8 8 8 9 0 9 4 3 7 0 6 *